Kirsten Hackenbroch

The Spatiality of Livelihoods –
Negotiations of Access to Public Space in Dhaka, Bangladesh

MEGACITIES AND GLOBAL CHANGE

MEGASTÄDTE UND GLOBALER WANDEL

herausgegeben von

Frauke Kraas, Jost Heintzenberg, Peter Herrle und Volker Kreibich

---

Band 7

Kirsten Hackenbroch

The Spatiality of Livelihoods – Negotiations of Access to Public Space in Dhaka, Bangladesh

Franz Steiner Verlag

Gedruckt mit freundlicher Unterstützung der Deutschen Forschungsgemeinschaft

Umschlagabbildung:
Ghat at the Buriganga River in Dhaka © Kirsten Hackenbroch

Bibliografische Information der Deutschen Nationalbibliothek:
Die Deutsche Nationalbibliothek verzeichnet diese Publikation in der Deutschen Nationalbibliografie; detaillierte bibliografische Daten sind im Internet über <http://dnb.d-nb.de> abrufbar.

Druck: AZ Druck und Datentechnik GmbH, Kempten
Gedruckt auf säurefreiem, alterungsbeständigem Papier.
Printed in Germany.
ISBN 978-3-515-10321-3

TABLE OF CONTENTS

LIST OF FIGURES

LIST OF PHOTOS

Unless otherwise quoted, all the photographs have been taken by me or by my field assistants, whom I instructed to do so.

LIST OF TABLES

LIST OF TEXTBOXES

LIST OF ACRONYMS

AL	Awami League (People's Party)
BBS	Bangladesh Bureau of Statistics
BIWTA	Bangladesh Inland Water Transport Authority
BNP	Bangladesh National Party
BRAC	Bangladesh Rural Advancement Committee (NGO founded in 1972)
BUET	Bangladesh University of Engineering and Technology
CI-sheet	Corrugated iron-sheet
CUS	Centre for Urban Studies
DCC	Dhaka City Corporation
DFG	German Research Foundation (Deutsche Forschungsgemeinschaft)
DMA	Dhaka Metropolitan Area
DMDP	Dhaka Metropolitan Development Plan
DSMA	Dhaka Statistical Metropolitan Area
GD	General Diary
ICG	International Crisis Group
IIED	International Institute for Environment and Development
JI	Jamaat-e-Islami (Islamist political party)
JP	Jatiyo Party
MP	Member of Parliament
NGO	Non-governmental organisation
NSDF	National Slum Dwellers Federation (India)
PRA	Participatory Rapid Appraisal
RAB	Rapid Action Battalion (special police force)
RAJUK	Rajdhani Unnayon Kortripokkho (Capital Development Authority)
SIP	Slum Improvement Project
SPARC	Society for the Promotion of Area Resource Centres
TI	Transparency International
Tk	Taka (literally money), currency of Bangladesh (BDT, Bangladeshi Taka), 100 BDT = 0,96 Euro (14.12.2011)
UNDP	United Nations Development Programme
WC	Ward Commissioner

ACKNOWLEDGEMENTS

My primary gratitude goes to the inhabitants of the study settlements of this research. I was welcomed in both settlements with utmost hospitality and it is difficult to express the warm welcome and close integration I experienced, especially from those whom I interviewed, but also from many others. A very affectionate thank you is reserved for the family with whom I lived, who accepted me as Auntie and *Apa* (sister) and made my stay a unique life-time experience. For my research they provided me with many of the insights that are discussed here. I hope that with my research I can contribute to a better understanding of the processes at work in the study settlements and similar urban areas, which in future could help those who shared their valuable time, experience and commitment with me.

Secondly I would like to thank my supervisors, Prof. Dr. Sabine Baumgart and Prof. Dr. Volker Kreibich, for their constant support and availability for discussions, and for encouraging me to go ahead with this rather unusual 'embedded planning research'. Furthermore, I would like to thank Prof. Dr. Roxana Hafiz and the faculty members of the Department of Urban and Regional Planning, Bangladesh University of Engineering and Technology. I was always welcome to join in discussion rounds between the faculty members and this provided me with valuable opportunities to discuss research issues. Similarly, discussion with the research staff of the Centre for Urban Studies was fruitful, providing me with further insights on planning issues in Bangladesh.

This research would not have been possible without the support of Bangladeshi field assistants. I would like to thank all those who supported me during the course of my research, from the very first field visits, to interviews, transcriptions and translations. My special thanks goes to Asif Rahman, Rifat Shams and Salimul Alam, who dedicated much of their time to joining me during long interview sessions and informal discussions in the study settlements in 2009 and 2010. They soon considered my research as their own and developed trustful relationships with my interviewees. Thank you for your commitment and the strength you showed during summer time electricity cuts and late hours of translating interviews in our 'Road 3 Office'. In Germany, I would like to thank my student assistants Alexander Hoba, Juliane Hagen, Ines Standfuß and Sonja Dieckmann for their great support, especially in executing the maps and figures I designed.

I would also like to express my gratitude to my friends in Germany and my PhD colleagues from the peer review group. Special thanks go to Dyah Widiyastuti, Genet Alem, Johanna Schoppengerd and Wolfgang Scholz for their support in commenting on my work. In the final phase, Michael Fink provided me with exceptional support by reading the final version and offering me a calm PhD-retreat in the Black Forest.

I would also like to thank the German Research Foundation (DFG) for funding this research within the priority programme 'Megacities – megachallenge: Informal dynamics of global change'. Being part of this programme provided me with the possibility for continuous exchange with other researchers and thus contributed considerably to the sharpening of my research ideas and approach.

Finally, let me extend my thanks to all those whom I could not mention individually, but who supported me in my research both in Bangladesh and in Germany.

PRELIMINARY REMARKS

Anonymity

The names of inhabitants of the study settlements are fictive and do not resemble the names of the research participants[1]. This is to ensure that no participant in my research will suffer harm due to her/his participation. Furthermore, the photographs depicted here have been chosen carefully so as not to divulge the participant's identity. Similarly, the study settlements have been re-named and materials which disclose their location have been removed in order to protect respondents' anonymity in front of a larger and locally aware audience.

Use of Bengali terms

In this research report I frequently use terms in Bengali, the language in which I conducted all my interviews. By using Bengali terms (in italics, except for place names) I aim to circumvent some of the problems emanating from translations, as specific Bengali terms carry a different meaning related to the social and cultural context of the language which an English translation would not be able to reflect adequately (see Chapter 4.2.5). In my explanations of the local terms used I seek to give an extensive English description and discussion of the inherent meanings.

While I am aware of the scientific transliteration used in the disciplines of Indology and South Asian Studies (Brandt 2010), I here prefer to use a simplified transliteration based on the Bengali alphabet and its pronunciation which will also enable non-Bengali speakers to read the terms correctly. Especially problematic in common simplified transliteration is the differentiation between the phonemes /a/ [আ], /ɔ/ [অ] and /o/ [অ and ও] (Brandt 2010: 9, 10). Often 'a' is used in transliterations to symbolise both /a/ and /ɔ/, leading to mispronunciation of the words by non-Bengali speakers due to the orientation on English pronunciation. Thus here I use 'a' only for the phoneme /a/ while I use 'o' for both /ɔ/ and /o/, which resembles the actual pronunciation more closely[2]. Only in cases of established English

1 Accordingly, if I quote directly in the text I indicate the source in parenthesis, e.g.: (Rohima, 18.01.2010). This then indicates an interview conducted with Rohima (her fictive name) on 18.01.2010. Additionally, a brief summary of information about Rohima is given in the side note next to the quote, so that the reader is able to understand the context of the quoted person. Furthermore, the details about the interviewee, including her/his occupation or in the case of key informant interviews her/his position that makes her/him a key informant, approximate age (calculated for 2010, the time of the last fieldwork), marriage status and number of children are compiled in the List of Interviews on page 377.

2 The limitations of this phonetic representation become obvious as the letter [অ] can be pronounced as either /ɔ/ or /o/. This can however not be represented by using the English language, and therefore this differentiation is not made here.

transliterations for place names do I use the established format, while indicating the pronunciation in square brackets. For example the local English transliteration for the neighbourhood Uttara in Dhaka fails to represent the pronunciation 'Uttora'. I however refrain from using transliterations of common words that have been deformed to meet English pronunciation. This is for example the case with the term *bosti* commonly used to refer to 'slum' (common variations in literature that do not resemble the Bengali alphabet are *bustee* and *bostee*) and *paka* for example used to refer to brick or concrete buildings, but used in other contexts as well (a common variation in the literature is *pucca*). In order not to complicate the transliteration further, I refrain from denoting the retroflex consonants. The simplified transliteration I am using in my work orients itself on van Schendel's system in his publication on the history of Bangladesh (van Schendel 2009).

All Bengali words used are explained when first introduced in the text. If they are used again, a reference is made to the explanation. All Bengali terms, and those English terms that are commonly used in Bengali but may have a slightly different or unfamiliar meaning compared to 'British English', are compiled in a glossary (p. 357) if used more than once. Here, the Bengali terms, their English translations as well as explanations concerning the socio-cultural meanings are given. The glossary also includes Arabic terms or terms of Arabic/Persian origin which are commonly used in Bengali, especially the names of religious holidays and events.

Perspective

In this research, I use both the personal pronoun 'I' and a more formal style of third person narrative perspective. As this may seem unusual to some readers, I would like to explain my approach. Given the embeddedness of my research in the local context, a 'formal style' suggests a distance that in reality did not exist. Thus I underline my personal involvement in the research process, on which I continuously reflected, by using the personal pronoun 'I' (see also the discussion on research ethics and positionality in Chapter 4.4). I do not limit this to the empirical chapters only, but also use 'I' where appropriate in other chapters to underline decisions I took during the research process or to underline main arguments and conclusions. The use of 'we' in Chapter 4 refers to work that I and my field assistants undertook.

Furthermore, the use of present and past tense in the empirical chapters is based on the differentiation between individual stories and generalising arguments and findings. Thus I strictly use the past tense whenever describing what a specific person did at a specific time. This does not mean that she/he is no longer doing this today, but given that the empirical research ended in 2010 the past tense seems appropriate – independent of the continuation of activities to date. I use the present tense when I formulate more general arguments, findings and conclusions based on the empirical study.

1 INTRODUCTION

Nasimgaon Khalabazar, Dhaka, April 2009

> One day in April 2009, Afsana as usual prepared to sell vegetables from a small market place in Nasimgaon, Dhaka, Bangladesh. Previously she used to put a wooden bed onto the sandy open field for display of the goods. But in recent months, after the election of a new government, this sandy field has been claimed by local leaders, affiliated to the now ruling political party, who subsequently established a concrete platform for market activities. They subdivided the platform into eleven units and distributed ownership of these units among themselves. The vendors who had previously used wooden beds to sell fruits or vegetables now had to pay rent to the new 'owners' of the public space. Because of her own affiliation with the ruling political party, Afsana was not made to pay rent but was given ownership of one unit. But the exact position of her unit remained contested due to contradicting oral and written agreements. Thus on this particular morning, she found that her neighbour had removed the vegetables she had planned to sell on that day. Seeing this she got angry and exclaimed: "You have removed the vegetables. If you can then remove me; I will not move from here. If you have power, then take possession of the *bhit* [shop unit[1]] by removing me." To underline her claim she called Roxana, a respected female political leader of the ruling party beyond the local level, to whom she was close. Roxana told one of the local leaders who had been involved in claiming the public space and establishing the concrete platform: "If Afsana files a case on the issue of woman torture then I will be the main witness". After this incident Afsana for a while did not get disturbed again in carrying on with her business. (This account is based on an interview with Afsana, conducted on 23.03.2010)

This was not the last encounter between her and the local leaders, and, along with the similar stories that could be told of other vendors in the same place, this narrative reveals the importance of the use of public space in the local economy as well as the continuous dynamics and contestations of access to public space in everyday life.

Setting the research problem

The above scene of everyday life is set in Dhaka, one of the densest cities worldwide, with an average density of 30,000 inhabitants per km^2 and low-income neighbourhood densities of 200,000 inhabitants per km^2, and a total population of 12 million inhabitants. Given these tremendous densities, public spaces are a scarce resource. At the same time, many urban livelihood activities are based on access to public space. In Dhaka, pavements, street spaces, vacant plots, public parks and city squares are used intensively for all kinds of economic activities, i.e.

1 The Bengali term *bhit* generally refers to a raised platform in a variety of contexts, for example in agricultural production or shop units in a market place. Here it refers to such shop unit demarcations on a concrete platform.

vending, production and provision of services. There has been considerable research on access to public space in the 'visible arenas' of cities, i.e. central areas that are to some degree controlled by state or municipal actors, even if these statutory actors apply informal modes of regulation. Bayat (2004) identifies the street (public space) as an "arena of politics" and draws on research experience in the Middle East, while Etzold et al. (2009) work on the contested spaces of street food vendors in the central localities of Dhaka. The above example, however, indicates how access to public space for livelihood activities is also contested on a neighbourhood scale and how inhabitants and local actors continuously negotiate and dispute access arrangements to use public space. These negotiations can, however, not be limited to economic livelihood activities only. Access arrangements to public space also include other uses, such as recreational activities or festivals as extra-everyday events.

The above narrative furthermore underlines how these 'negotiations' of access to public space are closely interwoven with political power and the ability to draw on sources of power, i.e. Afsana was able to make her spatial claim in opposition to a powerful political leader of the locality because she was able to get the support of Rizia, a politician active in higher level party organisations. This example thus also points at a local institutional setting shaped to a considerable degree by politics and political authorities. With the city ever-expanding, settlements of low-income groups only receive minimal attention from statutory authorities who are more engaged in serving the middle to high-income groups. While this does not mean a complete absence of statutory authorities, the low-income settlements are thus largely 'governed' by local leaders who, based on political and social definitions of leadership, form the local 'elites'. Besides the sources of power that became obvious in the introductory narratives, there exist other sources of power actors can draw upon. While the example showed two powerful women, the social gender relations of society in many cases suggest a different picture where women's access to space is considerably limited by gender norms.

In this context of low-income settlements and scarcity of public spaces, making them a contested resource, I want to investigate the importance of public spaces in everyday life and the underlying access arrangements negotiated among actors in a setting of diverse institutions and power relations.

Research objectives and methodological approach

The above discussion indicates the four topics that provide the main basis for this research, namely the spatiality of livelihoods, the negotiations of access to public space, the discussion surrounding the framing of the concept of urban informality and finally the question of spatial justice. Accordingly, the main objective in starting the research is to explore and analyse the importance of urban public space for the everyday life of urban dwellers and the mechanisms of how access to public space as a livelihood asset is negotiated among actors in an environment characterised by informality as a dominant mode of the production of space. A second objective is more normative: the discussion of how the findings relate to concepts of spatial justice, and what could be the role of planning in establishing and guaranteeing equal rights to the city and equal citizenship for all urban dwellers.

This research follows a qualitative approach of grounded theory, combined with elements of ethnographic research. The fieldwork thus provided the main input for the generation of findings with the help of emerging categories during data analysis. The methodological triangulation I applied included the methods of observation and participant observation, solicited photography, qualitative interviews and informal discussions. Especially the participant observations as an element of ethnographic research allowed me an in-depth investigation of the local negotiation processes of access to public space. This perspective was particularly useful to understand the power relations at work and the social norms and institutions defining access to public space, especially differentiated according to gender. Given this embedded research, I thoroughly reflected on my role as a researcher and the positionality I assumed and I was assigned by others.

The investigation focussed on two study settlements within the urban fabric of Dhaka. Nasimgaon is a low-income settlement that has been built without planning approval on government land since the 1990s. The settlement today is consolidated and of a high density, while its housing structures remain predominantly single-storey CI-sheet houses. However, the development pressures, which also endanger the existing public space, are made obvious by the internal transformation and densification processes, especially with encroachment on the last remaining public spaces and the introduction of two-storey houses. Furthermore, they are inherent in the continuous discussions about redeveloping the land by real estate developers due to its location adjacent to some of Dhaka's high-income settlements.

Manikpara is an older settlement founded in the 1970s adjacent to the old parts of Dhaka. Today the settlement is of a mixed socio-economic structure, with many low-income households but also an emerging middle class. Manikpara is consolidated, if not saturated, and public spaces in the internal area only exist in the form of the very limited and narrow road network. Accordingly, the public spaces adjacent to the settlement are used intensively. The development pressure on the area is also high due to its location adjacent to Old Dhaka and the plastic recycling industry as a main economic resource of the area. For the last couple of years Manikpara has been undergoing a highly dynamic transformation process into multi-storey buildings.

Structure of the research

This PhD thesis is structured into four parts and twelve chapters. An overview of the structure is provided in Figure 1.

Part I 'Situating the Research' begins with a discussion of the most relevant theoretical departures (Chapter 2). Based on these considerations, the research framework and methodological approach are outlined in Chapter 3, while Chapter

4 follows up with a detailed discussion of the application of research methods and reflections on my positionality. The concluding Chapter 5 sets the regional and local context of the research by reconsidering the specific preconditions in Bangladesh and Dhaka concerning the research topic, and introduces the study settlements.

Part II 'The Spatiality of Livelihoods' opens up the empirical parts of the research and consists of two chapters. In Chapter 1, the everyday and extra-everyday life activities taking place in public spaces are presented in a mainly descriptive style in order to provide the reader with an initial overview of spatial practices. These are then analysed and interpreted in the following Chapter 7 which focuses on the production and re-production of spatial practices and the specific public spaces created by social gender relations. Part II finishes with a conclusion of the main results.

Part III 'Negotiations of Access to Public Space' constitutes the second empirical part of the research and consists of two chapters. In Chapter 8, three narratives of the negotiations of access to public space are outlined, focussing on developments in three public spaces in Nasimgaon and Manikpara based on interviews and participant observations. These are then analysed and interpreted in Chapter 1 which analyses the underlying negotiation process in detail, especially with reference to the power sources drawn upon, strategies employed, outcomes achieved and dominant/resistance conceptualisations of space. Similarly to Part II, Part III closes with a conclusion summarising the main results of this empirical part.

Part IV 'Reconnecting the Research to the Theory Debate and Urban Planning' seeks to review the empirical findings of the research in two chapters. The first Chapter 10 discusses the findings of the empirical research in relation to the concept of urban informality and explores how the research can contribute to this theory debate. Chapter 11 then discusses the research findings in relation to spatial justice and provides entry points to utilise the outcomes of this research in urban planning.

Finally, in Chapter 12 I conclude this research by summarising the main findings, reflecting on the key issues of the research methodology and suggesting entry points for further research.

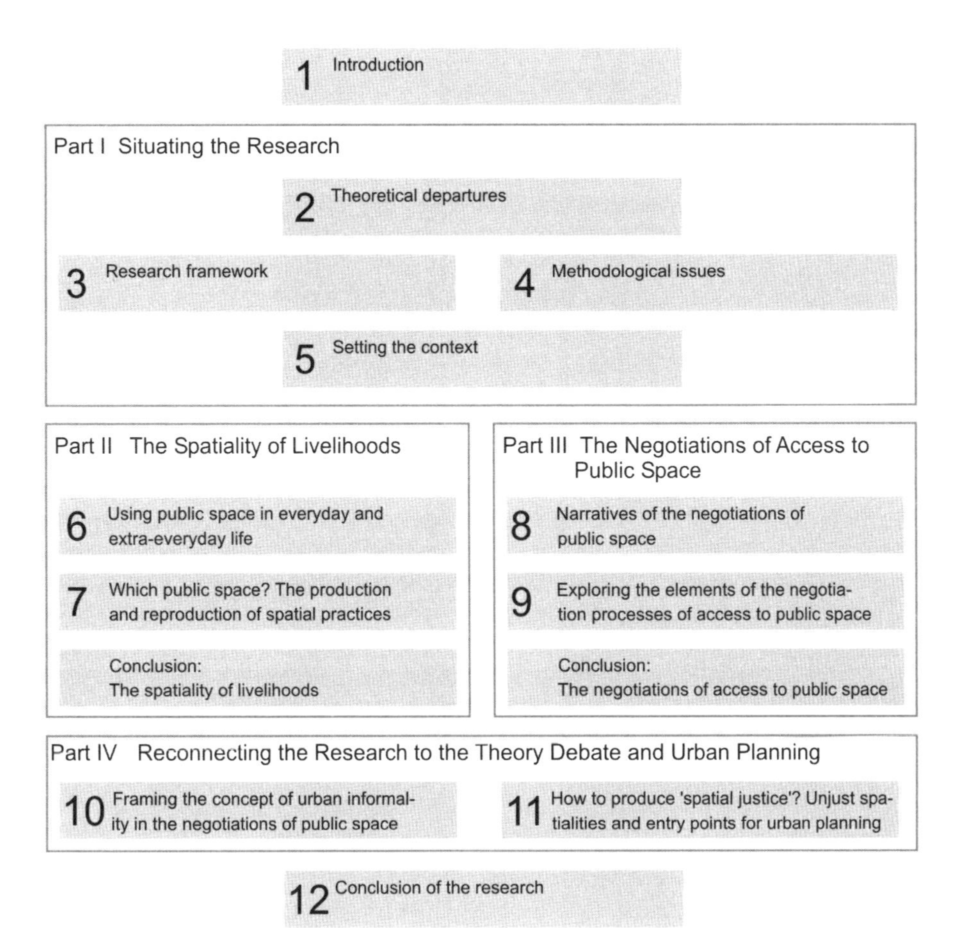

Figure 1: Structure of the thesis

PART I – SITUATING THE RESEARCH

The following chapters will lay the foundation for this research. In Chapter 2 I will outline my theoretical departures with regard to my understanding of public spaces, contested space and the negotiations of access to space, urban informality and spatial justice. Based on this, the short Chapter 3 elaborates the research framework and methodology applied in this research. This is followed by a detailed discussion of methodological issues in Chapter 4, which includes discussion of the selection processes, method application and data analysis, as well as reflections on research ethics and positionality which I consider highly relevant for this partly ethnographic research approach. Finally, Chapter 5 outlines the regional context of Bangladesh with its socio-cultural characteristics of relevance for this research and the urban context of the megacity of Dhaka. Furthermore, the study settlements are introduced and characterised in this chapter.

2 THEORETICAL DEPARTURES

In this chapter I aim to outline the theoretical departures of this research. Despite choosing an inductive grounded theory approach, this research was not void of any pre-knowledge and critical engagement in theoretical debates. The existence and usefulness of such a pre-knowledge and theoretical grounding has been emphasised by the later writings of the grounded-theorist Strauss (e.g. Strauss, Corbin 1994; Kelle 1994; see further discussion in Chapter 3.3). While in this chapter I outline the theoretical departures which informed the formulation and operationalisation of my research questions, I nonetheless consider this research a grounded theory research, for in the generation of findings I was guided by the empirical material and critically reflected my theoretical departures using the categories and theories evolving from the empirical data.

The first sub-chapter (2.1) discusses space as a (social) product and considers how public space constitutes an urban livelihood asset especially for the urban poor, starting with the definition of public space and arriving at the notion of 'contested space'. This notion of 'contested space' and the implications of negotiating access to public space in a setting of diverse institutions and differentiated power structures are discussed in the following sub-chapter (2.2). Subsequently, the usefulness of the concept of urban informality – seen as a mode of the production of space – to understand the constellations of negotiating access to public space will be discussed (2.3). Finally, the notion of spatial justice and the right to the city and possible implications for planning approaches will be outlined in the last sub-chapter (2.4).

2.1 THE SPATIALITY OF LIVELIHOODS

On my way home, I pass by a number of makeshift shops, erected at the Ghat [a small boat harbour] where the road crosses a lake and is not lined by buildings. There are two tea stalls, a few shops selling bamboo for construction sites and some mobile vendors who temporarily remain at that place. Jahid operates a stall selling tea, cigarettes, bethel nuts and leaves and cakes, while he lives in the low-income area on the other side of the lake. His tea stall is strategically well located in terms of customer flow – every day many people use this road to walk between the neighbourhoods, and there are a number of offices and a police checkpoint nearby. Furthermore, construction of a bus stop is going on and it is likely that this will draw more customers to his shop. Operating the tea stall is one of the few income generating activities that Jahid feels able to do. Since birth he is disabled, but the 300 Tk disability pension that he gets monthly is just sufficient to pay for the supply of electricity and water. Together with his wife, he started the tea stall, located in a public place which they claim for their livelihoods. The shop is their only possession, and he comments 'If this is removed, it will be very difficult for us to survive'. (summary of field notes, 24.03.2010, 30.03.2010; the location of this field note is not part of one of my study settlements)

The above example could be located anywhere in Dhaka, if not with local variations in every city of the world. Public space as a means to support an urban livelihood is a precious resource especially for those marginalised by society for whatever reasons. The following sub-chapters aim to arrive at an understanding of public space and to explore the importance of public space for urban livelihoods.

2.1.1 Arriving at an understanding of (public) space

In this sub-chapter I aim to outline the understanding of (public) space that guided me throughout this research, although during the course of this research I also had to question some of the understandings I departed from. I put 'public' in brackets to underline that I intend two understandings: that of space as a (social) product and that of space as a physical delineation of what constitutes public space in a city.

Understanding space as a product and part of dynamic processes

Central to the understanding of space in this research are the impacts of the 'spatial turn' on urban theory. The dominant paradigm of space as a container, a merely physical-material entity, was radically questioned by postmodern geography and hence space was seen as a social construct, a result of social processes and at the same time as an explanation for social relations (for example Massey 1984; Massey 2008 [1992]; Soja 1996; Gregory 1994; Löw 2001; for an overview see Bachmann-Medick 2009; Schmid 2005). Spatiality was reasserted ontologically as a central category of being, overcoming the binary of historicality (time) and sociality (Soja 1996: 71). This new and 'active' understanding of space has radically changed urban theory and led to re-engagement with and widespread reception of Lefebvre's work. Lefebvre, in his book 'The Production of Space', postulates that "([s]ocial) space is a (social) product" (Lefebvre 1991 [1974]: 26). Recent urban development trends in Dhaka illustrate the usefulness of this notion of societal relations being inscribed into space: the emergence of a real estate sector driven by the demand of elite groups leads to the production of housing estates that at best are not accessible to low-income groups, and at worst are implemented through the forced eviction of settlements of the urban poor. If the dominant discourse on housing in Dhaka is to serve the interests of the upper income groups, then the urban poor are denied a space in the city, and thus urban space is produced and reproduced according to the interests of a dominant part of society. Similarly, the access to and use of public space in everyday life can be seen as a result of negotiations between different actors – and the outcome always reflects the way society is organised (Madanipour 2010: 2). Drawing on the work of Lefebvre, among others, Massey (2006: 89–90) formulates three propositions concerning the conceptualisation of space. First, space is not merely a container or surface but "a product of practices, relations, connections and disconnections".

Second, "space is the dimension of multiplicity" and thus allows for a multiplicity of parallel experiences and trajectories. Third, "space is always in process" and "an on-going production".

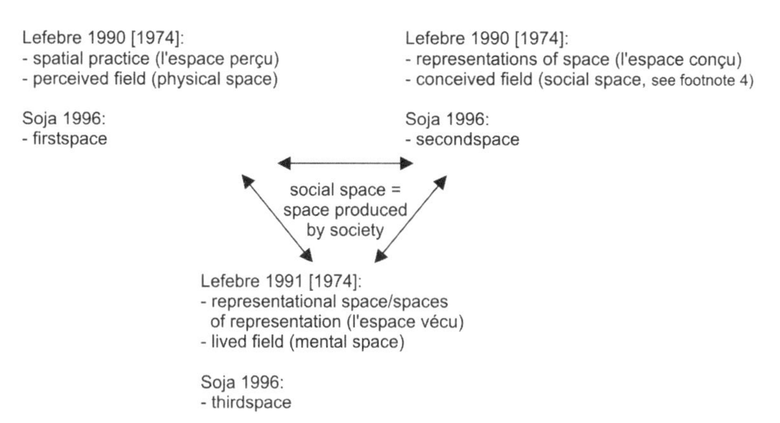

Figure 2: The dimensions of the production of space

For the ongoing production of (social) space Lefebvre suggests a triad of the production of space which he outlines in his 'Plan of the present work' (Lefebvre 1991 [1974]). It consists of the spatial practice, the representations of space and the spaces of representations (see Textbox 1), translating into perceived, conceived and lived fields, which form a trialetic relationship to produce 'social space'[1]. With this triad, Lefebvre moved beyond the 'classic dialectic' that is characterised by a starting point, the thesis, its opposition, the antithesis, and a third term that resolves the antagonism of the two, the synthesis (Schmid 2005: 312). Instead, Lefebvre considers three dialectically interwoven dimensions of the production of space that can each be understood as 'thesis', without one being the only starting point or privileged, and that have to be understood as equally relevant for the production of space (Schmid 2005: 313). In contrast, Soja reveals a preference for Lefebvre's 'spaces of representation' or what Soja himself terms 'thirdspace' (Soja 1996: 68), and he calls for always seeking a third possibility, "not just a simple combination or an "in between" position along some all-inclusive continuum" (Soja 2003: 60; see Figure 2 for an overview of Lefebvre's and Soja's terms). This critical thirding-as-Othering, as will be discussed later, can indeed contribute to an in-depth understanding of the processes at work in a society, and finally in the negotiations of access to public space for livelihoods.

1 Lefebvre's use of the term 'social space' is ambiguous – on the one hand he understands social space in a broader sense of the space produced by society, on the other hand social space is used to refer to the spaces of representations as opposed to physical and mental space (Schmid 2005: 208–210). In the following I adapt Schmid's interpretation of understanding social space in a broad sense as the space produced by society/social practice, and accordingly as the space containing the perceived, conceived and lived (Schmid 2005: 209–210).

Textbox 1: The dimensions of the production of space

In his conceptual triad of the production of space, Lefebvre differentiates the spatial practice (*l'espace perçu*), the representations of space (*l'espace conçu*) and the representational spaces/spaces of representation[2] (*l'espace vécu*) as follows:

"*Spatial practice* The spatial practice of a society secretes that society's space, it propounds and presupposes it, in a dialectical interaction; it produces it slowly and surely as it masters and appropriates it. From the analytic standpoint, the spatial practice of a society is revealed through the deciphering of its space. [...]

Representations of space: conceptualized space, the space of scientists, planners, urbanists, technocratic subdividers and social engineers, as of a certain type of artist with a scientific bent – all of whom identify what is lived with what is perceived with what is conceived. [...] This is the dominant space in any society (or mode of production). Conceptions of space tend, with certain exceptions to which I shall return, towards a system of verbal (and therefore intellectually worked out) signs.

Representational spaces [spaces of representation]: space as directly *lived* through its associated images and symbols, and hence the space of 'inhabitants' and 'users' [...]. This is the dominated – and hence passively experienced – space which the imagination seeks to change and appropriate. It overlays physical space, making symbolic use of its objects. Thus representational spaces may be said, though again with certain exceptions, to tend towards more or less coherent systems of nonverbal symbols and signs." (Lefebvre 1991 [1974]: 38–39, original emphasis)

Before moving on to a delineation of public space a closer investigation of the three dimensions of the production of space seems fruitful, for they might prove useful in understanding (public) space as a livelihood asset. These three dimensions, although together shaping the social space of a society at a given moment in time, are not necessarily equally distinctive of a certain period of the production of space and the relationships between them are not simple or stable (Lefebvre 1991 [1974]: 46). For example, the spaces of representation in the formation of 'abstract space' which emerged with globalisation in the 20th century are dominated by the other two dimensions: knowledge and power create a functional space, while the spatial practice is characterised by the implementation of the rationality of the state. The spaces of representations are thus limited to images and memories, which persist beyond the order of the state, but which again are influenced or 'emptied' by the dominance of male authority. Lefebvre here relates to feminist critique in underlining how male authority ('the phallic formant') has formed space, leaving few options for alternative expression in the representations of space. (Schmid 2005: 260pp.). For all the historical moments Lefebvre relates his conceptualisation to, he underlines how the dimensions stand in an unbalanced relationship to each other, forming specific modes of the production of space. On-

2 In the English translations of Lefebvre's 'The Production of Space' by Donald Nicholson-Smith *l'espace vécu* is translated as 'representational spaces'. However, other authors use the more accurate 'spaces of representation' (the equivalent *Räume der Repräsentation* is also used in the German translation printed in Dünne, Günzel 2006 [1967]) which is also adopted in this research.

ly for a short time in Venice during the 16th century, does Lefebvre analyse how the three dimensions achieved a consensus in everyday life (Lefebvre 1991 [1974]: 74).

In the above discussion on the production of space, space has not yet been operationalised to specific spatial levels. Lefebvre introduces three spatial levels where the production of space takes place, although these have to be understood as interlinked and interfering with each other. These are the global level referring to the broadest extension of the system and the most public spaces, the private realm at the level of residences, and the intermediate spaces consisting of avenues and squares leading to the private spaces (Lefebvre 1991 [1974]: 155). In the context of this work, the production of space investigated is at the intermediate level of the city and its neighbourhoods, while the interlinkages between these levels suggest that these production processes are at the same time influenced by the 'global' (e.g. global economic restructuring) or the 'private'.

For a basic understanding of (social) space the above explanations suffice. They are furthermore important prepositions for the further references to Lefebvre's work in the context of the spatiality of livelihoods and the negotiations of access to public space, especially concerning centres and peripheries, domination and resistance (see Chapter 2.2.1).

A delineation of public space

The above discussion of space as a social product outlines my understanding of the formation of (urban public) space. Nonetheless, for this research it is necessary to delineate what I understand as 'public space'. Public space in the context of this research is, in accordance with Brown's definition (2006b: 10), understood as outdoor spaces that are accessible to the general public. Brown uses the term urban public space

> "to mean all the physical space and social relations that determine the use of that space within the non-private realm of cities. 'Urban public space' includes formal squares, roads and streets, but also vacant land, verges and other 'edge-space'. It includes all space that has accepted communal access or use rights, whether in public, private, communal or unknown ownership; a common property resource, but one whose boundaries may change over time. Thus land that is privately owned but has been left vacant and is being used by traders would be considered as urban public space." (Brown 2006a: 10)

To a large extent I agree with this definition for delineating the 'physical' focus of this research; there are, however, a few clarifications to make. The definition includes both a physical delineation and a delineation referring to accessibility; the latter is inherent in the notion of social relations determining the use of space and by reference to accepted communal access or use rights. This notion of access rights is a central component of this research and needs further elaboration that goes beyond the definition provided above and is related to understanding space in the Lefebvrian sense as a product of social practice (the discussion continues in the subsequent sub-chapters). The above definition furthermore refers to the issue of ownership. In this research, public space can be in ownership of the state or private individuals – as long as the space is accessible to a wider public it remains a 'public space' regardless of ownership. Furthermore, public space in this research does not refer to the interior of public buildings, although these may be accessible by the general public. Thus the definition excludes the private spaces of dwelling units, but also the semi-private spaces of the courtyards which are accessible to all residents of a compound but restricted for outsiders.

Although not at the centre of this research, it is furthermore important to also define the 'other', i.e. private space and semi-private space. As Patel (2007: 47) understands it, private spaces are used by one household who defines the access or use rights, while semi-private spaces are shared among a number of actors who can impose restrictions to their usage. In order to clearly separate public from semi-private spaces, the imposition of restrictions to usage needs further elaboration. I suggest differentiating according to the range of actors able to restrict usage. For semi-private spaces, this range of actors is commonly limited to a specific group sharing a specific resource. For example, in Dhaka five families might share a central courtyard and these five families could define restrictions on its usage and access rules. Outside actors would only have a very limited ability to interfere in the usage of this space and it would not be accessible to a wider general public. On the contrary, the range of actors defining use rights and access restrictions to public space is diverse and open to a broader public.

The delineations of public, semi-private and private spaces outlined here are working definitions for this research and I am aware that these derive from a Western notion of a hierarchy of spaces. In her research on women's spatial experience in Kolkata, Bose (1998) doubts the applicability of the threefold Western differentiation of space in the South Asian context. She criticises especially the notion of an equal distribution of power once access is gained to one space: "[i]t [the threefold differentiation] perpetuates the notion that once women gain entry into public settings they obtain access to power and resources in society" (Bose 1998: 366). Such an assumption of equal access to resources and power is also at the base of Habermas' theory of communicative action (Habermas 1981). However, as Bose (1998: 266pp.) continues to analyse, such a starting point fails to consider differentiated power structures within a society. Accordingly, in this research I will critically assess the hierarchies of spaces and underlying power structures in the context of the empirical findings and the specific local setting of this research in Bangladesh (see Part II and Part III). The underlying assumption is that what constitutes public and private may differ between societies and even between sub-groups of a society depending on their socio-cultural values and traditions.

2.1.2 Public space and livelihoods

Public spaces as delineated above are at the heart of urban everyday life and in their usage much of a society's social relations are (re)produced. Public spaces, for example, contain expressions of gender relations in a society that can be read from the presence of men and women at different times as well as from their behaviour when in public. One of the most visible economic activities based on access to public space is street vending in open markets, along roadsides or on public squares. A rich body of literature is concerned with these activities and the underlying conditions for accessing such opportunities (for example Hansen 2004 on Lusaka; Anjaria 2006b on Mumbai; Etzold et al. 2009 on Dhaka; Brown et al. 2010 on Senegal, Ghana, Tanzania and Lesotho or Ha 2009 on Berlin to name just a few).

Access to public space for all sorts of economic activities is especially important for the livelihoods of the urban poor in many countries of the Global South. Using public space is a way to make a claim to participating in the city's economy, an economy that often does not provide a 'formalised space' for the urban poor. Many would not be able to afford to operate a shop in the municipal markets or in a shopping mall simply due to lack of financial resources. Furthermore, public spaces have increasingly become commercialised and this reduces the opportunities for 'access free of cost'. The most excluding of these developments is the *de facto* privatisation of public spaces as a result of the development of shopping malls, a phenomenon also shaping urban development in South Asia considerably. Access to public space thus has to be regarded as a key component of physical capital needed to perform any economic activities (Brown, Lloyd-Jones 2002: 191). Gehl (2007: 16) refers to these as "necessary activities" and analyses how public space in countries with less developed economies is still dominated by such indispensable activities necessary for a household from an economic point of view. In comparison he analyses how city life in Europe was transformed considerably during the 20th century. While around 1900 urban public space was dominated by work-related activities essential for the residents, i.e. necessary activities, the picture changed towards the end of the century with outdoor space being mainly used for recreational, i.e. optional, activities (Gehl 2007: 17).

Accessing public space, however, goes beyond understanding it as an arena for only economic activities. It is a fundamental resource also for social activities as well as for satisfying housing needs in the absence of sufficient 'formal' housing opportunities. The need of, especially, the urban poor to use public space for livelihood activities arises from different preconditions. Private space of the urban poor is often characterised by high densities and crowded living conditions and thus is "more restricted and fragmented than that of high income groups" (Brown, Lloyd-Jones 2002: 192). Public space, therefore, has a compensatory function for inadequate indoor spaces (Ghafur 2005: 36) and is of high social and economic importance as a place for trade and communal activities (Brown, Lloyd-Jones 2002: 192).

The space-dependency and spatial manifestation of livelihoods is what I refer to as the 'spatiality of livelihoods'. The origin of the term livelihoods goes back to the livelihood framework, a household-based approach for understanding the situation and actions of poor people which originates from research on poverty in rural areas. According to Chambers and Conway (1992) "a livelihood comprises the capabilities, assets [...] and activities required for a means of living. A livelihood is sustainable when it can cope with and recover from stresses and shocks and maintain or enhance its capabilities and assets both now and in the future while not undermining the natural resource base". In the past decade there has been a growing amount of research on livelihoods in urban areas (see for example the contributions collected in Rakodi, Lloyd-Jones 2002 and Brown 2006b; Sheuya 2004). Urban livelihoods differ considerably from rural livelihoods when considering the available, required and accessible range of assets – namely human, social, physical, financial and natural capital. Natural capital, e.g. land for agricultural production, is of eminent importance for rural livelihoods. On the contrary, access to infrastructure as well as to large concentrations of people and services is most important for urban livelihoods (Brown, Lloyd-Jones 2002: 191). However, while this serves as a base for using the term livelihoods, I do not use the livelihoods framework as an analytical concept in this research but 'livelihoods' is used to denote a range of activities that are carried out by an individual person – a perspective that in my eyes offers much more understanding of the 'spatiality of livelihoods' than the household perspective. The latter fails to recognise the social relations among family members, and especially when exploring the spatiality of livelihoods from a feminist perspective the household-based approach would conceal certain issues. Furthermore I am also critical of capitalising every household's asset, and thus refrain from measuring such capitals – which might even be extended to, for example, include political capital as the capacity to claim citizenship – in this research. What is important then in this research is to understand how the production of (social) space determines the spatiality of livelihoods. Let me explore this notion further.

The most important feature allowing for space-based livelihoods and shaping the spatiality of livelihoods is not the mere existence of public space, but access to public space. Public space is a resource over which the urban poor can rarely execute direct control because it is in the private ownership of affluent households or controlled by influential individuals or organisations. This means that access to this resource may be denied through regulation, cost and social exclusion (Brown, Lloyd-Jones 2002: 188). De Haan and Zoomers (2005) identify the inclusion of access as one of the main challenges for future livelihoods research. They further put emphasis on the 'conceptualised space' (in Lefebvre's sense the 'representations of space') and the notion of power in saying that "[a]ccess to livelihood opportunities is governed by social relations, institutions and organizations, and it includes power as an important explanatory variable" (de Haan, Zoomers 2005: 44). The notion of some additional order regulating access to public space and determining everyday life is also vividly expressed in Lefebvre's observation of a Parisian road junction 'Seen from the Window':

> "The succession of alterations, of differential repetitions, suggests that there is somewhere in this present an order, which comes from elsewhere. [...] Therefore, beside the present, a sort of presence-absence, badly localised and strong: the State, which is not seen from the window, but which looms over this present, the omnipresent State.
>
> Just as beyond the horizon, other horizons loom without being present, so beyond the sensible and visible order, which reveals political power, other orders suggest themselves: a logic, a division of labour, *leisure activities* are also **produced** (and productive), although they are proclaimed *free* and even 'free time'." (Lefebvre 2004 [1992]: 32, original emphasis)

Accordingly, what matters for the spatiality of livelihoods is the way a society is organising and producing space, and how this social space relates to what is experienced on the ground. The reality of access being governed by institutions and the notion of a social order beyond the scene leads to the notion of contested space. The governing institutions and social order could refer to the 'omnipresent state' as in Lefebvre's writings in a Marxist tradition, but it could also be a conglomerate of other institutions. For example, Bayat (2010: 139) refers to moral-political authority which could mean state power, but also the authority of individuals or socio-political movements with a particular doctrinal paradigm.

Different values attached to space – these may be economic or symbolic – by various actors frame contested space. Prominent examples for the contested spaces that entail the exclusion of the poor from parts of the city are the privatisation of public spaces in gated communities or private shopping malls, or large-scale transport projects such as highways and flyovers which leave no space for street vending activities. Different quarters of the city may be subject to different kinds of control – highly controlled areas often include the Central Business Districts, but despite a lower level of control in low-income areas, payments may still have to be made to local syndicates, police or other officials (Brown, Lloyd-Jones 2002: 192). Bayat (2010: 12) identifies how "outdoor spaces [...] serve as indispensable assets in the economic livelihood and social/cultural reproduction of a vast segment of the urban population, and, consequently, as fertile ground for the expression of street politics". In his investigation of the production of spaces in Mexico, Hernández-Bonilla (2008) analyses how public spaces in the *colonias populares* are a source of conflicts and contestations between established residents/communities, internal and external agents (other residents, political actors, local leaders or municipal authorities), who aim to privatise the collective resource of public space. Such contestations lie at the heart of the spatiality of livelihoods and accordingly the negotiations of use rights for public space.

A final note concerns the use of the terms 'social space' and 'public space' in this research. The term social space is used in an encompassing way to denote the space produced by society and at the same time shaping social relations (see footnote 1). Public space is the space (spatial unit) at the focus of this research. It can be understood as one product or result of the production of (social) space in a society, and at the same time as an explanation for social relations.

2.2 CONTESTED SPACE AND THE NEGOTIATIONS OF SPACE

> As I reach the Ghat today, I find the shopkeepers alarmed. They were informed by RAJUK [*Rajdhani Unnayon Kortripokkho*, the Capital Development Authority] that they occupied public land illegally and that their activities were prohibited, and thus that they were likely to be evicted in the near future. The shopkeepers believe that this is on the initiative of two RAJUK engineers who live in adjacent upmarket apartment blocks which were built by encroaching onto the lake. Presumably the RAJUK supervisor will inform the shopkeepers prior to eviction, so that they can remove their possessions in due time. However, since 31st March some of the shops remain closed and others have been deconstructed following the erection of a signboard proclaiming the illegality of their spatial claim. Jahid expects the RAJUK people to come and put barbed wire fencing along the pavement in order to keep the shops away. Until now he and the others do not know when the eviction will take place, but they heard that this morning evictions took place nearby and RAJUK people broke down some shops over there. (summary of field note, 04.04.2010)

The continuation of Jahid's story points at a classic example of contested space: while Jahid accessed public space in order to pursue his economic livelihood activities, his spatial claim was contested by an authority, here the statutory planning agency. The differentiated power structure between the state authority and the vendors who fear state action endangering their livelihoods becomes very obvious. This sub-chapter thus sets out to explore the role of power in terms of domination/resistance, institutions, and agency to make spatial claims in the negotiations of access to public space. It further discusses the implications of changing spatialities of power for vulnerability and insecurity.

2.2.1 The spatialities of power

The notion of 'contested space' identified above as determining the spatiality of livelihoods is a vivid example of how space is the product of society and the spatial practice of a society. According to Lefebvre, the contradictions of a society become visible in space (here he refers to space in general, thus public space may be one arena where the contradictions become visible):

> "Socio-political contradictions are realized spatially. The contradictions of space thus make the contradictions of social relations operative. In other words, spatial contradictions 'express' conflicts between socio-political interests and forces; it is only *in* space that such conflicts come effectively into play, and in so doing they become contradictions *of* space." (Lefebvre 1991 [1974]: 365, original emphasis)

Sharp et al. further note how "within such contradictions of space, particular places frequently become sites of contestation where the social structures and relations of power, domination and resistance are interwoven" (Sharp et al. 2000: 26). With reference to the urban poor's livelihoods carried out in public spaces, commonly without having explicit use rights, the questions of domination and resistance and the underlying spatialities of power contained in the above statement become central. Furthermore, urban public spaces can be understood as the most visible arenas for such contestation within a society. The public space is the one where resistance and domination are most visibly expressed and can develop momentum, as for example the recent uprisings/revolutionary movements in the Arab world indicate. Here, the street becomes political and it is this visibility that Bayat also refers to as an important dimension of "street politics": "While a state may be able to shut down colleges or to abolish political parties, it cannot easily stop the normal flow of life in streets, unless it resorts to normalizing violence, erecting walls and checkpoints, as a strategic element of everyday life." (Bayat 2010: 12). Although this research is primarily about using public spaces for livelihoods and not about social/political movements *per se*, the urban poor in accessing urban public space for their livelihoods do indeed make a political statement in many countries of the Global South (and presumably beyond).

Dominating and resisting power

Lefebvre's (and similarly Foucault's) conceptualisation of space refers to power which Lefebvre especially understands as domination (by the state), but he also attaches a notion of power to what he calls 'counter spaces'. In the context of the negotiations of access to public space for livelihoods, it is important to reflect how power is not only a means of the 'dominant' but is "dispersed throughout society" (Few 2002: 31), meaning that the poor and most vulnerable do indeed possess agency in that they "always seek to negotiate options that help to secure their livelihoods" (Bohle 2007: 130). Sharp et al. (2000: 2), referring to Foucault, explicitly underline the "positive and negative dimensions" of power where power not only refers to domination but also to the "ability to resist" – and hence "can be repressive and progressive, constraining and facilitative, to be condemned and to be celebrated". Thus, they further differentiate between 'resisting power' and 'dominating power' (or alternatively in this research dominant and resistance power): dominating power is "power that attempts to control or coerce others, impose its will upon others, or manipulate the consent of others" and resisting power is "power which attempts to set up situations, groupings and actions which resist the impositions of dominating power", ranging from subtle moments to pronounced forms of social organisation and social movement (Sharp et al. 2000: 3). Viewing the entanglements of power as inherently spatial and geographical, and power as emerging from spatial constellations (Sharp et al. 2000: 25), Sharp et al. continue by referring to Lefebvre's writing on the production of space (Lefebvre 1991 [1974], see quotation above):

> "Different social groups endow space [...] with amalgams of different meanings, uses and values. Such differences can give rise to various tensions and conflicts within society over the uses of space for individual and social purposes, and over the domination of space by the state and other forms of dominating social (and class) power" (Sharp et al. 2000: 25–26)

In the context of this research it is important to emphasise the last point, i.e. "other forms of dominating social (and class) power". In the case of the neighbourhoods investigated in Dhaka, state power or the state as an actor is rather peripheral, only present in very few instances. It is mostly local forces, often characterised as informal in their activities and legitimisation, that negotiate the social space and power relations. However, dominating power does not necessarily have to be exercised by the state, and it can be imagined how local constellations bring to the front forces that can be termed *de facto* state in these informal spheres (Roy, AlSayyad 2004b), commonly the local elite groups and authorities acting as middlemen, service providers and land brokers. It thus seems applicable to return to Lefebvre to deepen the understanding of dominating power in the production of space, despite his apparent focus on the state.

For Lefebvre, the contradictions of space are expressions of and reactions to abstract space, a space which he understands as simultaneously homogenous and fractured (Lefebvre 1991 [1974]: 355). This abstract space is homogenous on a global or national scale where the state aims at abolishing distinctions and differences and establishes 'centrality'. On the local level of everyday life this space is fractured, for example by the demands of the division of labour. However, the production of abstract space which he considers the dominant spatial conceptualisation of today's society, is dominated by the rationality of the state and experienced as a repressive space (Lefebvre 1991 [1974]: 352, 358; Schmid 2005: 269). For the negotiations of access to space, this would result in a dominant narrative which shapes negotiation results. Such "a body of constraints, stipulations and rules to be followed" results in a "normative and repressive efficacy" of social space (Lefebvre 1991 [1974]: 358).

But Lefebvre's understanding of abstract space as dominant does not mean that he understands it as exclusive in relation to other conceptualisations of space (Schmid 2005: 285). He thus goes on to analyse how this abstract space carries within it a new space, the differential space[3] (Lefebvre 1991 [1974]: 52). This differential space is Lefebvre's utopia of an urban society, a space of heterotopias, where social differences are recognised and mutually accepting of each other (Schmid 2005: 290). Although in Lefebvre's analysis of the periods of history this space is considered a utopia, it is nonetheless already partially existent today (Lefebvre 1991 [1974]: 381) and refers to the 'other', to the 'peripheries' and 'heterotopias' (Schmid 2005: 277). Similarly, Foucault in his essay *Des espaces autre* (Foucault 2006 [1967]; Schmid 2005: 278) understands heterotopias as

3 It is important to note that Lefebvre's conceptualisation of space, and especially his analysis of the periods of history, society and space, are Eurocentristic and draw on narratives of the occident (Schmid 2005: 282). The idea of "the rise of the West" thus has been criticised (Schmid 2005: 282, referring to Gregory 1994) and it should be questioned whether the linearity proposed by Lefebvre is accurate for understanding the state of the world. There is a strong notion of developmentalist approaches in this periodisation of history that seems to ignore alternative development trajectories. Bearing this criticism in mind, the conceptualisations of Lefebvre nonetheless provide important departure points for understanding urban spaces and thus the spatiality of livelihoods. Lefebvre himself, while primarily focusing on Western narratives, also brings his attention to the Latin American shanty towns which for him are acts of appropriation "greatly superior to the organization of space by specialists" resulting in a spatial duality, i.e. contradiction and conflict as well as weakening of dominated space (Lefebvre 1991 [1974]: 374).

'other spaces' that radically challenge the common spaces of everyday life. Lefebvre identifies counter space as a social practice, carrying with it the possibility of change:

> "When a community fights the construction of urban motorways or housing-developments, when it demands 'amenities' or empty spaces for play and encounter, we can see how a counter-space can insert itself into spatial reality; [...] against quantity and homogeneity, against power and the arrogance of power [...]." (Lefebvre 1991 [1974]: 382)

The conceptualisation of urban peripheries has also been considered central by Roy (2011: 232) for rethinking subaltern urbanism – "as a theorization of the megacity and its subaltern spaces and subaltern classes" – and decentering urban analysis (in the sense of recognising different narratives beyond the dominant inputs to urban theories). She particularly makes reference to Simone (2010) who understands peripheries both in relation to cities at the periphery of urban analysis and as a "space in-between [...] never really brought fully under the auspices of the logic and development trajectories that characterize a center" (Simone 2010: 40, quoted in Roy 2011: 232). This very much resembles the Lefebvrian notion of counter spaces, and Simone goes on further to understand the periphery as "potentially destabilizing of the center" (Simone 2010: 40, quoted in Roy 2011: 232).

Entanglements of power

What Roy ultimately proposes is an "itinerary of recognition that is dramatically different from that of the dominant map of global and world cities" and she draws attention to an understanding of resistance not merely as a "habitus of the dispossessed" but as being entangled with dominant power (Roy 2011: 232). This is an important point, indicating an understanding of power not as a static process but as highly dynamic and relational. Thus Sharp et al. understand the geography of domination/resistance:

> "[...] as a contingent and continuous bundle of relations; a geography that enacts a contested encounter within and between dominant and resistant practices which are themselves hybrid, rather than binary, and which are contingent upon and enmeshed within social networks, communication processes and economic relations." (Sharp et al. 2000: 27)

In their call for extending the livelihoods approach de Haan and Zoomers (2005: 37) also acknowledge that "[p]ower relations are re-created in interaction and thus constitute a dynamic process of 'wielding and yielding'". Similarly, Chambers (1995), with reference to the NGO sector, identifies 'uppers' and 'lowers' in terms of the positions of power. He goes on to argue that depending on context a person can belong to the 'upper' or the 'lower', and thus refers to multiple uppers and lowers (Chambers 1995: 208). This points at the entanglements of power, and Chambers goes on to describe a scenario of 'free to spin' which moves beyond existing and perpetuating hierarchies: "Democratic empowerment entails reversals which neutralise forces of dominance and liberate, allowing freedom for relationships in all directions" (Chambers 1995: 209). Another notion pointing in this direction is the 'metaculture' in Ipsen's model of three cultures (Ipsen 2002). With reference to claims to places by dominant and minority cultures, the 'metaculture' transcends these binary oppositions (Ipsen 2002: 240). Here a link can be established to the above discussion of thirdspace and differential space, as 'metaculture' in Ipsen's writings is very much related to the integration of immigrant communities in urban space and entails a notion of difference and heterogeneity.

Based on her research on feminist (project) groups in German cities and with references to Lefebvre and Foucault, Doderer (2003: 258) came up with a model of urban-societal topology. She differentiates hegemonial and subaltern strategies of claiming space which could be read, analogue to what has been identified above, as the spatialities of dominant and resistance power. Hegemonial strategies construct and build on a politics of order where space becomes both a platform and a means of control and regulation (Doderer 2003: 263). Here, traditional norms and dominant narratives are reiterated, for example in housing and urban politics. One extreme urban expression of this is the gated community which recreates (dominated) difference among the members of society. In contrast, subversive/subaltern strategies of claiming space are based on a relative formulation of difference that is continuously de- and reconstructed, as opposed to the hegemonic essentialist (i.e. not context-dependent and not allowing multiple readings) construction of difference (Doderer 2003: 265). These subaltern strategies for example encompass specific cultural expressions as developed by third generation migrants or the production of spaces beyond dualistic constructions of gender (Doderer 2003: 265). With reference to Fraser (1993), Doderer here postulates a "differential difference", where the construction of difference does not entail an imbalanced notion of power, i.e. one actor's power to claim space does not disempower another actor from claiming space (Doderer 2003: 266). This relates back to Lefebvre's utopian notion of differential space and the sporadic appearance of spaces of resistance as part of current modes of the production of space.

A careful note on the concepts of peripheries and resistance, however, also seems necessary, especially with regard to economic activities carried out in public spaces. While Foucault's de-centering of the notion of power has informed the idea of understanding resistance always in context to power and as opposed to dominating power (Sharp et al. 2000), and Scott in "Weapons of the Weak" (Scott 1985) attaches an act of resistance to every expression of the subaltern in everyday life, the usefulness of such an understanding of resistance is questioned by Bayat (2010: 51pp.). In his analysis of prevailing perspectives of marginality he elaborates how "resistance writers tend to confuse an *awareness* about oppression with *acts* of resistance against it" (Bayat 2010: 53, original emphasis). He furthermore points out that when understanding power as de-centred this notion "fails to see that although power circulates, it does so unevenly – in some places it is far weightier, more concentrated, and "thicker", so to speak, than in others" (Bayat 2010: 54), and demonstrates how the state matters in shaping this. Instead of reading an act of resistance into 'ordinary behaviours' and confusing, for example, self-help activities that in the end contribute to the stability of the state rather than challenging its domination with acts of resistance, he thus proposes the 'quiet encroachment of the ordinary' to provide a new perspective on the relation between 'marginalised' and dominant (state) power (Bayat 2004; Bayat 2010: 55p.). A similar understanding of resistance as not necessarily challenging a dominant order is provided by the analysis of 'legalism from below' which Eckert (2006) uses to refer to urban poor in Mumbai using the tools of state law against state actors.

With the notion of the 'quiet encroachment of the ordinary', Bayat refers to "noncollective, but prolonged, direct action by individuals and families to acquire the basic necessities of life (land for shelter, [...] business opportunities, and public space) in a quiet and unassuming, yet illegal fashion" (Bayat 2004: 81). It is thus an attempt to claim the opportunity for a livelihood, but it is not aiming at political transformation, and thus Bayat refers to this as a "nonmovement" (Bayat 2010: 64). With reference to the spatiality of livelihoods, this notion is particularly relevant because not every move to generate one's livelihoods by accessing public space is meant as inherently political and as an act of challenging authority. A critical notion, however, is necessary yet again. While this strategy of quiet encroachment may be employed widely by the urban poor, it also suggests the urban poor may be relative autonomous in their actions as long as their encroachments do not become too visible. This binary between the (state) authority and the quiet encroachers, characterised as the urban poor or a marginalised middle class, does not, however, take note of an in-between power structure and actors who employ encroachment strategies despite their better-off socio-economic status. This discussion of the logics of actors and their agency will be detailed under the concept of urban informality in Chapter 2.3.

2.2.2 Institutions governing access to public space

The above discussion has outlined some constitutive elements of negotiating access to public space, especially power. What remains blurred is the institutional set-up that lies beyond the spatiality of power and the agency of single actors. Accordingly, this sub-chapter aims to discuss the importance of institutions and social relations for the production of space.

Developing an understanding of institutions

In recent years, the New Institutional Economy discourse has shaped the debate on the relevance of institutions in human interaction. Accordingly, North has defined institutions as follows:

> "Institutions are the rules of the game in a society or, more formally, are the humanly devised constraints that shape human interaction. In consequence they structure incentives in human exchange, whether political, social, or economic. Institutional change shapes the way socie-

> ties evolve through time and hence is the key to understanding historical change." (North 1990: 3)

Institutions are further differentiated by North (1990: 47) into 'formal constraints', i.e. political and judicial rules, economic rules and contracts. On the other hand these may be 'informal constraints', including codes of behaviours and conventions (North 1990: 3). North understands the differentiation between such formal and informal constraints as a continuum, and refers to the complexity of societies to exemplify the move from "unwritten traditions and customs to written laws" (North 1990: 46). However, in this research the differentiation of formal and informal constraints moves beyond such a notion of a continuum. Instead, informality is understood as a mode of the production of space, and as such can both be inherent in formal or informal constraints (see Chapter 2.3 for the discussion of the understanding of urban informality). Besides this deviation in the understanding of informality, I nonetheless draw on North's general understanding of institutions as rules of the game. Speaking with Lefebvre's terminology they could be considered part of the representations of space or conceptualised space.

De Haan and Zoomers (2005: 36), while drawing on the above definitions in their call to include the governing of access to resources in the discussion of livelihoods, criticise how the New Institutional Economy debate has not progressed from institutions to power relations. With reference to the above discussion of dominant and resistance power, institutions can largely be understood as expressions of dominant power that are employed and defined by the dominating strata of society. However, movements of resistance and counter space narratives may set out to alter institutions and thus contest and challenge the way (social) space is organised and produced in a society. Accordingly, while a specific set of institutions governs access to public space, the outputs concerning the spatialities of power may differ considerably depending on which institutions are in place and enforced by which actors and organisations (see Chapter 2.2.1).

In this research I generally refer to the understanding of institutions as outlined by North. However, I do not adopt his delineation of formal and informal institutions, as this contradicts with my understanding of informality (see Chapter 2.3.1). Social norms, or codes of conduct and conventions according to North, are not necessarily informal in the sense of how informality is understood here, but they may be based on a large consensus of a society. In differentiating institutions, I rather refer to their origin. Here the delineation of de Haan and Zoomers (2005), based on Ellis (2000) and North (1990), proves useful as it differentiates between institutions, social relations and organisations. Accordingly, I refer to institutions based on statutory agreement, institutions based on social relations and socio-cultural traditions of society (also referred to as norms in the following), and finally to institutions (rules, behaviours, codes of conduct) introduced by specific actors or organisations.

Gender relations and public space

With regard to public space gender is an important social relation and, depending on gender relations within a society, institutions develop as constraints for human interaction. These especially take the form of norms, codes of conduct and behaviour, although they may also be more 'formal constraints', i.e. when unequal gender relations are translated into legal rules. This is for example the case with inheritance laws which in many countries remain patrilineal by law, leaving widows and daughters with only limited rights to property. Accordingly, feminist urban research understands social gender relations as being inscribed into spatial structures, while at the same time spaces are understood as gendered (Becker 2008: 653).

Since the late 1970s the discourse on space and feminism has changed considerably. The dualistic construct of public man/private woman had been a decisive socio-spatial institution creating gendered spaces within cities (Terlinden 2002), manifested in a "spatialization of patriarchal power" (Soja 1996: 110). But as gendered spaces became more hybrid, these dualistic approaches of early feminist geography were criticised by emerging postmodern feminists for falling short of recognising other categories of difference and identity. In his exploration of thirdspace from a feminist perspective, Soja then argues:

> "Cityspace is no longer just dichotomously gendered or sexed, it is literally and figuratively transgressed with an abundance of sexual possibilities and pleasures, dangers and opportunities, that are always both personal and political and, ultimately, never completely knowable from any singular discursive standpoint." (Soja 1996: 113)

Such a differential notion of gender relations results in a new spatiality, according to Soja (1996: 116) "one that cannot be so neatly categorized and mapped, where the very distinction between mind and body, private and public space, and between who is inside or outside the boundaries of community, is obliterated and diffracted in a new and different cultural politics or real-and-imagined everyday life". With reference to the above analysis of the spatialities of power, this then also impacts on dominant (hegemonic) and resistance power, and Rose identifies this contradictory geography:

> "The subject of feminism, then, depends on a paradoxical geography in order to acknowledge both the power of hegemonic discourses and to insist on the possibility of resistance. This geography describes that subjectivity as that of both prisoner and exile; it allows the subject of feminism to occupy both the centre and the margin, the inside and the outside. It is a geography structured by the dynamic tension between such poles, and it is also a multidimensional geography structured by the simultaneous contradictory diversity of social relations." (Rose 1993: 155, quoted in Soja 1996: 124)

With regard to gender relations and space, the above then evokes a differentiated analysis. For example, Breckner and Sturm (2002) analysed male and female mobility patterns in German public places and found how middle-class women used these places for social contacts, while women of low income 'inconspicuously' hustled across. Accessibility of public space for women and men depends on the "social and cultural norms and the acceptability of different types of work" (Brown, Lloyd-Jones 2002: 191). Accordingly, this differs significantly around the world. A study about slum dwellers in Kolkata, which shares a similar cultural background with Bangladesh, revealed that women in their child-bearing years experience the greatest spatial confinement. Many women preferred home-based activities while activities carried out in urban public space were seen as a 'last resort' (Bose 1998). Here, the notion of surveillance is at the core of gender relations (Bose 1998).

In her analysis of everyday life in Istanbul's *gecekondus*[4], Wedel (2004) identifies how the gendered spaces within *gecekondus* result in gendered political priorities and actions. The rejection of women's demand for a weekly market is justified with reference to their sexuality which would expose them to harassment, but yet another interpretation indicates a male unwillingness to create more public spaces for exchange among women (Wedel 2004). Furthermore, leaving the neighbourhood is not a matter of course for women and, accordingly, female spaces are reduced to 'being inside' of the *gecekondu* where the lanes serve as some kind of extension of private spaces. These internal spaces for female gatherings and information exchange are, however, not comparable to the male domain of the tea houses in terms of the (spatial) scale of political activities (Wedel 2004). In her account of street girls in Yogyakarta, Beazley (2002) analyses how, due to the dominant construction of femininity in Indonesia, street girls are double-marginalised and develop their own 'geographies of resistance'. Their claim to (the production of) space is, however, limited by the stigmatisation they experience from mainstream society and from the street boys who "use difference to divide and rule" (Beazley 2002: 1679). Both examples point at institutions that limit women's access to public space and female participation in the production of space. On the other hand, many NGOs in a developmentalist context nowadays refer to the empowerment of women in microcredit schemes. This aim can be understood as being rather narrowly defined, if considering a wider approach towards a female production of space. Given the narrow definitions, microcredit schemes have thus not led to a female production of space, as such economic strategies have done little to increase women's political agency (Batliwala, Dhanraj 2007: 25–26).

In a most intriguing account of clothing in Gujarat, India, Tarlo (1996) investigates how clothing is central to identity construction and can also be used to extend and challenge identities. Tarlo's ethnographic accounts vividly exemplify the social pressure to adhere to certain clothing norms in relation to spatial spheres, especially for young unmarried and recently married women, who are supposed to adequately follow the rules of *laj*, Hindi for shame, here referring to expressing shame by veiling oneself (Tarlo 1996: 161pp.). While non-adherence can be dramatically punished, her account furthermore tells the story of resistance against

4 *Gecekondu* is the Turkish term used for neighbourhoods built 'informally', i.e. without adhering to urban development plans and building regulations. The term directly refers to the process of claiming the space for housing, as it literally means 'built over night'.

these norms by young girls who identify themselves with urban society and, accordingly, when moving to the city challenge their rural society by wearing what is considered urban (Tarlo 1996: 194). While clothing is not deterministic for identity construction, the examples provided by Tarlo indicate how it on the one hand expresses the social norms of a society, while on the other hand presents a small window for resistance, although involving the need to carefully negotiate such action.

2.2.3 Agency in negotiating access to public space

With the above discussion of power and institutions in mind, how do actors actually make claims to public space, and who is able to do so? There has been considerable research on how the urban poor strive to secure their share of city space and claim resources they are not provided with by the state. Two narratives contradict each other: the narrative of achievements by 'deep democracy' (Appadurai 2001) or in 'political society' (Chatterjee 2004) and the structuralist narrative of dependency whereby the poor are only assigned very limited agency, like in Lewis' 'Culture of Poverty' (Lewis 1959). The latter narrative has been criticised heavily and this notion does not provide any insights to this research. Instead, it is worth paying attention to the other narratives, and specifically asking which kind of agency the urban poor have, and how does it differ from those of other groups of society.

In his article 'Deep Democracy', Appadurai (2001) exemplifies how through the 'Alliance', made up of the three organisations SPARC, the National Slum Dwellers Federation and Mahila Milan[5], the urban poor in Mumbai negotiate access to land and infrastructure by producing knowledge through self-survey, organising housing exhibitions and 'toilet festivals'. By these forms of "counter-governmentality" the slum dwellers claim political citizenship and recognition at local and international levels, as the following two examples read:

> "[…] the Alliance is also keenly aware of the power that this kind of knowledge [generated through self-surveying] – and ability – gives it in its dealings with local and central state organizations (as well as with multilateral agencies and other regulatory bodies). The leverage bestowed by such information is particularly acute in places such as Mumbai, where a host of local, state-level and federal entities exist with a mandate to rehabilitate or ameliorate slum life. But none of them knows exactly who the slum dwellers are, where they live or how they are to be identified. This fact is of central relevance to the politics of knowledge in which the Alliance is perennially engaged." (Appadurai 2001: 35)

> "This is nothing less than a politics of recognition from below. When a World Bank official has to examine the virtues of a public toilet and discuss the merits of faeces management with

5 SPARC (Society for the Promotion of Area Resource Centres) is an NGO working on urban poverty issues in Mumbai. NSDF (National Slum Dwellers Federation) is a grassroots organisation of slum dwellers across India. Mahila Milan is an organisation of poor women focusing on women related issues and urban poverty in India.

> the defecators themselves, the poor are no longer object victims, they become speaking subjects, they become political actors". (Appadurai 2001: 37)

However, this celebration of the poor's agency should also be looked at with care, for such practices are not non-exclusionary and may not result in equal citizenship for everyone within a community or society.

Accordingly, Chatterjee (2004) makes the distinction between civil society and political society and it becomes clear that this is also a question of agency. Civil society, he argues, appears "as the closed association of modern elite groups, sequestered from the wider popular life of the communities, walled up within enclaves of civic freedom and rational law" (Chatterjee 2004: 4). Political society is then what is available to the poor or marginalised groups for making their claims, and, with reference to several Indian examples, Chatterjee (2004) demonstrates that not every group is able to succeed in political society. With reference to the example of the rural poor in West Bengal he states:

> "The rural poor who mobilize to claim the benefits of various governmental programs do not do so as members of civil society. To effectively direct those benefits toward them, they must succeed in applying the right pressure at the right places in the governmental machinery. This would frequently mean the bending or stretching of rules, because existing procedures have historically worked to exclude or marginalize them. They must, therefore, succeed in mobilizing population groups to produce a local political consensus that can effectively work *against* the distribution of power in society as a whole. This possibility is opened up by the working of political society. When school teachers gain the trust of the rural community to plead the case of the poor and secure the confidence of the administrators to find a local consensus that will stick, they do not embody the trust generated among equal members of a civic community. On the contrary, they mediate between domains that are differentiated by deep and historically entrenched inequalities of power. They mediate between those who govern and those who are governed." (Chatterjee 2004: 66, original emphasis)

Inherent in Chatterjee's account of political society is thus a notion of unequal citizenship, and thus a differentiated agency for claiming civic rights depending on whether one has access to civil society or political society. Baud and Nainan (2008) investigated who can participate in and make use of 'invited space', i.e. different kinds of platforms where local government in Mumbai 'invites' discussion and joint elaboration of citizens' needs for services. The findings point at a more limited agency of the poor, who negotiate in political space and are in a dependency relationship with Councillors:

> "Finally, we find two models of what we prefer to call "negotiated spaces". For low-income vulnerable groups of citizens, the "political space" remains one through which they are able – to some extent – to negotiate rights. For middle-class citizens, an "executive space" is opening up, which increases their direct negotiating power with local government and provides a basis for collective organization, expanding their rights at the city level. To what extent do we find a "democratic deficit" in these contrasting models? It is clear that different groups of citizens have different rights and [are, sic] dependent on the political will of representatives working for them. Vertical and horizontal accountability has increased for the middle class, as citizens can now hold their government to account to some extent." (Baud, Nainan 2008: 498)

Thus the setting of power relations influences the agency of the urban poor and sets limits, especially in societies based on patron-client traditions (see Chapter

5.1 for an analysis of the persisting patron-client structures of power in Bangladesh).

As indicated, being restricted to operating in 'political society' or 'political space' means that long-term gains are more difficult to achieve, however, the urban poor retain agency in certain negotiations. Especially Chatterjee's notion of 'political society', however, points at the temporality of achievements (Chatterjee 2004: 60–62; see also Chapter 2.2.4). Abourahme's (2011) account of everyday life in Palestine between refugee camp and checkpoint uncovers how, despite constraining conditions, the inhabitants of the camp continue in their everyday lives to claim and appropriate spaces by negotiating "seemingly endless liminalities" (Abourahme 2011: 459). By resorting to strategies that make use of the specific space in-between camp and checkpoint, e.g. hawking, organising traffic, organising the queue in return for payments, and demonstrating, the inhabitants display "a flexibility, a readiness to take risks, an ability to maneuver through different temporal orders and instrumentalize spatial fragmentations" (Abourahme 2011: 459) – and thus they demonstrate how they possess agency despite their marginalisation from society. Similarly, insurgent citizenship can be understood as agency of the subaltern challenging dominant orders (Holston 2009; Holston 1998; Meth 2010). The above account of domination and resistance power had already indicated the direction in which these arguments point: that those marginalised do nonetheless possess agency, whether expressed in subaltern activities, resistance power or counter spaces, or merely through strategies of coping with dominant society while not challenging it.

2.2.4 Risk and uncertainty in negotiation processes

Based on the notion of contested space and the above discussion on power, institutions and agency, I will now elaborate on how to conceptualise the negotiations of public space considering issues of risk and uncertainty. The notions of risk and uncertainty are understood here as inherent in the lack of fixed regulations and boundaries. These result in continuous contestations and changing spatialities of power, i.e. newly emerging spatial configurations and constellations as a result of changing power structures.

Inherent in the dependence on access to public space for livelihoods are the risks of being dislocated from a specific place and thus the uncertainty about whether the current livelihood strategy can be maintained, as the opening story of Jahid has demonstrated. The notion of risk here refers to the question of access rights, and thus to the conceptualised space which is produced through the various organisations and actors claiming space in the context of social norms and values. In the context of violent conflict in Sri Lanka, Bohle (2007: 130) refers to "geographies of violence and vulnerability as social spaces that have to be mapped according to the relative positions of vulnerable actors within shifting fields of power that deeply influence their abilities to live with violence". This research, however, does not focus on vulnerability to violent conflict, but to negotiations, claims and contestations of access to public space. While the subject matter differs, it can be argued that the basic characteristics of the spatiality of risk and uncertainty remain similar.

Risk and uncertainty emanating from 'shifting fields of power' are also at the heart of urban informality, here understood along with Roy and AlSayyad (2004b: 5) as an "organising urban logic" that, contrary to formality, "operates through the constant negotiability of value" (for a detailed discussion of urban informality see Chapter 2.3). One such example is the "unmapping of space" that Roy (2003; Roy 2004) refers to in relation to the territorial politics and ever-shifting spatial boundaries employed strategically by the West Bengal Left Front government. This territorial politics was based on contradicting layers of property mappings at Kolkata's urban fringes. In consequence, the vesting of land (confiscation and acquisition of privately owned land by the state based on an urban land ceiling act or in the 'public interest') resulted in high flexibility for the state in its decisions to commercialise colonies of squatters. For the inhabitants of the colonies this translated into a temporary condition and a continuous risk of dislocation. This notion of temporality is also inherent in other works. Chatterjee (2004: 60, 62), in his writings on the achievement of political society in West Bengal, underlines the temporality and contextuality of political society in being successful in claim-making. This non-permanence is also reflected in the identification of "grey areas" that lack stability in the conflict regions of Sri Lanka and that Bohle (2007: 137) sees as being the most unstable and insecure due to overlapping protection regimes. Furthermore Yiftachel (2009: 89) in the context of unrecognised Bedouin Arab settlements in Beer Sheva, Israel, refers to "gray space" as being positioned "between the 'whiteness' of legality/approval/safety, and the 'blackness' of eviction/destruction/death". Accepting "gray spaces" means that activities and populations are branded as "permanent temporariness" (Yiftachel 2009: 90), thus the concept indicates the pending character of the status-quo at any point in time and essentially uncertainty and insecurity.

2.2.5 Subsuming the characteristics of negotiations of space

Sub-chapter 2.2 has presented an array of aspects that are inherent in the negotiations of (public) space as a contested resource. Terming public space a contested resource especially refers to the different values attached to space by different groups of actors and to their competing claims to use this space (see also Chapter 0). Diverging interests necessitate negotiations of space which in consequence determine access rules and access limitations for user groups.

Both in Lefebvre's and Foucault's writings there is a notion of power that refers to domination/hegemony and resistance/counter spaces or a subaltern sphere. Such a notion of power points at the entanglements between different strata of society and the idea of counter spaces and insurgent citizenship carries with it an opportunity for change. While not all actions beyond dominant power can automatically be understood as resistance in the sense of challenging a dominant form of power, everyday life strategies of claiming spaces, e.g. the quiet encroachment of the ordinary, can be important components of the urban poor's livelihoods. Furthermore, this underlines the agency of the urban poor in negotiations – although at the same time the literature points out the limited agency of the marginalised as opposed to what is referred to as elite groups. While this research thus builds on an understanding of negotiations where each actor has options to follow an individual strategy, there are nonetheless institutions that limit each actor's strategy considerably. The gender relations of a society determine how space can be accessed by the different genders, while the space that is produced by society again influences its gender relations. Thus, beyond individual strategic choices, actors in negotiation processes are always embedded in such an institutional context that ultimately defines their range of strategies. Finally, the notion of risk and uncertainty has been identified as being ever-present in the production and reproduction of public spaces. Especially where actors make claims outside dominant spheres or outside legal frameworks, as for example the claim-making in 'political society', these are marked by a 'permanent temporariness' and require permanent re-negotiations. The next sub-chapter will provide further insights by discussing the concept of urban informality and its usefulness in understanding the production of space and the underlying negotiation processes.

2.3 FRAMING THE NOTION OF URBAN INFORMALITY IN THE NEGOTIATIONS OF (PUBLIC) SPACE

Jahid and the other shop owners at the Ghat claim this 'left-over' space in order to pursue their livelihoods. They do not have a right to this space in any formal, statutory sense – they are neither owners nor leaseholders nor tenants of this plot. Their presence may be termed informal as it is not backed by any statutory regulations, nor by a general agreement of society. Their strategy of claiming space informally is instead contested by residents of the surrounding high-income areas and RAJUK as the executing statutory institution. Part of this contestation is the signpost erected next to the shop of Jahid and others informing about a High Court verdict against encroachment:

"This is to inform the public that for the purpose of the preservation and protection of [the] Lake from illegal encroachment, [the area's] Society submitted a writ petition [...] to the High Court against Bangladesh and others [indicates any citizen of the country and the state]. The honourable High Court has given a verdict on 14 July 2009 that those people who have grabbed the land plots within the territory of [the] Lake are supposed to be evicted [will be evicted]. Moreover, the honourable High Court has ordered that it will be considered as a punishable crime to build any kind of structure on the territory of [the] Lake. Proper legal measures will be taken against those who will be involved in such an offense."

Interestingly, the same residents who submitted the writ petition to the High Court and the planning authority RAJUK have accepted encroachment onto the lake by up-market residential buildings in the vicinity. Two such ten-storey apartment blocks violate existing planning regulations and building codes, but this kind of informality, with the backing of capital, has gone uncontested. Informality has different faces and is involved in the production of various spaces – and while up-market informality is accepted, informal livelihood strategies by the urban poor are criminalised, despite their comparatively minor and less permanent encroachment. (reflections[6] on Jahid's case, 12.07.2011)

The concept of urban informality, especially if understood as an "organizing logic" and as a "mode of the production of space" (Roy, AlSayyad 2004a: 5), appears particularly useful in framing the discussion on the negotiation of access to public space. An understanding of urban informality that goes beyond seeing it as separate from 'formality' and instead focuses on the entanglements of state, civil society and the market in the continuous production and reproduction of space can significantly contribute to an analysis and explanation of urban everyday life realities. Furthermore, as the example of Jahid indicates, informality as a mode of the production of space may be employed by both the marginalised or ordinary as well as elite groups, with different implications in terms of outcomes and depending on the available sources of power.

2.3.1 From a dualistic concept to a hybrid understanding

The discussion of informality goes back to the structuralist approach of urban labour market research conducted in the early 1970s that distinguished between formal and informal economies (International Labour Office (ILO) 1972); Hart 1973). The informal economy was defined as being outside state regulation and thus used to refer to those who were not employed in the organised and regulated labour market but rather resorted to other forms of wage labour and self-employment. This structuralist notion of informality as marginality was critically taken up by dependency theorists who saw an expression of economic globalisation and global capitalism in the informal sector. This was famously analysed by Perlman in 'The Myth of Marginality', where she concluded that the *favelas* of Rio de Janeiro were not:

> "*marginal* but in fact integrated into the society, albeit in a manner detrimental to their own interests. They are not separate from, or on the margins of the system, but are tightly bound into it in a severely asymmetrical form. They contribute their hard work, their high hopes, and their loyalties, but do not benefit from the goods and services of the system. *It is my contention that the favela residents are not economically and politically marginal, but are excluded and repressed, that they are not socially and culturally marginal, but stigmatized and excluded from a closed class system*" (Perlman 1976: 251, original emphasis).

In Perlman's recent writings (Perlman 2004) and, for example, Davis' 'The Planet of Slums' (Davis 2007), marginality has re-emerged as a category of urban informality and globalisation. Such narratives of marginality, however, draw a picture of despair that does not match the dynamics of many urban processes and denies the possibility that the marginalised groups possess agency with which to change their situation (Roy 2012: 691–693; see also Chapter 2.2.3).

6 I wrote these reflections, and similarly the ones at the beginning of Chapter 2.4, when I started reviewing the story of Jahid to frame this chapter on theoretical departures.

On the other hand, neo-liberal legalist approaches celebrate the informal economy for the urban poor's creative and heroic entrepreneurialism (see especially De Soto 1989; De Soto 2000). In 'The Mystery of Capital' De Soto analyses how the urban poor, excluded from the legal property markets, resort to the extra-legal sector (De Soto 2000). Underlying this is the assumption that if the informal sector formalised, the urban poor could capitalise their assets and participate effectively in the economy (De Soto 2000; Bromley 2004). De Soto's concept of 'overcoming informality' has been heavily criticised for simplifying real world complexities of formality and informality (Bromley 2004).

A transformation of the structuralist and legalist conceptualisations of informality has been observed in the last decade. The conceptual dualism dominating the definitions of informality became blurred when it was recognised that reality does not present any fixed boundary between 'formal' and 'informal'. Informality can no longer be understood as being separate from formality, but can be seen as "a series of transactions that connect different economies and spaces to one another" (Roy 2005: 148). But this research also moves beyond the notion of a formal-informal continuum. The seeming boundary between the two spheres is fluid in nature, transforms continuously and is contested in response to the changing patterns of economic potentials, social norms, power structures, cultural customs and political arrangements. Thus, drawing on Roy and AlSayyad (2004b: 5), informality is understood as "a process of structuration that constitutes the rules of the game, determining the nature of transactions between individuals and institutions and within institutions".

2.3.2 Beyond poverty: from the 'quiet encroachment of the ordinary' to informalisation of the state and 'elite informality'

In the above sub-chapter, the understanding of informality as marginality has been criticised. Indeed this is what has dominated the urban discourse for a long time. The picture of the slum was at the heart of understanding informality. However, informality moves far beyond the picture of slums and involves not only the urban poor, as this sub-chapter aims to argue. While informality can indeed be understood as a practice of marginalised groups to make claims to entitlements and citizenship, it can at the same time be used by elite groups to make their claims to urban amenities and produce 'dominant spaces', referred to as 'elite informality'.

In the discussion on resistance power (see Chapter 2.2.1), I have already outlined Bayat's 'quiet encroachment of the ordinary' (Bayat 2004; Bayat 2010) as a strategy of claiming urban resources. Access to urban facilities, according to Bayat, is not claimed by visible confrontation or direct challenging of state authority, but through informal negotiations with single authorities. The strategies of 'deep democracy' (Appadurai 2001) or claim-making in 'political society' (Chatterjee 2004) can also be understood as informal practices of those who are rarely in possession of dominant power. Chatterjee also underlines how political society operates through para-legal arrangements (Chatterjee 2004), indicating how cer-

tain claims "could only be made on a political terrain, where rules may be bent or stretched, and not on the terrain of established law or administrative procedure" (Chatterjee 2004: 60). Altrock refers to such a notion of informality as 'supplementary informality':

> "[It is supplementary] to the registered performance of state and non-state actors in the formal world and can contribute to the well-being of a particular group that has different reasons to violate rules and to establish their own ones [...]. Informal institutions in this sense are set up when the state is too weak to implement its formal rules but there is a need for social order anyways (for instance informal land development and subdivision)." (Altrock 2012: 176)

The failure of Global South states to cope with the constant stream of rural migrants to the cities and the subsequent informal provision of housing and services is at the base of this understanding of informality. It is at the heart of megaurban narratives of slums, for example in Neuwirth's 'Shadow Cities' (Neuwirth 2006), where he documents his experiences in four settlements in Nairobi, Mumbai, Istanbul and Rio de Janeiro. It is also evident in the processes of social regulation emerging in response to fragile states as in the research of Kombe and Kreibich in Tanzania (Kombe, Kreibich 2006; Kreibich 2012). As Roy (2011) analyses, the image of the slum has become the dominant narrative of the megacity of the Global South, and 'supplementary informality' is the type of informality associated with it. But informality is not synonymous with poverty, nor is it necessarily characterised by the absence of the state, as Roy continues to argue:

> "Informal urbanization is as much the purview of wealthy urbanites as it is of slum-dwellers. [...] But they [the forms of informality adopted by wealthy urbanites] are expressions of class power and can therefore command infrastructure, services and legitimacy. Most importantly, they come to be designated as 'formal' by the state while other forms of informality remain criminalized." (Roy 2011: 233)

Here, then, is a different notion of informality, one triggered by elite groups, and one where the state is actively involved. Besides informality as an expression of the subaltern, of marginalised groups and of periphery, there has been a growing body of literature considering the active role of the state. Meagher (1995: 279), in her reassessment of the informal sector in Sub-Saharan Africa, concludes how the expansion of informality is not a process occurring "outside the state" but the result of an environment of "state complicity". In her analysis she thus underlines the usefulness of moving beyond understanding informality as a marginalised sector and towards the structuralist approach of informalisation, "conceived as a wider economic response to crisis" (Meagher 1995: 259). Roy has taken up this conceptualisation and proposes understanding urban informality as a deregulated system:

> "Deregulation indicates a calculated informality, one that involves purposive action and planning, and one where the seemingly withdrawal of regulatory power creates a logic of resource allocation, accumulation and authority." (Roy 2009b: 83)

Here, the state is not understood as an entity characterised by a limited ability to cater for its population, but as an entity using 'calculated informality' to select whom to service. Accordingly, the state formalises or criminalises population groups and spaces, ultimately producing "an uneven urban geography of spatial value" (Roy 2011: 233). Such a notion of informality as a strategy of the powerful tolerated or even supported by the state is vividly expressed in Yiftachel's (2009) conceptualisation of gray spaces undergoing dynamics of blackening or whitening, which he developed in reference to Bedouin settlements in Beer Sheva (see also Chapter 2.2.4):

> "The understanding of gray space as stretching over the entire spectrum, from powerful developers to landless and homeless 'invaders', helps us conceptualize two associated dynamics we may term here 'whitening' and 'blackening'. The former alludes to the tendency of the system to 'launder' gray spaces created 'from above' by powerful or favorable interests. The latter denotes the process of 'solving' the problem of marginalized gray space by destruction, expulsion or elimination. The state's violent power is put into action, turning gray into black." (Yiftachel 2009: 92)

State informality, however, is not only used to support elite groups. Anjaria (2006a) in his research on street hawkers in Mumbai refers to the 'predatory state'. Here, state officials gain from not legitimising street vendors' status and instead keep them in "a constant state of flux" in order to exercise their power and exploit the vendors financially, collecting *hafta*, informal payments outside official rules (Anjaria 2006a: 2145–2146). This is clearly a misuse of state power by the statutory agents which goes on unhindered where power imbalances exist. The person acting for the state uses the legitimacy of the state to continue his/her activities. Similarly, politicians often build on patron-client relationships and in return for votes and political support supply infrastructure – another informal relationship which is based on differentiated power structures (de Smedt 2009; Banks 2008). In research on Dhaka, Etzold (2012) has also noted how regular state-led evictions of street vendors in Dhaka are not carried out with a view to 'once and for all' clear the respective spaces, but rather serve as a reminder to the ordinary to continue their regular payments to the powerful, and thus as a means to reconfirm existing power relations.

In understanding urban informality as a mode of the production of space, on the one hand I include informality as a practice which refers to the struggle of the ordinary – e.g. street vendors, newly arriving rural-urban migrants seeking a chance in the city, established inhabitants of low-income settlements – in search of a share of city space, civic and political citizenship or livelihood opportunities. At the same time, I include informality as a practice of the elite or powerful, for example those backed by political parties or those who occupy prominent positions in civil society, and the state, referring to the circumvention of formal modes. Both practices of informality can involve the state, whether in active roles of state complicity or in passive roles of toleration and benign neglect. Based on the above discussion of the different 'faces' of urban informality, it is also useful to re-consider the widely used terms of 'slum' and 'informal settlement' in the context of this research (see Textbox 2). Urban informality is not understood as being attached to the spatial category of slum or informal settlement *per se*, but rather moves beyond such spatial delineations of urban neighbourhoods to refer to the processes and modes of the production of such and other spaces.

Textbox 2: Defining 'slum' and 'informal settlements' in the context of this research

This research is set within the low-income settlements of Dhaka. Often these are referred to as 'slums', for example in the regular survey by the research institute Centre for Urban Studies (CUS) in Dhaka (CUS et al. 2006).

The word **slum** has become popular both among urban researchers and practitioners, a development that is criticised by Gilbert as it transports a rather dangerous message given the numerous negative connotations it bears (Gilbert 2007). Referring to a 'slum' always involves a process of 'othering', and it seems likely that creating such differentiations by deeming some housing areas 'decent' while others are branded sub-standard will continue in every society (Gilbert 2007: 707). Furthermore, Roy has criticised how the slum has become the dominant narrative of the megacity, limiting discussion of other contributions to urban theory (Roy 2011).

My experience in Dhaka's low-income areas made me seriously reconsider the term slum, which in my eyes does not do justice to a large share of the population of this city. It labels almost 40% of the urban population and settlements as a homogenous mass, although "most 'slums' are anything but homogenous and contain both a mixture of housing conditions and a wide diversity of people" (Gilbert 2007: 704). The Bengali term *bosti* does not help much either – it originates from Sanskrit and once referred simply to neighbourhood without negative connotations, but over time it has come to be used synonymously to the English word 'slum' by urban researchers in Bangladesh[7]. I also found inhabitants of some areas that were labelled 'slums' by local urban researchers talking disparagingly about the adjacent *bosti*.

The term **informal settlement** is also misleading, for it does not specify what kind of settlement it is and what is informal and what is formal. As Altrock (2012: 179) notes: "To speak of 'formal' or 'informal' settlements obscures the fact that people living in them live in hybrid modes of more formal or informal status, actions and interactions". The term informal also does not mean that these settlements have developed completely outside of state influence, for the state is often actively involved in producing informality (Roy 2009a). Nor is the term – although this is often confused – used synonymously to slums and settlements developed under preconditions of poverty, as informal urban development is used as a strategy by all strata of society (Roy 2010: 99). This can also be observed in Dhaka, where large up-market real estate projects are developed despite being in violation of the Master Plan which had defined their sites as flood retention areas (see Chapter 5.2.2).

Accordingly, I refrain from using the terms slum or informal settlement to describe my study areas. Instead I choose the neutral terms settlement and neighbourhood, which neither carry a stigmatisation nor confuse the heterogeneity of informality. Despite my choice, it is unavoidable to use the terms slum/*bosti* in Chapter 5 when discussing the current conditions of planning in Dhaka, as these terms are widely used in the literature and surveys I refer to. Whenever I do not directly refer to literature in Chapter 5, I use the term 'low-income settlement without planning approval' to differentiate these from settlements with planning approval as well as from middle or high-income settlements without comprehensive planning approval ('elite informality').

7 In a Bengali-Bengali Lexicon (Sahityo Academy 1988 [1966]) two meanings are attached to *bosti*. The first is living area, locality and village. The second meaning translates as "filthy dirty living area of the low class" (Sahityo Academy 1988 [1966]: 1476). The Sanskrit origin *vasati*, however, does not carry this negative connotation (Turner 1966).

2.3.3 Informality as a mode of the production of space

The above discussion has revealed how space is negotiated in a hybrid institutional setting of entanglements between actors and institutions both rooted in statutory and informal spheres. If, along with Roy and AlSayyad (2004b), I understand urban informality as a mode of the production of space which is employed by a variety of actors, the question remains how this relates to the negotiations of space and spatialities of power discussed above. Two possible styles of urban informality as a mode of the production of space will be discussed here, namely informality as an expression of resistance, as a counter space or space of insurgency; and informality as an expression of domination. Although given the hybridity and entanglements of actors and power sources these are not always clearly separable, the distinction between these two styles nonetheless provides a basis for the following discussion.

Urban informality as a mode of the production of spaces of resistance can be related to the 'counter spaces' Lefebvre saw as existing in 'heterotopias' or 'other spaces' parallel to a dominant spatial order (see Chapter 2.2.1). This embraces the informality of the marginalised as in Bayat's 'quiet encroachment of the ordinary' (Bayat 2004), alternative forms of economic participation (Altvater, Mahnkopf 2003: 28), and the claims made in 'political society' (Chatterjee 2004) by which certain groups seek to appropriate a share of city space. Visibility of these modes of the production of space varies considerably, depending on whether they present a challenge to a dominant order of organising and producing space. As Jachnow analyses, if informality consists of numerous individual fates of self-organisation it can be attractive for the state because the risks of informality, such as the failure of a street vendor or the collapse of a non-authorised building, are carried by individuals (Jachnow 2003: 90–91) and thus do not necessarily affect and challenge the organisation and order of society. Roy refers to "urban informality as a way of life" and states with reference to Simone:

> "Such conceptualizations of "urban informality as a way of life" pay special attention to how the informal emerges as a response to the lack of "stable articulations" of "infrastructure, territory, and urban resources" and becomes a "generalized practice" of "countering marginalization" (Simone 2006). In doing so, they signal that the informal is then an alternative urban order, a different way of organizing space and negotiating citizenship" (Roy 2012: 695)

Secondly, informality as an expression of an alternative urban order can be a form of insurgent citizenship challenging dominant authorities, as for example the criminal gangs in Brazil claiming access to urban space and resources (Holston 2009), but also the attempts at 'deep democracy' of slum dwellers in Mumbai (Appadurai 2001)[8]. Such movements of insurgent citizenship claim their legitimacy from a discourse on the right to the city and the right to equal citizenship.

8 Insurgent citizenship, however, does not necessarily have to operate through informality as a mode of the production of space. In an interesting account of how the urban poor claim citizenship in India, Eckert (2006) analyses how marginalised groups make use of the law

Resistance and the production of counter spaces is however only one possible expression of urban informality. At the same time, urban informality can be a mode of the production of space characterised by dominant power, operating via the 'representations of space', i.e. the 'conceptualised space' in Lefebvre's triad. Such dominant power may include the state and state actors, but it can also include other actors able to draw upon similar resources of power and legitimation. In Chapter 2.3.2 I already introduced the state as an actor of urban informality. Such informality by the state can become an expression of domination, or as Eckert puts it in relation to police activities in Mumbai, "[t]he state's patterns of domination are often shaped by extralegal (or illegal) practices" (Eckert 2006: 45). In her research on the food production trade in Bissau, Guinea-Bissau, Lourenço-Lindell also observes how state officials operated "in both the formal and the informal economy and thrived on the manipulation of the two" (Lourenço-Lindell 2004: 89).

Urban informality as an expression of domination is, however, also employed by criminal cartels and in patron-client relationships. In Mexico, the *cacique*, leaders of political parties or mafia-like organisations, are involved in resource allocation and commonly work for political parties or other interest groups (Jachnow 2003: 84), while in Bangladesh the *mastan* is in a similar position of power (Banks 2008; Siddiqui et al. 2010).

While the above discussion has outlined the styles of urban informality as a mode of the production of space, it has not yet addressed the impacts these have on the production of space. Although informality allows the marginalised to make certain spatial claims, spatial arrangements tend to be built on rather insecure conditions (see also the discussion on risk and uncertainty in Chapter 2.2.4) and especially so if urban informality is applied by dominant groups and the state. For example this has been referred to as a condition of 'permanent temporariness' (Yiftachel 2009: 90) and 'unfixed values' (Roy 2004). In her research on Kolkata, Roy finds how the government operates through an 'unmapping of urban space' and thus produces an "uneven geography of spatial value, the fractal geometry of regulated and deregulated space, that is the landscape of urban informality" (Roy 2012; see also Chapter 2.2.4). Altvater and Mahnkopf also point out the insecurity of informal arrangements by saying that human security is especially endangered when universal values of statutory administration are no longer in place. Security is then achieved in an environment of corruption and clientelism, determined by inclusion in vertical networks, leading the authors to ask whether existential insecurity and reduced human security are elements of a new form of ruling (Altvater, Mahnkopf 2003): 18p.).

On a concluding note, impacts of urban informality as a mode of the production of space are characterised by a flexibility of arrangements, while the effects this flexibility has depend on the accessibility of power and the influence actors

against state institutions acting extra-legally in what she refers to as "citizenship as resistance" (see further discussion in Chapter 2.4).

can exert in the 'representations of space' as the determining dimension of the production of space.

2.4 SPATIAL JUSTICE: ACCESS TO PUBLIC SPACE AND THE RIGHT TO THE CITY

> Coming back yet again to Jahid's story, what then is the 'right to the city', and who can actively claim full citizenship and who is denied full citizenship? The narrative vividly illustrates the differentiated citizenship existing in society, which allows some groups to make their claims to already designated urban space and resources, while others are denied similar access. Here, the question emerges as to what constitutes spatial justice, and how can spatial justice be achieved in an arena characterised by deeply differentiated power structures? (reflections on Jahid's case, 12.07.2011)

As the above reflections indicate, this sub-chapter ends on a more normative note concerning the potential role of politics and urban planning for the production of 'spatial justice'. Based on the previous discussion of dominant and resistance power and resulting exclusions and inclusions, especially with regard to whose spatial claims and informality are accepted and appreciated and whose are condemned and criminalised, the question arises as to what constitutes spatial justice and how it can be achieved. The notion of spatial justice is closely tied to the debate about the right to the city and the right to equal citizenship. But before I discuss these action-oriented movements and the potential contributions of politics and planning, I aim to discuss the meaning of spatial justice, with specific emphasis on the spatial component.

The spatiality of (in)justice

In his recent publication 'Seeking Spatial Justice' Soja (2010) attempts to theorise spatial justice. He starts with the example of the successful court case of the Bus Riders Union in Los Angeles. The Bus Riders Union (used synonymously for a coalition of actors) won the case against the Metropolitan Transit Authority, with the outcome that the bus network had to be extended to improve services in the inner city and for the urban poor and ethnic minorities, instead of investments being made in a fixed-rail network that would serve the already wealthy and the marketing of the city from a global perspective. Transport planning in the metropolitan region had been discriminatory for decades, and Soja describes the results accordingly:

> "The outcome of this socially and spatially discriminatory process was an unjust metropolitan transit geography, favoring the wealthier, multicar-owning population in the suburban rings over the massive agglomeration of the immigrant and more urgently transit-dependent working poor in the inner core of the urban region." (Soja 2010: x)

The example on the one hand indicates the spatiality of (in)justice and on the other hand presents "an exemplary model of successful urban insurgency in the search for racial, environmental, and spatial justice" (Soja 2010: xviii). While insurgency will be discussed below in the context of the right to the city and equal citizenship, it seems necessary to go into further detail about the spatiality of (in)justice. As has been analysed above (see Chapter 2.1.1), space in this research is understood as a social product and as shaping social relations. Soja's understanding of spatial justice is also founded on the reassertion of spatiality and thus he argues that:

> "[...] the spatiality of (in)justice [...] affects society and social life just as much as social processes shape the spatiality or specific geography of (in)justice. (Soja 2010: 5)

Soja's understanding of spatial justice thus includes an explicitly spatial process-oriented component, where unjust spatialities/geographies are produced and reproduced over time. The commercialisation and privatisation of public spaces and the primacy of property rights can be understood as expressions of power and 'boundary making' that produce unjust geographies (Soja 2010: 44–45). The critical geography perspective that understands processes of the production and reproduction of (in)justice as inherently spatial also means that spatial injustices will always exist, that "location in space will always have attached to it some degree of relative advantage or disadvantage" (Soja 2010: 73). Accordingly, what Soja aims at in 'Seeking Spatial Justice' is not simply redistributive justice, but a process-oriented acceptance of difference, which I have also discussed in Chapter 2.2.1 especially with reference to 'differential space' as the utopia of an urban society where social differences are recognised and accepted (Lefebvre 1991 [1974]) and 'differential difference' (Doderer 2003). In contrast the liberal egalitarian theory of justice as developed by Rawls in 1971 focused on distributive justice or social justice as an outcome and without an assertive spatial component concerning the processes of how justice is produced and reproduced over time (Soja 2010: 76pp.).

The process-oriented concept of justice and differential space is also inherent in Harvey's contributions on the right to the city, understood not only in terms of "a right of access to what the property speculators and state planners define", but as "an active right to make the city different" (Harvey 2003: 941). As Soja points out, the struggle over the right to the city "aimed in part at a fair and equitable distribution of urban resources but even more so at obtaining the power over the processes producing unjust urban geographies" (Soja 2010: 83).

Claiming the right to the city and citizenship

In the past years movements claiming the right to the city have been ever-increasing on all scales, following the formulation of the World Charter on the Right to the City in 2004. But even earlier Lefebvre's original idea, *le droit à la ville*, had been taken up and provided the base for rights to the city movements and the claims to citizenship. The Bus Rider Union's case is an example of claiming the right to the city in a strategic coalition of various actors and by legal means; however, many rights to the city initiatives are rather localised movements of resistance to oppressive regimes and institutions. In his account of blogger's spatial practices in pre-Arab Spring Cairo, Fahmi (2009) analysed how blogger activists created spaces of resistance on central public spaces and thus made claims to their right to the city. These spatial claims were then followed by the ordinary who increasingly started to express their claims for rights, for example to housing and salaries in the public sphere.

Claims to the right to the city and equal citizenship have been analysed by Holston (2009) in the context of rights talks by criminal gangs in Brazil as expressions of insurgent citizenship. According to him the 'peace of the street' was only possible in a system which sought to maintain a discriminatory and differentiated citizenship (Holston 2009: 20). Political democratisation, however, has resulted in increased everyday violence and injustice against citizens, a continuation of differentiated citizenship, and he is critical of democracy's effects to create a 'just society' in Brazil:

> "Brazil's democracy has thus far produced a dangerous, hybrid space of citizenship as a sphere of social change in which the legal and illegal, legitimate and criminal, just and unjust, and civil and uncivil claim the same moral ground of citizen rights and respect by way of contradictory social practices. This conjunction of opposites is certainly perverse. Nevertheless, it also indicates a fundamental characteristic of democratization in Brazil and elsewhere: the equalities of democratic citizenship always produce new inequalities, vulnerabilities, and destabilizations, as well as the means to contest them." (Holston 2009: 16–17)

The statement furthermore underlines the process-oriented component of (spatial) justice which needs to be continuously negotiated, as spatialities of (injustice) are continuously produced and re-produced.

Equal citizenship does not necessarily have to be claimed by extra-legal acts of resistance. In an interesting account of how the urban poor claim citizenship in India, Eckert (2006) analyses how marginalised groups make use of the law against state institutions in what she refers to as "citizenship as resistance". In one of the cases, she analyses a group of local youths who fight against illegal liquor dens in their neighbourhood. Their approaching the police as a statutory actor brought no results, as the police themselves were involved in the businesses. Subsequently, the youths made use of the law and the police reacted in some cases, while at the same time charging the youths in several cases causing some of them to be brought to jail. Over time, the group of local youths extended their repertoire of legal instruments and were able to differentiate between illegal and legal grounds for arrest. Consequently, they extended their knowledge of legal issues and became legal advisors in many cases within the neighbourhood concerning, among other things, land rights and land tax. Here, the claim to citizenship is approached via legal institutions against state actors, or as "legalism from below" (Eckert 2006: 54). While a claim to citizenship and a resistance strategy, in making use of statutory institutions this initiative can be understood to reproduce the dominant order (Eckert 2006: 68) rather than challenging it in an insurgent claim of citizenship. In the following I will discuss how urban planning can contribute to spatial justice, which includes the notion of equal citizenship and equal opportunities to access common resources.

The (potential) role of planning in producing spatial justice

Above I have discussed the idea of spatial justice and elaborated the need to understand this in the context of both time (process) and space. Any process of urban planning produces a specific (public) city space, and this distributive effect of planning (Davy 2009: 234) evokes the question of how planning can influence the production of spatial justice. As the above discussion on urban informality and especially on 'elite informality' has indicated, urban planning can widen the gaps between those included in society and those excluded from society.

In his analyses of the production of spaces of wealth and spaces of poverty and the potential effects of a socio-ecological land policy to reduce injustices, Davy (2009) concludes that spatial planning cannot end poverty. He states, however, that "planners at least can avoid creating spaces of wealth at the cost of the poor [...] [and] can increase the opportunities of the poor to profit from rural or urban land uses" (Davy 2009: 254). With reference to a diversity of property rights not limited only to private property but also including spatial commons, Davy calls for a polyrational approach to planning. In another article, Davy had criticised the monorationality of planning in ignoring differences, and instead called for polyrational planning which "like consensus building, rather emphasizes different perceptions of urban land and nurtures a culture of difference" (Davy 2008: 309).

I would like to close this discussion of the possible contributions of planning with Healey's article 'On creating the 'city' as a collective resource' (Healey 2002), which rounds up the above discussion on the need for the right to difference as part of the right to the city and spatial justice from an urban planner's point of view. In her article, Healey calls for an exploration of 'multiple readings' of the city in the context of strategic urban governance and recognises the risk of domination by powerful groups:

> "[...] the effort of encouraging multiple readings of the city has the potential to generate not only knowledge resources but integrative processes in the diffused power context of urban governance, strengthening consciousness of an urban public realm. Such efforts may thus provide the basis for mobilising a plurality of actors to act 'for' particular qualities and attributes of 'the city'. But given the plurality, and the risk of dominance by a narrow conception promoted by a powerful interest-group, such processes of reading, sharing and mobilising should keep alive the multiple readings, the 'many cities in one city', and should acknowledge the many institutional sites where such readings and mobilisings may happen. In this context, the city planning office, or the mayor's strategic planning group, should not expect to produce the definitive 'reading' and plan. But they may have an important role in sustaining debates about the city and helping participants in governance arenas across the area to articulate their view and debate its implications with others." (Healey 2002: 1786)

In her further statement the relation of space and time is most explicit, and thus also the processes that I analysed above with reference to Soja (2010) as producing and re-producing spatialities of (in)justice:

> "Such a governance activity needs to have a knowledgeability to grasp multiple spatialities and temporalities, and an imaginative capacity to appreciate the complex ways in which actions taken now may play out through space and time" (Healey 2002: 1786).

Healey's strategic urban governance to create the city as a collective resource thus suggests a way how different actors, among them urban planners, can contribute to creating spatial justice, namely by "mobilising attention to the diverse and contested nature of meanings and representations of the city" (Healey 2002: 1786).

3 RESEARCH FRAMEWORK

Based on the theoretical considerations outlined in the previous chapter, this chapter sets the framework of the research. It presents the research objectives and underlying central assumptions (3.1) as well as the research questions (3.2). Furthermore it sets the epistemological and methodological framework (3.3) before the subsequent chapter focuses on the details of the research methodology and methods.

3.1 RESEARCH OBJECTIVES

In the previous chapter I identified a number of research gaps as well as ongoing debates on the production of public space and urban informality, with a specific focus on rapidly urbanising economies, to which this research can potentially contribute. These are:

- the limited research on the importance of the spatial component of urban livelihoods and specifically the notion of access to public space, i.e. the **spatiality of livelihoods**;
- a growing body of literature on the contested nature of (access to) public space that acknowledges how various actors engage in **negotiation**s to maintain, secure and/or claim access to public space as a livelihood asset, and how both dominant and resistance power are employed throughout negotiation processes;
- an ongoing debate on the conceptualisation of urban **informality**, not as distinct from the state but rather as an 'organising logic' and a 'mode of the production of space' (Roy, AlSayyad 2004a) intertwined with 'formality' and the state in a hybrid relationship of 'state complicity' (Meagher 1995); and
- an ongoing debate on **spatial justice,** the notion of citizenship and the right to the city, especially with regard to urban groups at the 'margins', whose interests tend to be neglected or excluded by urban planning and the 'dominant' strata of society.

These four topics provide the main basis for this research and thus inform the main objective: to explore, analyse and understand the importance of urban public space for the everyday life of urban dwellers (spatiality of livelihoods) and the mechanisms of how access to public space as a livelihood asset is negotiated among the actors (negotiations) in an environment characterised by informality as a dominant mode of the production of space (informality). This objective is pursued through consideration of two predominantly low-income settlements in the megacity Dhaka, Bangladesh, as study areas, which have not been planned according to the existing planning legislation but have developed 'informally', i.e.

regulated mainly by community actors and institutions, but not necessarily without the involvement of the state. A second objective that results from the previous analysis of research gaps and relevant theoretical debates is more normative and will be discussed briefly, mainly in order to outline important issues for constitutive research: the discussion of how the findings relate to concepts of spatial justice, and what could be the role of planning in establishing and guaranteeing equal rights to the city and equal citizenship for all urban dwellers.

3.2 RESEARCH QUESTIONS

The objectives can be operationalised into research questions that relate to the four main topics identified above, namely the spatiality of livelihoods, negotiations, informality and spatial justice. Given the grounded theory approach to the research (see Chapter 3.3), these research questions have to be read and understood as consecutive. They became more focused during the research process, as a result of the continuous analysis of the empirical results and inclusion of emerging categories in the investigation.

Spatiality of livelihoods

- How 'space-based' are urban livelihoods?
- For which purposes are public spaces needed and used, and by whom – with specific regard to gender and socio-economic status?
- Which importance/meaning/value is attached to public space in everyday life?
- What makes space a contested resource?

These research questions take up recent research on urban livelihoods (for example Rakodi, Lloyd-Jones 2002) and ask about the importance of access to public space for urban dwellers. Given the high densities prevalent in many low-income settlements in Dhaka, the availability and accessibility of open spaces seem to be of crucial importance for the urban poor's livelihoods, e.g. for street vending, small market stalls or production activities. As first empirical investigations on the importance of space, analogue to the literature review, soon pointed to the contested nature of public space, this triggered development of the research questions on 'negotiations'.

Negotiation

- How is access to public spaces negotiated among actors with diverging interests?
- Who are the actors and what are their specific interests, sources of power and legitimisations in the negotiation process?

- Which access arrangements exist concerning use of/access to public space and what are the consequences of these for the inhabitant's livelihoods and access to public space?
- How are the negotiations influenced by social relations and institutions in the perception of the negotiating actors?

The contested nature of public space points to the necessity of negotiating access among actors. Only recently has livelihood research turned to the question of access, which is seen as being "governed by social relations, institutions and organizations" and includes power as an "explanatory variable" (de Haan, Zoomers 2005: 27). Accordingly, this research sets out to investigate how actors negotiate access to space, including their strategies, sources of power, agency and legitimation in negotiation processes as well as the transparency and visibility of such processes. Access arrangements – whether formal, informal or 'hybrid' – can reserve street space for traffic use based on police orders or deny hawking in the interest of shop owners; monetary and non-monetary costs may be fees payable to the municipality, a land lord or 'muscle man', or extra time to guard a place against invasion; social norms and practices can restrict access to public spaces for poor women or reserve street sections or crossroads for weddings or religious celebrations. Negotiations and the strategies of actors are furthermore always shaped by the social relations and institutions acknowledged and developed by a specific society. The interpretation of negotiations draws on the notion of the 'production of space' (Lefebvre 1991 [1974]) to elaborate how a society produces its own (social) space.

Informality

- How can informality be conceptualised within the negotiation processes?
- How is the state involved in informal local negotiation processes?
- Which spatial patterns are produced by the logics of informality, with specific regard to land use transformations and urban densities?

Dhaka's low-income areas are seldom recognised by official authorities and thus the statutory planning system does not provide for open spaces such as markets, playgrounds or community squares and parks. Instead, access to space is negotiated largely among actors operating outside the statutory framework. Statutory authorities are nonetheless involved in local negotiation processes – but they do not always follow formally recognised and accepted objectives. Thus this research sets out to conceptualise the logics of informality and informalisation in the negotiations of public space and to explore the relevance of informality as "an important analytic tool in understanding the practice of planning and the production of space" (Roy 2010: 87). Furthermore the aim is to understand the outcomes produced by such 'informal' negotiation processes, especially with regard to emerging patterns of spatial and social exclusion/inclusion and uneven geographies of power.

Spatial justice

- Which 'spatialities of (in)justice' emerge as a result of the negotiations of access to public space?
- What are potential entry points for urban planning to achieve spatial justice?

The negotiation processes of access to public space hardly seem to create just spatialities, neither on a city level nor on a local (settlement) level. The current framework of an inadequate statutory planning system and a highly politicised 'informal sphere' does not enable the urban poor to claim their 'right to the city'. This research thus aims to briefly discuss potential entry points for spatial planning with regard to the findings on urban informality as a mode of the production of space.

3.3 RESEARCH METHODOLOGY

This research is epistemologically grounded in hermeneutic or interpretist tradition, where "the emphasis is on *understanding* rather than *explanation"* (Marsh, Furlong 2002: 20, emphasis in original). It is thus characterised by a predominantly inductive research approach, where knowledge in the form of theories and concepts is generated on the basis of empirical data, and thus in an exploratory process (Dick 2008: 115; Kelle 1994: 284). Consequently, in order to explore, explain and understand the everyday life realities in urban public space I chose a grounded theory approach. This approach, introduced first by Glaser and Strauss, rejected the dominant deductive approach of theory testing and instead called for a "discovery of theory from data" (Glaser, Strauss 1967: 1).

In my research, however, I employed my pre-knowledge of the topic, which explicitly includes my everyday life experiences both in Germany and during my first visits to Bangladesh, as well as existing theories, concepts and analytical frameworks to develop a profound understanding of the research problem and research gaps. The research questions were informed both by this background and by the observations made during the first exploratory field visits (see Figure 3). Thus, I did not follow a strict inductive approach as originally formulated and called for by Glaser and Strauss (1967). In the past there was indeed a strict dualism between deductive and inductive research approaches. But soon there were critiques that the grounded theory approach tended to "overstress the extent to which existing theory can be completely ignored" (Turner 1981: 228). Later especially Strauss (e.g. Strauss, Corbin 1994) acknowledged the over-emphasis on induction in his original writings. To overcome the duality of approaches, the idea of *abduction* or *retroduction* was introduced, led by Peirce and Hanson (see for example in Reichertz 2007; Kelle 1994: 308). Here, existing theories and existing knowledge, combined with knowledge generated from empirical investigations, leads to the development of hypotheses (Kelle 1994: 308) which will eventually evolve into theories. In fact, Glaser and Strauss in their research on dying hospital patients that led to 'The Discovery of Grounded Theory' (Glaser, Strauss 1967) did employ a broad theoretical framework before enhancing and expanding it based on their empirical findings (Kelle 1994: 308). Furthermore, for the data analysis they explicitly considered the theoretical pre-knowledge they had acquired (Kelle 1994: 318pp.). Accordingly, my research questions and theoretical considerations informed but did not strictly determine the categories to generate and analyse my empirical data. In an open approach towards the empirical data, I searched for the categories emerging from the material in order to contribute to the aim of grounded theory – the development of mid-range theories (Glaser, Strauss 1967: 32; also see Figure 3).

Elements from ethnographic research also influenced the methodology. Ethnographic research is a form of social research that aims to explore the nature of social phenomena with the help of 'unstructured' empirical data and thus allows development of an open set of categories (Aktinson, Hammersley 1994: 248). In this regard, it is congruent to the logic of knowledge generation involved in a grounded theory approach. Bringing an ethnographic component into research based in the urban planning discipline is a concept not widely applied. Participant observation as an ethnographic method (see Chapter 4.2.2) enabled me to generate a deeper understanding of the implications of access to public space in everyday life. While to some degree I was able to employ an 'insider perspective' to generate knowledge, it has to be noted that social research of the kind carried out here, and ethnographic research in particular, requires a comprehensive reflection of one's positionality. The objectivity of social research has been widely questioned and it has been pointed out that "the accounts produced by researchers are constructions, and as such they reflect the presuppositions and socio-historical circumstances of their production" (Aktinson, Hammersley 1994: 252). This is all the more relevant when considering that I performed this research as a woman from a Western European background and thus from a monetarily rich society, in a South Asian setting in low-income areas under conditions of monetary poverty (see Chapter 4.4 for a more detailed critical reflection of my role as a researcher).

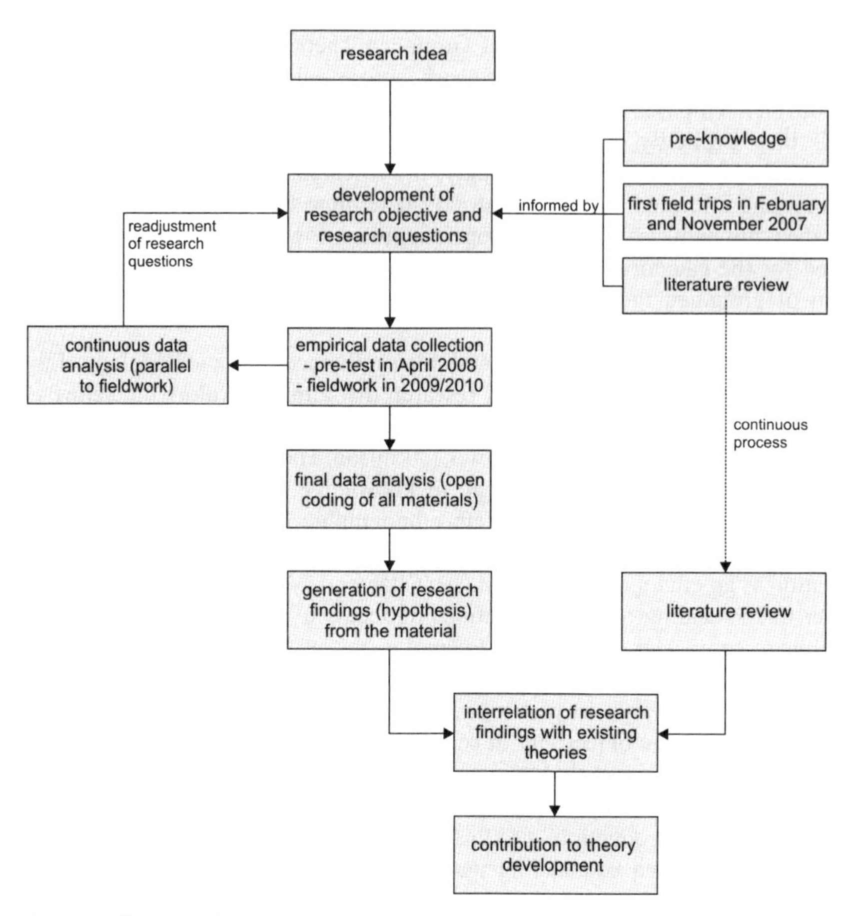

Figure 3: The research process

4 METHODOLOGICAL ISSUES

Based on the research methodology outlined in the previous chapter, this chapter presents the underlying methodological issues of this grounded theory research. In the first sub-chapter (4.1) the sampling process for this research is discussed, including the selection of two study settlements and the sampling of research participants according to the principles of theoretical sampling. Sub-chapter 4.2 discusses the research methods and tools that have been applied in the field, while sub-chapter 4.3 outlines the approach to data analysis. Finally, sub-chapter 4.4 reflects on the methodology applied, especially with regard to research ethics and positionality.

4.1 NARROWING DOWN: THE MULTI-LAYERED RESEARCH PROCESS

This research is based on a three-layered sampling process, narrowing down towards the focus of the research (Figure 4). To answer the research questions I first selected the research sites, i.e. the megacity of Dhaka as the geographical setting and the two study settlements Manikpara and Nasimgaon (see Chapter 4.1.1). Then, while conducting the fieldwork, I made choices concerning the public spaces investigated as well as the persons to be interviewed in order to explore their everyday life and the study settlements' institutional set-ups. Here I applied the method of 'theoretical sampling' (Glaser, Strauss 1967; Flick 2009, see Chapter 4.1.2).

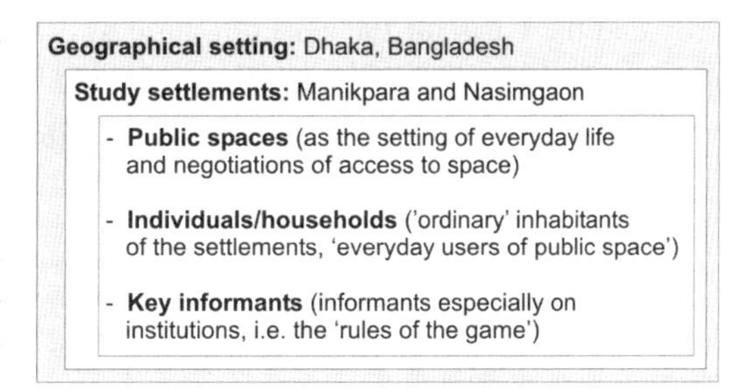

Figure 4: Layers of the research process

4.1.1 Selecting the research sites

Geographical setting: Dhaka

Geographically this research is located in the megacity of Dhaka. Dhaka was selected as the study location for the German Research Foundation's priority programme "Megacities – megachallenge: Informal dynamics of global change" which this research was part of. Being the primate urban centre of Bangladesh,

Dhaka is a rather unique case in the country with dynamics in both economics and land development that are not experienced in other cities of Bangladesh. Furthermore, Dhaka differs considerably from other megacities of the region, such as Kolkata, Delhi, Mumbai and Karachi, in terms of culture, urban growth patterns, economic dynamics and political organisation. Nonetheless, parallels can be recognised especially with regard to the behaviour of non-state actors in the production of space (see for example Roy 2003 and Roy 2009b for Kolkata and Delhi, Appadurai 2001 for Mumbai, Hasan 2009 for Karachi). Thus, while acknowledging the context-specificity of knowledge produced by this research on Dhaka, the categories emerging from research in this South Asian megacity may well contribute to theory building beyond its local context. The specific urban context of Dhaka will be discussed in Chapter 5.2 to set the stage of this research.

Selection of study settlements

For the selection of study settlements the focus stemming from the research questions was on low-income areas. Thus the 'Slums of Urban Bangladesh' mapping and census conducted by the Centre for Urban Studies in 2005 (CUS et al. 2006) provided the basis for the selection process. This survey identified 5,000 clusters of slums and squatter settlements according to the following definition:

> "[...] slums were defined for the purposes of the present study as settlements with a minimum of 10 households or a mess unit with a minimum of 25 members and:
>
> - predominantly very poor housing;
>
> - very high population density and room crowding;
>
> - very poor environmental services, especially water and sanitation;
>
> - very low socio-economic status;
>
> - lack of security of tenure.
>
> To qualify as a slum, an urban community had to meet at least four of these criteria." (CUS et al. 2006: 14).

A pre-selection based on size of the clusters using the information provided by the census of CUS (CUS et al. 2006) and Quickbird image analysis minimised the potential study settlements to 14. The two study settlements to be selected were then further required to match the following criteria:

- relatively self-dependent urban settlement, i.e. the area should reveal its own internal urban structure with a network of open spaces including streets and footpaths and areas for supply with goods for (at least) daily consumption,
- own internal organisational and institutional set-up,
- advanced stage of settlement development, i.e. not settlements in infancy stage but rather consolidated ones with established urban structures (location within the built-up city and not at the fringes) and organisational/institutional set-up to be able to investigate the research questions, and

- minimum degree of security against sudden and forced eviction to ensure the continuity of the research process[1].

Apart from these basic similarities, the selection process for study settlements followed the logic of contrasting cases (Yin 2003: 54; maximum variation cases according to Flyvjerg 2004: 426) with respect to the integration into the urban fabric of Dhaka, the integration into formal institutional frameworks and the focus of livelihood activities (see Table 1). After site visits to 13 of the potential settlements[2] Manikpara was selected for type A. Subsequently, Nasimgaon was selected as the second study settlement, type B, based on a maximisation of contrasts compared to Manikpara (type A) (further descriptions of the study settlements in Chapters 5.3 and 5.4).

Contrasting criteria	*Study settlement A*	*Study settlement B*
Integration into the urban fabric of Dhaka	– integrated through network of streets, footpaths, public spaces – building structures integrated (no immediate perception of 'slum settlement')	– development distinct from the urban fabric of the surroundings ('island effect', perception of 'slum settlement')
Integration into formal institutional frameworks	– formal institutions showing their presence in the area, involved in decision-making	– dominance of informal institutions, area largely governed outside formal decision-making systems
Focus of livelihood activities	– centred around employment in one dominant industry (e.g. leather, plastic recycling, garments)	– rather small-scale, self-employed and service-oriented livelihoods (e.g. vending, small production units, rickshaw pulling, household works, etc.)

Table 1: Contrasts for study settlement selection

After selection of the two study settlements, I familiarised myself with the local conditions through a number of exploratory site visits. Furthermore, the exploratory field visits went along with preparing base maps of the areas including the network of roads and footpaths, and the building structure. This mapping of street networks and land uses conducted jointly with researchers of the Humboldt-

1 At the time of selecting the settlements (February 2007) this criteria seemed to be fulfilled. But soon after its introduction to office in January 2007 the caretaker government started to evict illegal structures and slum areas on government land. While Manikpara is safe from eviction, Nasimgaon suddenly became threatened just before the elections were held in December 2008. Despite continuous threats since that time Nasimgaon settlement was still in place at the time of writing. For more information on the insecurity of tenure in Nasimgaon see Chapter 5.4.1.

2 One area was not accessible to foreigners because the only route led via the military cantonment.

University in Berlin in November/December 2007 provided accurate base maps for identifying public spaces of relevance for further detailed study.

4.1.2 Investigating the research questions 'on the ground'

For investigating the research questions on the ground, different methods were applied. Below, I will discuss how the combination of methods contributed to answering the research questions on spatiality of livelihoods, negotiations, urban informality and spatial justice. Furthermore I will discuss the sampling process for the data collection.

Combination of methods for answering the research questions

In order to investigate the research questions within the study settlements, a multiple method approach was employed, i.e. 'methodological triangulation' (Denzin 1970). Triangulation according to Denzin (1970: 297) refers to the "combination of methodologies in the study of the same phenomena". He differentiates four types: the use of different data sources (data triangulation), the use of different researcher groups, observers and interviewers (investigator triangulation), the use of multiple perspectives and hypotheses (theory triangulation) and the application and combination of multiple methods (methodological triangulation) (Denzin 1970: 301pp.; Flick 2008: 13pp.). When triangulation was developed as a research strategy it was largely understood as a means to reduce biases originating from single methodologies and to thus increase the validity of the research (Denzin 1970). The critiques of this understanding emanating from a positivist research tradition caused Denzin to reframe the goal of triangulation:

> "The goal of multiple triangulation is a fully grounded interpretive research approach. Objective reality will never be captured. In-depth understanding, not validity, is sought in any interpretive study." (Denzin 1989: 246, quoted in Flick 2008: 20)

In this research, triangulation of different qualitative research methods was employed to provide complementary views on the research topic (Dick 2008: 125) and to investigate the research questions from different perspectives in order to achieve more in-depth answers and contributions to theory building (see Figure 5 for an overview of the methods applied in relation to the research topics).

The spatiality of livelihoods was explored by

- observations of public spaces (semi-standardised)
  - revealing the range of everyday life activities carried out in the selected public spaces, including the dynamics throughout the day and on weekdays/holidays,
  - revealing the impacts these activities have on the public spaces (e.g. appearance, conflicting uses) as inputs for the further research on 'negotiations'
- solicited photography

  - revealing the everyday life activities and mobility patterns from the perspective of the 'everyday users of space',
  - capturing their perceptions and conceptualisations of public space,

- qualitative interviews I (follow-up/recurring interviews of the solicited photographers, informal discussions and focus group discussion),
  - deepening the findings revealed by the solicited photography,
  - discussing institutions (norms, especially social gender relations) defining access to public space for different user groups (continued below)
- participant observation
  - complementing and confirming the findings generated by employing the above methods using a '24 hours-perspective'.

The negotiations of access to space were explored by a similar set of methods, however, here the contributions of each method differed from above:

- qualitative interviews I (solicited photography, follow-up/recurring interviews of the solicited photographers, informal discussions and focus group discussion),
  - discussing access rules to public space and power sources/structures within the negotiation of access from the perspective of the 'everyday users of space',
  - discussing institutions (norms, especially social gender relations) defining access to public space for different user groups (continued from above)
- qualitative interviews II (key informants, i.e. with community members who had specific knowledge about the institutional setting of the settlements)
  - understanding the 'rules of the game' which regulate access to public space,
  - understanding the role of (political) organisations and actors in the negotiations of access to public space, including their power sources and ways of legitimising claims to space
- participant observation
  - complementing and confirming the findings generated by employing the above methods,
  - observing the everyday negotiations of access to space and the resulting transformations of public space and everyday life spatial practices in case of changing access rules.

For the two topics of urban informality and spatial justice no specific fieldwork methods were designed. Rather, the results of the fieldwork concerning the spatiality of livelihoods and the negotiations of access to public space were interpreted with a view to contributing to the discussions on urban informality and spatial justice.

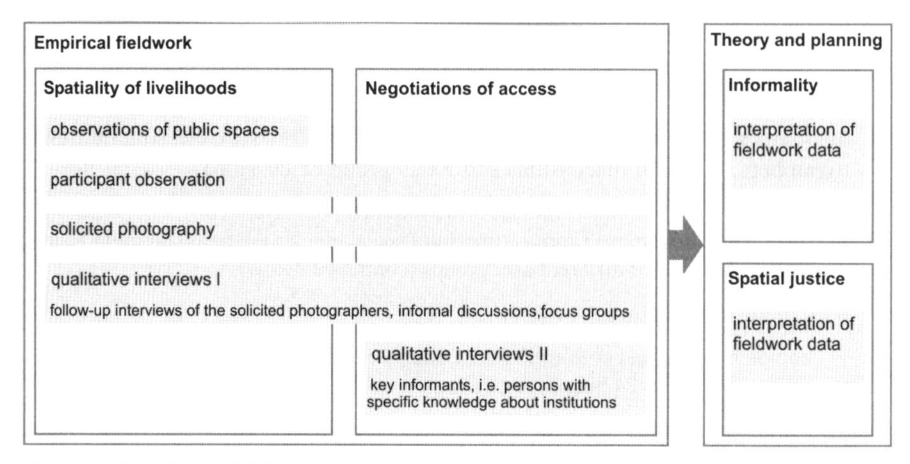

Figure 5: Overview of fieldwork methods in relation to the four research topics

Sampling within the study settlements

Within the study settlements, the research focussed on three dimensions as the starting points for data collection (see also Figure 4), namely:

- public spaces,
- individuals/household members as 'everyday users of space', and
- key informants for the institutional dimension, i.e. 'the rules of the game'.

The selection of public spaces, providing the starting point for observations, was based on the availability of public spaces in the two study settlements. For the introduction of these spaces, see Chapter 5.3.2 (Manikpara) and Chapter 5.4.2 (Nasimgaon). For selecting the 'everyday users of public space' as participants of solicited photography (and additional/recurring qualitative interviews), the method of theoretical sampling was used, as described below. The selection process of key informants was then based on the number of actors and organisations involved in negotiating access to public spaces and on the availability of such informants.

The selection of individuals as participants for solicited photography followed the method of 'theoretical sampling' (Glaser, Strauss 1967; Flick 2009). Here, the selection of cases is stepwise and analysis of the already collected data and knowledge gained determines what data to collect next (Glaser, Strauss 1967: 45). Thus theoretical sampling considerably differs from random sampling techniques which aim to create representativity. Instead it sets out to discover new contributions to theory by selecting cases that promise to be most relevant (Flick 2009: 159). Or as Glaser and Strauss put it:

> "The basic criterion governing the selection of comparison groups for discovering theory is their theoretical relevance for furthering the development of emerging categories. The re-

searcher chooses any groups that will help generate, to the fullest extent, as many properties of the categories as possible." (Glaser, Strauss 1967: 49).

The data collection phase ends only when 'theoretical saturation' has been achieved, which means that "no additional data are being found whereby the sociologist can develop properties of the category" (Glaser, Strauss 1967: 61). The principles of theoretical sampling and saturation apply to both the process of data collection as well as the process of data analysis where they inform the amount of the material processed (see Chapter 4.3 on data analysis).

Accordingly, the 'everyday users of space' were selected with a view to achieve a wide range of responses and perceptions on the research topic. The most important criteria for selecting participants were sex, occupation, location of the vending unit or workplace and general availability and willingness to participate in solicited photography and follow-up interviews (see Table 2 for the tentative selection target). A stepwise procedure was followed with three main rounds of selecting participants (April 2009, May 2009 and March 2010), each round informed by the analysis of the previous interviews and the identification of participants who promised to reveal the most relevant additions to the evolving categories.

Area	*Sex*	*Occupation*	*Location of vending unit/workplace*
Manikpara	balance of female/male participants	– plastic drying on the slopes – plastic sorting (male/female) – plastic business owner – women staying at home/working from home	– embankment slopes – Embankment Road – other
Nasimgaon	balance of female/male participants	– semi-permanent/permanent vendor/production unit in public spaces (different products) – semi-mobile/mobile vendor in public spaces (different products) – operator/owner of shops (stable vending/production unit) adjacent to public space – rickshaw puller, boat man – garment worker	– Nasimgaon Math – Khalabazar – other (living at different places, livelihoods not based on the two spaces named above)

Table 2: Selection criteria for participants of solicited photography

Despite these informed criteria, practical experience showed that availability and willingness to participate became equally important. Furthermore, access especially to two groups was limited. The first group was women in Manikpara working from home or not involved in any work for economic purposes. They refused to accept cameras because they did not leave the house and visit public spaces in their everyday life. Instead of solicited photography, a focus group was held with three women who were not engaged in work outside of their houses. The second group was the women in 'regular' employment both in Manikpara and Nasimgaon. The working times in the plastic sorting shops and factories (8am to 8pm) and the working time in the garment factories (from 8am often up to 10 or 11pm) did not leave much time for taking pictures. Although finally some women in regular employment participated, the low quality of results – relatives took a considerable amount of pictures or even all pictures – made it necessary to conduct another two focus groups (see Chapter 4.2.4).

The selection of informants on institutions was informed by the actors and organisations mentioned during the interviews with individuals in specific places, as well as by informal discussions during participant observation about the management of the public spaces under focus. The aim was to speak with as many representatives of organisations or persons with specific knowledge about the local organisations and institutions as possible. However, coverage depended largely on availability and willingness to talk. Furthermore, many of the other individuals who participated in the research also contributed information regarding institutions. Among the key informant interviews are also interviews that followed-up a specific event. For example, there was a conflict about the relocation of rickshaw garages on Nasimgaon Math and thus the rickshaw garage owners and rickshaw pullers became key informants about this conflict and the institutions at work.

4.2 FIELDWORK METHODS

The following sub-chapters will explain the research methods employed for investigating the research questions (see Figure 5). At the end of each sub-chapter the limitations experienced during the fieldwork are considered, first for each individual method and then in a general overview at the end of the chapter (4.2.5).

Figure 5 provides an overview of the field work schedule. A pre-test and exploratory period was conducted in April/May 2008. While the main field work took place from January to June 2009 and February to April 2010, the periods in between and until December 2010 were covered by regular field visits conducted by my research assistants according to my instructions and based on the progress of analysis.

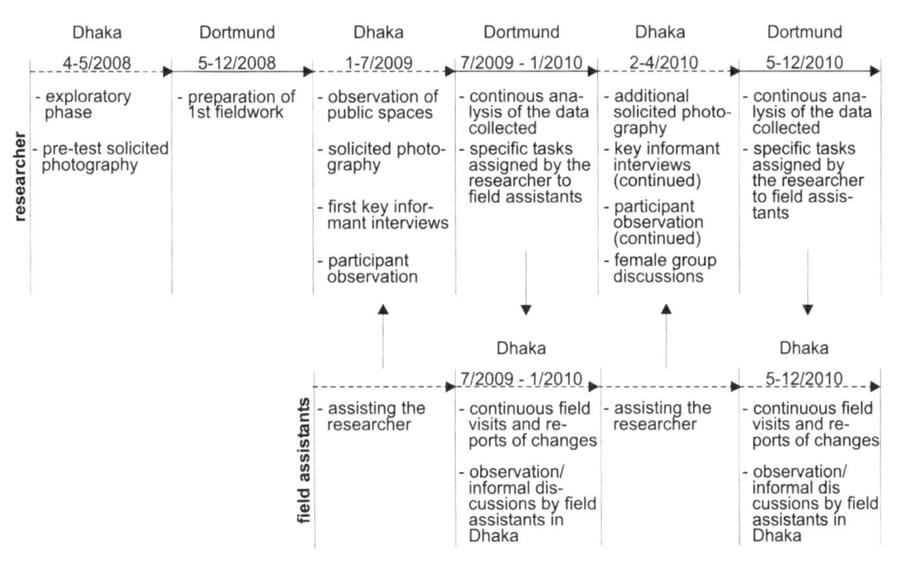

Figure 6: The fieldwork schedule of the researcher and field assistants

4.2.1 Observation of public spaces

At the beginning of my 2009 fieldwork, when I started to familiarise myself with the study settlements, I began with the observation of selected public spaces. In observing public places before going on to participant observation, solicited photography and interviews, I made use of the strength of the observational method that Adler and Adler (1994: 382) identified as the "ease through which the researchers can gain *entrée* to settings". The observations were qualitative, without pre-defined categories of measurement, leaving the researcher to search for evolving and relevant concepts and categories, and naturalistic, meaning that they followed the actors and their interactions in the natural stream of everyday life (Adler, Adler 1994: 378). However, the observation of public spaces, in contrast to the participant observations (see Chapter 4.2.2), focussed mainly on depicting the activities and physical items in a public space at a certain moment in time rather than on a continuous account of interaction and behaviours among space users. Thus they can be categorised as 'descriptive observations' for orientation in the public places investigated (Flick 2009: 288). As such they constituted the basis for more 'focused observations' carried out as participant observations once the observer became more familiar with the settings, processes and social groups involved (Adler, Adler 1994: 381, see Chapter 4.2.2).

In both study settlements the observations were carried out on a weekday and a Friday – the weekly holiday in Bangladesh due to Islam being the dominant religion. The weekday observation went on from 7am to 10pm in three-hour intervals (i.e. 7am, 10am, 1pm, 4pm, 7pm and 10pm). On Fridays, the most interesting differences were expected during prayer time, thus the early morning and late evening observations were not considered to be necessary. In Nasimgaon observations were carried out in two public places, in Manikpara in three. Each round of observation started in the same place at the preset hour and continued along exactly the same route. Thus once the activities and physical items in a specific place had been mapped and taken note of, e.g. from 7am to 7:25am, the researcher moved on to the next spot in the same settlement, where observation took place with a little time shift, e.g. from 7:25am to 7:50am.

The observation involved locating the observed activities and items on the base map (developed previously, during exploratory visits), noting down the characteristics in a table and taking photographs of activities and items. Categories only evolved during the data entry, while during observations an open approach was followed in order to capture the diversity of activities without narrowing it down to pre-defined category systems. These detailed notes about the nature of activities and number of persons involved together with the photographs informed the sampling of individuals as participants for solicited photography and provided the researcher with a vivid picture of everyday life in the observed public places. Although variation was sought with respect to weekdays and timing, the researcher is aware of concerns over validity and reliability of observations as a method (Adler, Adler 1994: 381). The observation of public places was, therefore, only one method of gathering information on the physical public space, and triangulation with other methods was employed to allow an in-depth understanding of the observed processes (Flick 2009: 285).

During observation rounds many residents were curious about what we, i.e. the research team consisting of me and two field assistants, were doing with our clipboards with maps and tables, and some expressed anxiety about eviction, immediately connecting us to the government. Once we explained about the research purpose and underlined our detachment from any government organisation we received positive reactions. These observation rounds thus provided a valuable opportunity for me to get into contact with the inhabitants more closely, as they kept seeing me continuously for two whole days in each settlement and thus got used to my presence in the area. From this point onwards, I was able to transform first encounters into consolidating relationships between the researcher and the researched (see also below, especially on identification of participants for solicited photography and informal discussions).

4.2.2 Participant observation

Participant observation –the ethnographic element – became an important aspect of methodological triangulation. After the descriptive observations detailed above I shifted to 'focussed observations' that are usually employed once "observers become more familiar with their settings and grasp the key social groups and processes in operation" and attention is directed to "a deeper and narrower portion of the people, behaviors, times, spaces, feelings, structures, and/or processes" (Adler, Adler 1994: 381).

Participant observation took place as a continuous process: whenever I went to the field I engaged in participant observation, ranging from simply observing what is taking place to informal discussions and extensive *adda*, a Bengali word that refers to "the practise of friends getting together for long, informal, and unrigorous conversations" and is often seen as something quintessentially Bengali (Chakrabarty 2001: 124). Especially in the phase of building up relationships of trust and rapport I used every opportunity to sit down with people who reacted positively to my presence and have tea while 'doing *adda*'. On some of these days of 'getting familiar with each other' I had numerous cups of the strong and sweet tea that is available for two or three Taka per cup. Taking tea is one of the social activities people engage in throughout the day, and buying tea for someone also is a gesture often employed to keep a good relationship with someone, for example with local leaders or the police. So once I had tea with one person it was common that not only one glass was bought but a whole tin can, so that tea could also be given to neighbouring shop keepers, colleagues, neighbours in the compound or friends. These informal small-talks provided valuable insights to everyday life and at the same time gave me an opportunity to find suitable participants for solicited photography and key informant interviews.

I also decided to rent a room and stay in the study settlements in order to experience a holistic perspective of everyday life. This enabled me to overcome the restriction of my presence to the daytime and early evening hours and offered me an insider perspective on the socio-spatial practices throughout a 24-hour day. This intensive exposure to the field also meant that I was able to witness the more subtle negotiation processes and make contacts with those who were only active in the evenings, e.g. the night guards and water vendors.

Fortunately, there was a family in Nasimgaon that I had become familiar with during my first exploratory visits. When I came back to the field at the beginning of 2009 I was determined to find a place to stay in Nasimgaon, but was still wondering how to realise it, especially as it was not easy to judge security if not familiar with at least the landlords. Thus I was very surprised when I found out that the family I knew had actually extended their living space during the eight months that I had been in Germany and instead of being owner-occupiers were now landlords, renting out three rooms at the back of their own. So when I visited them and asked them how they had managed to extend their compound, they offered me the opportunity to move into one of the rooms which was going to be vacated from 1st of March 2009. This was exactly what I was looking for, and thus soon after this meeting we agreed that I would move in at the beginning of March.

While this was possible in Nasimgaon, no such opportunity arose in Manikpara. I had established very good relations with some families, but the local conditions considerably differed from Nasimgaon. Manikpara is much less of a 'socially familiar' community but is characterised by a higher degree of 'strangers' publicness' (see also the analysis in Chapter 7.2), and thus women's lives are more confined to the house than to the neighbourhood. In my few visits to Manikpara in the evening or night time, I noticed that after sundown the public space became even more male-dominated than during daytime, while in Nasimgaon women still moved around at such times. Furthermore, I was told repeatedly about places that were unsafe at night where even male interviewees felt uncomfortable. The security situation and the different community structure would not have allowed me as a woman to spend my time outdoors at different times of day and evening. Even during the daytime moving around alone as a woman without a clear purpose (e.g. going to work) is not considered acceptable and honourable behaviour by the local society. Not adhering to the rules of this relatively conservative society would have resulted in a bad reputation, which I could not afford for my research. Thus my participant observations in Manikpara were all carried out during daytime visits, sometimes extending to the evening hours, but I was always accompanied by at least one of my field assistants.

When I started renting a room in Nasimgaon this was perceived as an event in the neighbourhood. At first it seemed that the whole neighbourhood came to visit me to verify the rumours they had heard about a foreigner living there. At the same time, I myself had to adapt to this new everyday life. In her participant observation during research among the Garo ethnic community in Bangladesh, Bal (2007: 17) refers to this overcoming of an "initial clumsiness" and acquiring of skills for a different everyday life as no longer "being the baby". Of course the first bucket shower on the open space behind the house, next to the latrine, fully dressed, and then needing to change right on the spot, where there was none of the comfort of seclusion a bathroom offers, was a challenge. So was the lack of privacy even at night, when I shared my bed with another two people. Although I adapted fast, I cannot deny that the possibility of having alternatives in the end, of having a choice due to financial resources, was what made me accept these living conditions rather easily. And as an outsider I was indeed 'always observed' (Forsslund 1995: 59), so much so that at the beginning of my stay I had to tell the ladies of our compound that I would like to take my bucket-shower without their watchful (female) eyes. Furthermore I had to adapt to a very different understanding of privacy, which was by far the most difficult adaptation for me with my Western European background, where multi-generational households and the existence of extended families are exceptions. My landlady found it only natural to share my bed at night to keep me company and was really surprised when I mentioned that I was accustomed to sleeping alone (however, I had my room to myself for only a few nights as the eldest daughter came back from the village because of troubles with her husband and then joined me in my room, a situation that was simply taken for granted by the family of my landlord). Similarly, when I was 'home' during daytime or in the evening, the other women and children staying in the compound dropped in for a chat, making it difficult for me to reserve enough time for documenting my experiences.

I documented the participant observations in field notes (Emerson et al. 2010), and also trained my field assistants in writing field notes when they visited the study areas during my absence. In analysing these field note accounts, I reflected my own positionality (see Chapter 4.4), considering that accounts by researchers are social constructs (Adler, Adler 1994: 387) and thus my own identity was inevitably interwoven with these field notes. Being aware how observational research can be ethically sensitive and vulnerable to ethical malpractice (Adler, Adler 1994: 387), I continuously and openly communicated my research aims and objectives, but nevertheless I had to experience the limitations of estimating the impact of a participant observer's presence in the field (see Chapter 4.4.2).

4.2.3 Solicited photography

I applied the visual method of solicited photography oriented on the application by Dodman (2003) with a view to generating an alternative insight into individuals' perceptions and use of public space for livelihood activities. The method enables circumvention of some of the problems that may be experienced when only relying on 'language-focused' methods such as interviews. By putting the perception of participants at the centre of the research and giving them the opportunity to express themselves through the means of photography, they are being considered experts of their own socio-spatial environment (Krisch 2002: 137). For implementation disposable cameras were distributed to selected participants (see Chapter 4.1.2 for the sampling of participants, and Figure 7 for the process applied) who were given the task of taking pictures of public space.

The sampling process included a preparatory phase with informal discussions – literally meaning drinking tea and 'doing *adda*' – during which I could find out whether the person considered actually met the criteria for selection (see Chapter 4.1.2, Table 2). Initially I had developed a questionnaire for these discussions, however, in reality this questionnaire only served as a reminder for me to touch the relevant points during informal discussions, while I did not fill it in during these discussions. Furthermore, these discussions served to establish a mutual relationship of trust between me and the participants, which proved important for the interview phase that followed the picture-taking. Once I considered someone's case 'relevant' for my research I handed over the camera with instructions to take pictures of public space. However, I needed to be more specific in my explanation, especially considering the need to translate a term like 'public space' into Bengali, for which there are several options. Furthermore, during a pre-test of the method in April 2008 I had experienced that although I did not

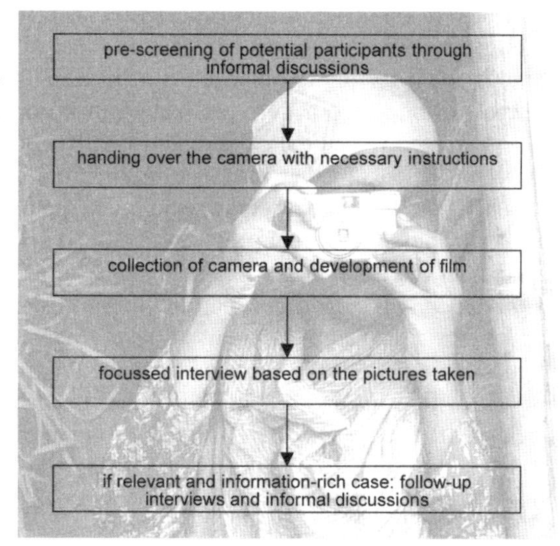

Figure 7: Process of solicited photography

want to give very detailed explanations in order to avoid participants only taking the pictures I had in my mind, it was necessary to provide some details and ideas. I thus referred to 'outdoor' spaces in order to avoid indoor photographs less relevant to the research, and to 'locations' and 'places' that the participants considered important for their livelihoods and/or that they visit regularly in their everyday lives. Furthermore, I explained that these could be photographs of spaces with both negative and positive connotations for the participant.

The resulting photographs varied considerably. Many participants were keen to take 'beautiful photographs' for me, with the aim of proudly showing the nicest places in Dhaka or even Bangladesh. I was, however, looking for both positive and negative perceptions and everyday life places. Nonetheless, the 'beautiful' images taken to show me this beauty, or the images taken by someone who travelled in order to show me a specific location, also symbolise a certain spatial perception, although often beyond everyday life spatial practices. But in quite a few cases they indicated participants' desires and aspirations, especially about the rural homesteads.

Although I had promised the participants to take the family pictures desired with my digital camera, so that the camera handed out could be used for the research-related pictures only, I had to experience that the power and force exerted by relatives, friends and local authorities who wanted their pictures taken were much stronger than anticipated. Thus in most of the films developed there were pictures taken for the reason that 'She/he asked me to/insisted that I take her/his picture'. In one case a lady was called to a political office because one of the leaders wanted to see the camera and finally also wanted his picture taken. This was very difficult for her to refuse. Apart from that, it was impossible to ensure that only the selected participant would take pictures – even if they were very enthusiastic and serious about the activity I had asked them to do. In a few cases I handed out the camera to a female member of a household, but during discussion of the pictures it turned out that the husband, or in one case the male employer, had taken a considerable amount of them[3]. 'Taking away' cameras and asking for one's picture to be taken revealed a lot about power relations both within families but also within neighbourhoods. It further indicated the inscription of gender difference in the society: when I asked the women about this they did not contest this 'male claim of space' but had given the cameras to their husbands as a matter of course.

3 In one extreme case, slightly different from the above, the *jamai* (son-in-law) of a female participant in Manikpara was present when I handed over the camera to her. He immediately got fascinated by the camera and asked me whether I could also give one to him, telling me about a marriage ceremony he was invited to next week where he would like to take pictures. I explained to him that he should not take the camera from his mother-in-law, because I had assigned her with a task. My 'bad feeling' was confirmed when I came back to collect the camera and she told me that the *jamai* had taken the camera forcefully from her, and this had even led to her husband fighting with the *jamai*. Concerning my research, the majority of the pictures were taken during the wedding and the discussion was thus not very fruitful.

Another case triggered the very interesting opportunity of having an 'internal observer' in the study settlement. One participant of solicited photography who was very enthusiastic about taking pictures especially while observing community activities and conflicts on the public space in front of his shop, asked me for another camera after finishing the first. During the course of almost one year, and especially during my absence from the field between August and December 2009, he received six additional cameras[4] which proved a very efficient way of being continuously updated about the happenings in this specific public place. This 'internal observer' method was not replicated in other public places due to a lack of adequate participants.

Once the films of solicited photography had been developed, a focussed interview about the pictures took place with each participant (see Table 3 for the number of participants and Chapter 4.2.4 for more details on the interviews). Furthermore, I entered into a long-term relationship with the participants – while often I conducted only one long interview about the solicited photographs, I kept visiting the same person during my field visits and engaged in numerous informal discussions with her/him (see Chapter 4.2.4). After the interview about the photos I handed over the set of photographs to the interviewee. Before I left, I thanked each participant for their participation and commitment by handing over a gift.

4.2.4 Qualitative interviews

Although participatory elements like the solicited photography and the ethnographic approach of participant observation played an important role during the fieldwork, qualitative interviews remained the core fieldwork tool. The following different kinds of interviews were conducted, depending on the context and type of interview partner (see also Table 3 for an overview of numbers and the List of Interviews on page):

- focussed interviews with the participants of solicited photography about the pictures taken, including narrative/episodic elements,
- problem-centred interviews including narrative/episodic elements mainly with key informants on local organisations and institutions, but also as additional interviews with participants of solicited photography,
- focus group discussions with groups of women, and
- informal discussions/unstructured interviews as part of participant observation.

4 Of these six cameras one got lost, one was mainly used by his son in the village during the *Eid-ul-Adha* festival and another one was mainly used by his friend, a political leader.

	Manikpara		*Nasimgaon*	
	persons interviewed	*number of interviews*	*persons interviewed*	*number of interviews*
focussed interviews (solicited photogr.)	10	10	17	23[5]
problem-centred interviews (solicited photography)	1	1 (incl. one Venn diagram)	4	7 (incl. three Mobility Maps)
problem-centred interviews (key informants)	8 plus friends in one case	10 (incl. one Venn diagram)	15	17 (incl. one Venn diagram)
focus group discussions	7	2	4	1
informal discussions	uncounted (field notes)		uncounted (field notes)	

Table 3: Overview of interviews conducted

If the number of interviews is higher than the number of persons interviewed some persons have been interviewed more than once. Additional to this overview, three interviews have been conducted with experts.

All interviews followed an inherently qualitative and non-standardised approach guided by the grounded theory based methodology and interpretist paradigm. For interviews this implies that the researcher does not enter the field with a predetermined theoretical structure of categories but that types and categories for theory building develop mainly during interviews with the research participants (Lamnek 2005: 348). Interviews are further characterised by flexibility and openness: the interviewer uses interview guides – except in the case of informal discussions – without the necessity of following a strict sequence of topics and questions and thus the course of interviews is largely determined by the interviewees (Lamnek 2005: 350, 352).

Focussed interviews with the participants of solicited photography

The interviews with participants of the solicited photography about the photographs they had taken followed the interview technique of focussed interviews (Flick 2009: 195pp.; Lamnek 2005: 368pp). Focussed interviews are characterised by an initial stimulation: “The persons interviewed are known to be involved in a

5 This is because two participants received a second camera, and another participant, the ‘internal observer’ mentioned above, received a total of seven cameras. Of the seven cameras only five of the developed films were discussed. One of the cameras was stolen and another one could not be discussed because of my long absence from the field and a parallel transition period among my field assistants. Thus once my new field assistants had developed a good relationship with the participant the photographs were no longer a ‘fresh memory’ and a discussion did not seem fruitful.

particular situation: they have seen a film, heard a radio program (or) read a pamphlet" (Merton, Kendall 1956: 3, quoted in Lamnek 2005: 369). The solicited photography represented such a stimulus, although it differs as participants decided on the pictures taken instead of having a pre-defined set of pictures presented to them. The interview guide focused on understanding why the interviewee had taken the picture and what the space shown on the picture meant for her/his everyday life. Thus the questions concerned the everyday life activity patterns including the purpose for which a specific place was visited, at what times, and in whose company, as well as the values the interviewee attached to these places. Besides, specific pictures provided an entry point into a discussion of access rules to the public spaces that provided a livelihood to the interviewees and thus explored the institutional setting within the study areas from the perspective of the ordinary users of space. In 2010 the additional solicited photography interviews in Nasimgaon were also extended to include the matter of women's clothing, in addition to the one women's focus group held in Nasimgaon[6] (see below).

Contrary to the discussion of focussed interviews in the literature, the interview technique here was not applied for testing hypotheses in a positivist research process (especially Lamnek 2005: 369) but was embedded into the grounded theory and interpretist approach. But focussed interviews were hardly ever applied in 'pure form', and mixing them with other interview techniques increases their potentials. Thus here the discussion about pictures also aimed at providing impulses for narrative or episodic interview passages. The interview guide was handled flexibly and the development of the interview was to a large extent governed by the interviewee her-/himself. The underlying principles of non-manipulation of the interviewee, specification of her/his point of view and definition of situations, collection of a wide spectrum of meanings with regard to the stimulus, and depth/profundity (Flick 2009: 195) were adhered to.

Problem-centred interviews

The interviews conducted with representatives of organisations and other persons with 'special knowledge' on local institutions, i.e. key informants and experts, were based on the technique of problem-centred interviews (Flick 2009: 363pp; Lamnek 2005: 368pp), but also included narrative and episodic elements. The key informants were members and employees of local committees, established long-term residents or people who had started a specific project in public space, for example established a garden in Manikpara. The interview guides were developed individually for each interviewee or group of interviewees. Topics commonly covered the responsibilities of management of public spaces, the specific access rules existing, the nature of conflicts about the use of public spaces and the ways

6 In Manikpara, the clothing style and mobility patterns were instead covered by two focus group discussions, which – given the problems experienced with women participants of solicited photography in Manikpara – seemed more appropriate to adequately cover the topic.

of solving those, the institutional setting within the study settlements as well as the history of development with specific reference to the development of public spaces and the involvement of (political) community organisations. Often the discussions followed the narrations of the interviewee, who recalled specific events and thus triggered the interviewer to ask questions to improve understanding of the situation, the motives of the interviewee (and her/his organisation) and her/his perception of the other actors involved.

Additionally, follow-up interviews with participants of solicited photography were held. Here the interview either centred on a recent event or specific topic, for example a new regulation that had been introduced and how the interviewee had dealt with it, or it was supported by the PRA (participatory rapid appraisal) methods of Mobility Maps or Venn diagramming (Kumar 2002). Three interviews were conducted with Mobility Maps with a view to identifying participants' spatial mobility patterns. Venn diagramming, a technique commonly used for studying and understanding local people's perceptions about local institutions and individuals as well as power structures (Kumar 2002: 234), in three cases helped to identify the institutions and individuals exercising power in specific places and to reveal the power relations at work. Both Mobility Maps and Venn diagramming in the first place provided stimuli for conduction of a problem-centred interview.

Focus group discussions

In three cases focus group discussions were held to capture women's perspectives on the spatiality of livelihoods and the norms and values existing about gendered access to public space. Two focus groups were held in Manikpara to compensate for the problem of finding female participants for solicited photography (see Chapter 4.1.2). For each focus group I selected one woman I knew and asked her to invite two or three other women, friends or colleagues, with whom she would feel comfortable in discussion. One focus group was held with women from low-income households who were all employees in the plastic recycling industry. Another focus group was held with women from slightly better-off families who did not need to work to support their families and thus mostly stayed at home. The third focus group was held in Nasimgaon with four women who all had worked in the garments industry, but only one did so actively at the time of interview. The interview guide for all focus groups included the discussion of daily activity patterns, perceptions of working at home/working outside home and working indoors/outdoors, mobility patterns and dress codes, the perception of the neighbourhood and identification with Dhaka and/or the rural home especially from a women's perspective.

For these women's focus groups I excluded men from being present in order to create an atmosphere for open exchange. In one case, which involved the woman whose husband was dominating in providing answers (see below) and who had fought with the *jamai* for taking away his wife's camera (see footnote 3), I and my field assistants had to be very insistent to keep the room 'female only'. Her husband was very eager to join the group, but the women supported us in our claim for a 'female only forum'. However, during the discussion he found several excuses to enter the room, so that once he wanted to eat his dinner I had to postpone the second part of the focus group to the following week.

Informal discussions/unstructured interviews as part of participant observation

Apart from the three interview types characterised above I held numerous informal discussions during my field visits. Quite often these took the form of the Bengali *adda*. During my field visits I always passed by a few places where discussions developed with the owner of the shop and often others who stopped by on their way or who were already passing their time at the same spot. In Manikpara there was one place on the embankment slopes where a person I had known and who became a key informant had a small shop for sorting and storing plastic materials. His friends, most of them engaged in politics in the locality in some way, gathered at this spot two times a day as it provided the necessary privacy to smoke cannabis while sitting outdoors[7]. In Nasimgaon there was another shop that became my most common '*adda* place'. The owner of the shop was sitting in his shop all day, and while sometimes he had to deliver or sell goods to customers, this was most of the time done by an employee. Thus he spent a lot of time observing what was going on in the public space adjacent to his shop while eating bethel nuts with bethel leaves. I spent many hours sitting with him, and often other people that joined in, on a bench in his shop with a view of the public place, drinking tea, chatting about recent happenings in the locality and philosophising about life.

While these two locations described above were the main *adda* places they were not the only ones, and I had numerous other informal discussion rounds at tea stalls and at the places where my interviewees lived or worked (within the limitations for me as a female researcher see Chapter 4.4.2). Visiting the participants of solicited photography and the key informants regularly before and after the first interview and engaging in informal discussions with them enabled me to follow their livelihood and spatial strategies and the negotiations on access to public spaces over a long period of time. Furthermore, through this I was able to build the trust and rapport for a more detailed account. I also experienced that it took a long while for people to speak openly to me about the politics 'hidden behind' everyday life in public spaces[8].

7 While I was always welcome for a cup of tea and 'doing *adda*', smoking only took place once I had left the round and was never openly displayed in front of me, a female foreign researcher.

8 For example, I got to know Rohima in March 2009 who together with her husband Shahin sold fruits in a small open market space. I handed over a camera to her in April and we did two interviews in April and July 2009. Then, during a third interview in January 2010 she told me about how the place had changed due to local politics and how they struggled to adapt to the continuously changing environment. Her statement during this interview indi-

Documentation of interviews

The focussed and problem-centred interviews, all except one, were voice recorded and later on transcribed in the Bengali language by field assistants. At the beginning of an interview, after introducing the aims and objectives of the research, the interviewees were asked for their consent to record the interviews. It was pointed out that the data would be made anonymous and their names would not appear in the research products.

Objections to this type of documentation often concern constraints to the naturalness of answers given once a recorder is present, and a researcher hopes that in the course of the interview the recorder will soon be forgotten (Flick 2009: 373). All the interviewees agreed to having the interviews recorded[9], but it could be observed from the flow of the interview and extent of the answers given that not everyone felt completely comfortable with the recording. Often the comments and additions made after the recorder had been switched off provided very valuable additional input[10].

The informal discussions were not recorded, as they normally evolved naturally and asking for permission to record would have interrupted the flow of discussion and stifled the conversation. Instead, after such rounds of informal discussions I sat down with my field assistants and journalised the contents and arguments from memory to field notes (see also Chapter 4.2.2 on documentation of participant observation).

Interview situations

During the interviews a number of challenges arose, specifically with regard to the place and setting of interviews and time availability.

To establish a convenient, relaxing and familiar interview situation interviews were mostly held at the participants' homes and in a few cases in their shops. In

cates how it took time to build up the necessary trust between us in order to give detailed accounts of what was going on: "*Apa* [Bengali for sister, referring to me] was here at that time, but we could not tell anything about this since we were afraid of telling. They [members of the local committee] used to threaten us always" (Rohima, 18.01.2010).

9 There were only two interviews that I did not record on purpose. These were expert interviews with government officials. Especially in one of these interviews the interviewee did not respond to any of my questions but kept telling me general information that was largely not new to me. I had expected this and in both cases had decided prior to the interviews that recording would make the interviewees even more suspicious of talking to me.

10 This was extreme with one interviewee. We interviewed him at his house because the market place was too noisy for a discussion. However, he was not comfortable taking us to his house where we had never been before, and although he had agreed to recording he was also not comfortable with the tape. After the interview we went back to his shop and he invited us to stay with him for tea and banana. Once back in his shop he became very talkative again – obviously this was much more his environment for meeting us, and he felt more comfortable just chatting without a recorder.

both cases it was not always possible to create a setting where the interviewers and the interviewee were undisturbed by others and thus 'free to talk'. Instead, other members of the household or neighbours kept joining in, some just as 'listeners', others to display their knowledge and even correcting the words of the interviewee. It was often not possible to change the situation on the spot, and thus I accepted it and tried to capture the dynamics evolving from these new situations.

In many cases the new constellations evolving during the interview reflected the social position of the interviewees and the persons joining. At the same time, I was often not able to keep out others due to their position in society. For example, one interview with a lady became distorted once her husband joined, but because of his dominating role – in fact he was giving answers in her place, not giving her the space to express her opinion and telling us that she did not know about these things – I could not ask him to leave in any polite way. Thus a different setting had to be found, in this case a women focus group where it was easier to exclude male participants from the beginning. In another women's focus group I wanted to hold with three women, a fourth woman was present, the mother of one of the participants. She was by far the eldest in the round and it would have been very rude to tell her to leave the room. In another case a colleague and best friend of the interviewee came for a visit during the interview and quietened the interviewee when he was about to talk about the area's *mastans* (musclemen) and politics. At other times, however, interviews that transformed into group discussions provided additional insights, for example in another interview where the leader of a political youth organisation joined the interview with his uncle and provided additional information and perspectives, causing the two of them to discuss local politics in a joking mood which revealed a lot about their perceptions. The boundaries between interviews and participant observation were thus always fluid (see also the experiences of Woiwode 2007: 122).

My experience of participant observation also helped me to understand that interrupting one's discussion or coming to listen is not perceived as a matter of violation of privacy – quite contrary to what is commonly accepted in Germany. Furthermore, everyday life in a low-income settlement, where families occupy one room and share other facilities and a common courtyard area with their neighbours, is not based on privacy but on company (see my experiences during participant observation above, also Woiwode 2007 121).

In addition, a large number of interviewees were rather busy and it was very difficult to find time for a long interview. In Manikpara the workers of the plastic recycling industry normally work from 8am to 8pm with only a short lunch break granted by their bosses. The only time slot available for interviews was on Fridays when working hours end at lunch time, but then especially the women were engaged in domestic work. Similarly, shopkeepers could not easily leave their businesses for a long time, and people commonly operated their shops or stalls from 8 or 9am to 11 or 12pm. 'Not being present' because of an hour-long interview would mean a loss in business, a sacrifice I could not expect someone to make. Others, especially political leaders and persons with a rather organised 'business emporium' like the leaseholder in Manikpara, were regularly called for other meetings during the interviews. For example, community members came to seek their advice as members of the *shalish* (traditional judgement system from rural Bangladesh, see also Chapter 5.1.4)[11], or someone important called whom they had to attend to.

Instead of seeing all these incidences only as hindrances for my interviews I took them as an additional experience of people's everyday life, as they revealed very openly what matters in people's everyday life in terms of social contact and livelihood strategies. Often spontaneous changes of my plans, guided by what happened in the field, gave me the opportunity to observe events that I had not planned for in my field schedule.

4.2.5 Challenges and limitations

The challenges and limitations discussed here relate to the issues of language, seasonality and my inability to be constantly present in the field. Other challenges related to the embeddedness of the research will be discussed in Chapter 4.4.2.

Language and translation

A challenge to interviews as a method of oral communication arose from the language barrier. Once my research focus on the everyday life negotiations of public spaces in urban settlements of Dhaka was defined, it was obvious that this qualitative research would involve close communication with the local inhabitants. Thus I started learning Bengali, first with a view to be able to exchange greetings and introduce myself and the aims of my research in order to build up relationships. Later, when my research took on a more ethnographic character and involved my staying in the field for participant observation, being able to speak Bengali became indispensible as a means of communication. Thus the level of Bengali I achieved to speak helped me in keeping contacts, living in the area, moving around independently and explaining the purpose of my stay. Furthermore, it helped during interviews, when I no longer had to rely on the parallel translations given by my field assistants but could follow the basic flow of the conversation.

But despite my acquired language skills, I had to rely on field assistants during all my interviews and most informal discussions. Involving translators especially in such qualitative research bears certain risks and challenges. Translators are "actively participant intermediaries making judgements which may transform the message received" (Bujra 2006: 175). This is both valid for asking questions and translating answers. The interview method chosen required close, trustful and empathic interaction with the interviewees and skills in listening which largely

11 Once the political leaders I was interviewing were called for a *shalish* and invited me to join them, giving me the chance to observe the negotiations on the spot (see Textbox 6 in Chapter 9.1.3).

depend on personality and behaviour (Lamnek 2005: 354). Although I was careful in selecting translators whom I considered to have the appropriate skills of communication, it still makes a difference whether one relates with the interviewee directly or through third persons, i.e. translators. Furthermore, due to the other obligations of interviewees[12], I could not always pursue the lengthy process of translating the full answers into English and posing every question in English first before its translation into Bengali. So I had to trust that I had trained my translators/interviewers well enough in order to carry out large parts of the interviews by themselves with me only following the basic line of arguments according to my abilities in Bengali. Translating back and forth during an interview can also lead to a rather stiff atmosphere (Bujra 2006: 175), which may be less encouraging for the interviewee to talk freely – thus advantages of time and flow of speech had to be weighed against the disadvantages of not being able to check everything.

Furthermore, translated words often carry different meanings than the 'original', and awareness of this is needed both when posing questions and when carrying out translations (Crane et al. 2009: 41). There were a considerable number of cases when I was not happy about how a question had been asked in Bengali, for example when it was formulated in a suggestive manner or in a way that did not trigger the narrative answers desired. At other times, the question I had in mind did not make sense in the local social context, illustrating that ideas and concepts from one language cannot always be translated into another (Bujra 2006: 176). As Bujra (2006: 172) points out, translation is a "social relationship involving power, status and the imperfect mediation of cultures". She continues to note:

> "The problem with dependence on local translators is that one can be restricted and trapped within their perspective on their own society. It is not until one begins to speak the language and enlarge the canvas of relationships that one can decipher how social location or political position might affect a translator's interpretation." (Bujra 2006: 174).

The translation problems experienced during the interviews continued in the translation of the Bengali transcripts. While the aim was to preserve the translated text as authentically as possible (Crane et al. 2009: 43), this raised questions about the extent to which inept speech or grammar should be improved, or whether the translation should be literal or colloquial. As language is always an expression of socio-cultural values and practices, I decided to keep many Bengali words that carried a broader meaning than any English translation would have been able to reflect (see the glossary with detailed explanations on pages 357). Besides the problems inherent in translations discussing different cultural concepts and meanings, working with my translators also presented opportunities. In discussing the meanings of words and sentences with my translators I was able to reflect on my own positionality and to question implicit 'ethno-centristic' theories and concepts (Crane et al. 2009: 44) more thoroughly than if I had taken language 'for granted'.

12 The persons I interviewed were often involved in work or household chores, and it was not easy to get time, for example, from a woman whose children expected her to prepare food at a set hour and whose husband expected her to sit in the shop at the other times selling tea and snacks to their customers.

Seasonality and presence in the field

Most of the fieldwork was conducted between February and June. From March onwards the temperatures increase and go as high as 40 degrees in May, while cooling rains occur only sporadically. Both in 2009 and 2010 power was cut every alternate hour or every two hours to cope with the electricity crisis in the country. Conducting interviews in a CI-sheet (corrugated iron-sheets) room heated up by sunlight throughout the day without electricity to operate the ceiling fan and without proper ventilation due to lack of windows constituted a challenge to concentration both for the interviewee and the interviewer's team. As during my stay up to the beginning of July 2009 the monsoon was late, I experienced only small events of rainfall not comparable to the rains common in the monsoon.

While I am aware of the seasonal differences, especially between the dry and rainy seasons, in the use of public spaces, it was not possible to include a research period during the rainy season in the research schedule. However, I compensated for my absence in the field during these times with the help of my field assistants. Through their continuous field visits and corresponding field notes and photographs, I was able to understand the seasonal changes in the research participants' livelihood strategies occurring between August and December. Furthermore, this continuous presence in the field enabled me to closely follow conflicts that occurred about public space and the emergence of new or re-framing of existing organisations and institutions without interruptions for almost two years. A further drawback of the seasonally limited field periods was that I have not experienced the fasting month of *Ramadan*, the *Eid-ul-Fitr* celebrating the end of Ramadan and the *Eid-ul-Adha* or *Kurbanir Eid,* commemorating the willingness of Abraham/Ibrahim to sacrifice his son Ishmael, in the study settlements. Here again I had to rely on accounts provided and photographs taken by my field assistants.

4.3 ANALYSIS OF EMPIRICAL DATA

In line with the grounded theory approach to this research, the analysis of empirical data was conducted in an open and iterative process. This was informed by the approaches to coding developed by Glaser as well as Strauss and Corbin (Kelle 1994: 313–333). Both suggested a first round of open coding by looking at the material without a pre-defined set of categories:

> "The analyst codes for as many categories that might fit; he codes different incidents onto as many categories as possible. New categories emerge and new incidences fit existing categories" (Glaser 1978, quoted in Kelle 1994).

Strauss and Corbin, in line with their recognition of the implications of pre-knowledge and existing theories in grounded theory research, explicitly include these in the coding process, as long as they are not imposed on the material (Kelle 1994: 324pp.). In a second step Strauss and Corbin (Kelle 1994) propose conducting axial coding, which aims to investigate and establish interrelations between the emerging codes and categories. Here they make use of existing theories in order to decide which linkages are meaningful for the theory building process. The last step of selective coding then subsumes the interrelated codes and categories into core categories which then inform the theoretical concept and contribute to theory building (Kelle 1994: 331).

Accordingly, in my research I started the analysis of the materials with a very open mind and without a pre-defined set of codes. While going through the text passages of interviews and field notes, I coded sentences, paragraphs and longer conversations with as many codes as seemed adequate to capture the meaning of the text from different perspectives. Soon the number of codes increased and it was necessary to group them under main headings. Informed by my pre-knowledge as well as the codes that I had already applied, I established 12 main codes, namely:

- imagined spaces/spaces of representation,
- perceptions of urban life,
- spatiality of livelihoods,
- management of public spaces,
- access rules and arrangements,
- claiming space,
- legitimation of using/claiming public space,
- contestations,
- institutions/norms/social relations,
- power,
- negotiations and strategies,
- livelihoods.

These broad main categories were of course not fixed, but also developed in an iterative process of re-organising and re-considering the codings I had assigned at regular intervals. Similarly, once I had gone through all the materials once, I compared the text passages coded into one sub-code. I then decided whether I needed to re-group these text passages by establishing different sub-codes, or whether a text passage I had coded earlier should also be included in another code that had evolved later on. Throughout the coding process, and especially once I had gone through all my materials and started re-organising them in a second round of coding, I wrote memos explaining why I considered a text passage relevant within a sub-code and which new categories had evolved. I also documented the interlinkages between codes with the help of memos (axial coding). From there I was able to write code-summaries, where I discussed potential findings and contributions to theory building. These code-summaries then laid the basis for formulating the empirical chapters, where I again re-arranged and interlinked these results and finalised the process with selective coding.

To support the data analysis I used the qualitative data analysis software MaxQDA. I inserted all written documents of my research, i.e. the translated interviews and field notes, into the database provided by the software. Subsequently, I used the programme's coding function for organising, coding and categorising the material. For the interpretations and setting-up of linkages between codes, I used the function of writing memos for each code and its characteristics. The retrieval function enabled me to quickly establish links between coded text passages in different texts and on specific topics, which then provided the main input for generating the research result. Using the software does not change the process of data analysis as such, as the method would have been the same in a 'manual' approach. However, given the amount of data inputted into the programme (160 texts of different length and depth), it eased the process of organising and re-organising the data into codes without losing track of things.

The non-text documents, especially the solicited photographs and maps from observation of public spaces, could not be managed with this software. Instead both the photographs and the maps were printed on posters and hung on the walls so that during the analysis of interviews I was able to always include the visual interpretation. I then included memos in the Max QDA file on the interpretation of these maps and photos wherever appropriate in order to integrate these with the text-based data. Interestingly, some patterns of the maps and solicited photographs only became obvious to me when I had generated first findings from the written materials. Especially concerning the main codes of 'spatiality of livelihoods' and 'imagined spaces/representations of space' the visual material re-confirmed many of the consolidating findings and results.

4.4 REFLECTIONS ON RESEARCH ETHICS AND POSITIONALITY

As I was not culturally familiar with the environment in which I conducted my research, I thoroughly reflected on research ethics and my positionality in the work with the research participants. Thus in this sub-chapter, I will reflect on the application and adherence to standards of research ethics. While a number of critical issues arising during my field work have already been addressed in Chapter 4.2, I will discuss the (changing) position I had while conducting my research in terms of positionality and negotiations of power.

4.4.1 Research ethics

Ethical practice in research – previously mainly discussed in biomedical sciences – has become a growing concern for social sciences, and various ethical guidelines have been developed by research institutes' umbrella organisations and universities. In their book 'Research Ethics for Social Scientists' Israel and Hay (2006) discuss the development and evaluation of social scientists' research practices around four concepts: informed consent, confidentiality, beneficence and non-maleficence, and the problems relating to research relationships. I will briefly discuss all the four concepts, but will put most emphasis on 'beneficence and non-maleficence' while the problems relating to research relationships and the dimension of 'power' will be discussed in the next sub-chapter on positionality.

'Informed consent' requires that research only be conducted after explaining the intentions of the research, including the intended outcomes, to the research participants (Brydon 2006: 26). Prior to the first fieldwork in each study settlement, I visited the relevant local authorities – the Ward Commissioner in Manikpara and members of the local committees in Nasimgaon – to inform them about my research and get their approval of my continuous presence in the areas. Thereafter I informed each research participant about the aims and purpose of my work and asked for their consent to participate. Accordingly during my field visits, and especially during observation and participant observation, I explained my presence to everyone whom I met and who questioned me about my motivation in being there.

Secondly I granted 'confidentiality' to all research participants. This involved making all data sets anonymous and keeping the information provided during any interview confidential (see paragraph on anonymity in the preliminary remarks to this research). Of course, my interviewees often came to know each other because of my recurring visits to the same place and informal discussions held. Especially the participants of solicited photography were often based in the same public space, and thus it was easy for them to identify whom else I had given cameras. While I had guaranteed confidentiality and never reproduced what someone had told me in an interview in front of others, knowledge about who else was participating in my research on some occasions triggered conflicts with deepening relationships. The conflicts manifested themselves in speaking ill of the other in front of me, and I interpreted them as competition for 'the best relationship' with me, the *Bideshini Apa* (Bengali, meaning 'foreign sister'; the word *bideshini* refers to 'white' foreigners, see more in Chapter 4.4.2). To handle this situation, I took to explaining again about the purposes of my research and the relevance of having a variety of participants, and took care not to 'favour' one 'competing' participant over the other in terms of time spent at their places.

The concept of 'beneficence and non-maleficence' refers to the expectation that (social) researchers should "minimize risks of harm or discomfort to participants" and, under certain circumstances, should also seek to "promote the well-being of participants" (Israel, Hay 2006: 95). In my research I consider 'minimising risk' both as an issue of guaranteeing confidentiality (see above) in order to avoid distress arising from interview participation, and of preventing any misinterpretations of the research and its results by others (Israel, Hay 2006: 97, Brydon 2006: 28–29). In conceptualising ethical sensitivity as being developed by researchers as part of a social practice Lo Picollo and Thomas (2008) put emphasis on the significance of social context in structuring our moral perception and judgments. They argue that social research, and thus planning research, can be part of different social practices – that of scholarship, i.e. knowledge production in pursuit of truth and the development of an understanding of the world, and that of politics and governance (Lo Piccolo, Thomas 2008: 13). Locating one's research in both of the practices at the same time is not possible without creating tensions or changing the practices; one action might have a different significance depending on the underlying social practice (Lo Piccolo, Thomas 2008: 13).

Although this research is considered as 'scholarly research' and not research undertaken as part of a planning process, commissioned by policy-makers, it is nevertheless research on planning issues, and thus might entail a reaction and intervention into people's lives. Furthermore, ethnographic research always involves a 'representation of others' even if it does not claim to speak for or on behalf of them – an ethical and political responsibility is evident in this regard (Aktinson, Hammersley 1994: 253). This research's findings might be used by others following a different 'social practice' in a way that had not been anticipated by me or that I would strongly disapprove of (Lo Piccolo, Thomas 2008: 17). This point becomes particularly relevant considering the conceptualisation of 'slums' commonly employed by city governments. Instead of including 'slums' as integral parts of cities, they are often stigmatised, and government reactions range from open confrontations in the form of eviction to benign neglect (see Textbox 2 and Chapter 5.2 on the implications of using the word 'slum').

I here would like to point to the difficulties thus arising with regard to 'avoiding harm' and 'good' ethical practice. Like Weinberg in her research on the use of a planning document in a maternity home (Weinberg 2002, in Israel, Hay 2006: 105) I do not want to harm the residential unit 'slum' which, despite giving rise to exclusionary practices by elite groups, does provide a home and a livelihood, including the internal support structures offered by the community, to a large proportion of city dwellers in Dhaka and in the Global South in general. Considering this, I seek to minimise risks by articulating clearly the implications of my research and the possible impacts on planning. Here, my research becomes political and normative and I take seriously the responsibility of helping address injustice (Brydon 2006: 37) or of 'doing good' (Israel, Hay 2006), but at the same time I am aware of the limited abilities of researchers to achieve "a meaningful change in the lives of the groups they study" (Israel, Hay 2006: 110).

4.4.2 Positionality

> "Any research context is riddled and cross-cut by relationships of power, from those between the sponsors of the research and the researcher, and between the researcher and the researched, to power relationships within the culture of the research setting, relationships between classes and clans, landholders and landless, educated and illiterate, elders and juniors, women and men, rich and poor." (Brydon 2006: 27)

This intercultural research, built on a qualitative methodology and including an ethnographic component of participant observation, necessitated a thorough and continuous reflection on the issue of positionality. The power relations between me and the communities/research participants were manifold, and they were constituted both by my autobiography and by the profile of the communities (see also Butler 2001: 272). Often one's own autobiography is only reflected on briefly by the researcher her-/himself, but can provide a very vivid account of one's 'cultural luggage' (see for example Forsslund 1995: 4pp.). Furthermore, as Brydon (2006:

36) points out, "the researcher's positions and roles are negotiated by both the researcher and the researched community throughout the research process".

In the following I will set out to discuss the power relations I experienced during my research and how the identity ascribed to me changed during a process of continuous negotiation of my role. I will set out to exemplify the different categories and expressions of my positionality which will give a vivid picture of the challenges and the opportunities that arose from my intercultural research. While this reflexivity is crucial in order to adequately assess the research data – both interviews and participant observations – there, however, remains considerable doubt about the extent to which one's positionality in an intercultural research setting can be understood fully (Butler 2001: 273).

Autobiographic reflections

> "Race, ethnicity, class, gender, religion, marital status and other non-demographic characteristics, including one's worldview, often define the position and identity of the researcher in relation to the researched community." (Apentiik, Parpart 2006: 34)

Looking back at what I wrote about in my personal diary and communicated 'home' to my family and friends reveals how different everyday life in Dhaka is compared to where I come from:

- Dhaka is a restless, bustling city full of people with never-ending traffic jams and a dense network of activities, at first sight suggesting chaos, but a closer look reveals the existence of a hidden order, there are rules about how traffic is organised, how people move around.
- There are manifold activities spilling out onto the streets, creating an atmosphere of hustling and bustling which is 'completed' by constant noise and air pollution. Poverty is ever-present, at a scale I have not witnessed before.
- The first visit to a number of potential study settlements left me with an overwhelming impression of the friendliness of people. What impressed me is the high degree of security I felt at any time, despite being in a new environment.

Even though I had both research and working experience in Sub-Saharan African cities and low-income settlements, I was overwhelmed by Dhaka, this first megacity I had visited in the Global South. I was amazed by the hustle and bustle of street life, the countless mobile vendors and small entrepreneurs. At the same time, I was surprised to see so few women in public spaces, and I was taken aback by the images of poverty I faced. The research I commenced then enabled me to listen to the stories of the poor attentively and relate my first impressions to the life stories of Nasimgaon and Manikpara's inhabitants.

My degree in urban planning, accomplished in Germany, entailed a specific understanding of everyday life organisation, and the role of the state as a regulator. Further, my upbringing in a middle-class family meant that I had always experienced a certain standard of spacious living conditions – although when travelling I always quickly became accustomed to local conditions. In Dhaka everyday life is organised very differently from what I have experienced and learned in Western Europe. Rules which in Germany are very often written, exist rather in a 'mental space' produced by society instead of the state. While with sufficient financial resources one can afford very good living conditions in Dhaka with regard to housing, the permanent water and electricity crisis, traffic jams and air pollution are outside of one's own influence. A further challenge was to adapt to the living conditions in my study settlement Nasimgaon (see Chapter 4.2.2). My relatively fast adaptation was only possible because my landlords integrated me into their family life from the very first day of my stay, and both my landlords and the tenants of the compound always signalled that I was most welcome.

Bideshini – about 'losing colour' and common ascriptions to foreigners

As a female foreigner from a different cultural background I was a *Bideshini*[13], which also tends to suggest 'whiteness' and a European-North American cultural background. Being 'white' is rated very highly in Bangladesh's society and it is associated with relatively high socio-economic status while 'being black' carries a low-income, working class image. Both among Bangladeshi friends from middle-income families and within the communities of my study settlements, I repeatedly listened to accounts of how 'blackness', especially for a girl, was considered a negative point for 'match-making' in marriages. Using skin whitening creams is thus very common in all strata of society and my landlady in Nasimgaon regularly applied 'Tibet snow' both to herself and to her two young daughters.

I experienced this 'identity-creation by colour' extensively for I conducted my fieldwork usually during daylight and in the dry season with a high number of sunshine hours. Almost every time I was out on the streets people were concerned about me 'losing colour', i.e. losing my white complexion and 'becoming black', and recommended carrying an umbrella. 'Losing colour' ('becoming black') and 'gaining colour' ('becoming white') here stand in contrast to the Western European conceptualisation, where 'gaining colour' refers to acquiring a healthy complexion from exposure to sunshine[14]. After several months of fieldwork almost all the participants in my research expressed their concern about how I had 'become black' and thus lost in 'good looks'. This anecdote quite clearly points to the identity I had been assigned – I was identified as the 'white' *Bideshini*, which automatically entailed an upper socio-economic background. The concerns expressed about me 'losing colour' point to the perception of me as a stranger not fit for 'rough everyday life' and furthermore as someone who contested certain norms of

13 The word *Bideshini* was actually never used to address me when present, but my identity was perceived as *Bideshini*.

14 It has to be noted that, additionally, being exposed to heat in Bangladesh is also understood as a health risk, which brings a health component into the socio-cultural aspect of 'becoming black'.

the society: it is not common for someone 'white', i.e. from the high-income strata, and especially a female, to expose herself to this outdoor environment as continuously as I did. If I were a Bangladeshi woman I think many people would have disapproved of this behaviour, but being a *Bideshini* gave me a certain freedom in this regard.

Being *Bideshini* immediately raised another expectation. When working in development research and in the context of urban poverty, it is unavoidable that inhabitants perceive a foreign researcher as someone from a donor agency or NGO who is going to implement projects in the area and bring benefits to the community (Apentiik, Parpart 2006: 37). This was particularly the case in Nasimgaon where a large number of NGOs have been involved for a long time period and foreigners visiting the respective NGO projects are a common sight. I reacted to this by communicating clearly about my research aims and objectives. That I conducted my research without an introduction from elsewhere, with my only attachment being to a university, and that my contacts were not channelled through NGO or government support presented an advantage in this regard, because I was able to distance myself from other organisations' agendas.

It became obvious that I was perceived and had been accepted as someone from outside the donor and NGO world when Rizia, a local political leader I had become familiar with, invited me to join a meeting of a local committee of an NGO involved in water supply. The following field note of the discussion indicates no fear of discussing critical issues despite me being present as an outside observer:

> "In between some more members, the cashier and the secretary came to join the meeting one by one. The cashier said that one person wants to rent their office space for 4,500 Tk and will give an advance of 10,000–20,000 Tk. The cashier proposed that they could rent out their office to that person and rent a nice room for 1,500 Tk for their office. Then some members discussed about the matter and one person said 'No it's not necessary'. Then another person said 'If it is rented out then we have to make sure that we can still use this office at the time when some donors will come so that they can sit'." (field note, 24.04.2010)

This incident shows some degree of 'allowing me in'. This, however, was not valid for all research participants as becomes obvious from some interviews, especially with key informants who were involved in party politics. These interviewees often followed a specific narrative in their responses that I would interpret as 'ideal answer types' which tried to avoid issues of conflict and rather produced an account of events aiming at pleasing donors. With recurring meetings and the building of rapport however, it was possible to get 'less filtered' responses even from these politicians.

Apa – Being a sister

While the above points at a very 'outsider' identity, this represents only one part of the multiple and often fluid identities and roles that were attributed to me. At the same time I was *Apa*, a very common Bengali form of addressing someone as 'sister'. Accordingly, for the children of my interviewees I was *auntie*, while the elder generation, who due to their age in relation to mine could not address me as *Apa*, addressed me using the personal pronoun *tumi* instead of *apni* – a distinction resembling the difference between the German personal pronouns of *Du* and *Sie*, although this comparison does not fully do justice to the very complex set-up of personal pronouns in Bengali. In general, it is very common in Bengali to refer to non-family members using kinship terms:

> "After being introduced to somebody it is customary for the people of Bengal to give up using the formal formulas of address such as /babu/ and /mošae/ for Hindus, and /šaheb/ and /mia/ for the Muslims, and use some appropriate forms of kinterms for address. These kinterms are generally determined on the basis of age or status. [...] Any male person, who is little older in age or is higher in status will be regarded as an elder brother and will be addressed as such." (Dil 1972:113–114)

While certainly being called *Apa* especially by elder women was an indication of respect towards me as a foreigner with a high level of education, I experienced that 'becoming *Apa*' soon entailed a sense of familiarity and a relationship of trust between me and the inhabitants of the study settlements. While this opened up opportunities for very good researcher-researched relationships and very frank *adda* rounds, identifying me as quasi-kin also entailed expectations and obligations. Whenever I visited the study settlements, most of my interviewees expected a visit and considerable time to be spent with them, telling me how they 'felt good' in gossiping and doing *adda* with me. I became a listener from outside for a number of mainly female interviewees who repeatedly sought my advice for decisions concerning their household, or asked my opinion on what they had decided. Aiming to give back as much as possible, I assumed this additional role of *bhandobi*, female friend, and tried to give my interviewees the opportunities to talk while listening sympathetically (Israel, Hay 2006: 108).

Making me the *Bideshini* and *Apa* furthermore prescribed me the role of a relatively rich family member and consequently I was repeatedly approached for financial support, especially in phases of hardship. This attribution is very common in Bangladesh, where the help of richer persons to 'their own poor', here referring either to direct relatives or fictive kin, is deeply enrooted (see Gardner, Ahmed 2009 for the case of London-based Bangladeshis and their relation to their home villages). Whenever I could, I contributed with some extra tea or fruits both at my compound and during interviews. I further decided to celebrate my birthday in Nasimgaon before my departure from Bangladesh, as a means to say thank you to everyone and fulfil the expectations of 'being *Apa*'. According to the local custom, I distributed meal packs to people I had become familiar with or whom my landlords considered important for their own social network – in this way I was able to support my landlords in re-confirming their own social relations. On the other hand my landlords had made it very clear that I should only pay the usual monthly rent, without any financial expectations beyond this.

Bideshini, Apa, bhandobi – Being an unmarried female

Bideshini, Apa and *bhandobi* point at my female identity that indeed was quite crucial for my research. On the one hand, only being female opened up the everyday private life in the housing compound to me during my participant observation. This private and semi-private space especially during the day is a female space, where men are only present for a short while, i.e. to have food or take a nap. No man, unless sick, could be found staying in a housing compound throughout the day. Even if not at work, men spend their days in the public sphere, meeting for *adda* or watching movies at the numerous tea stalls. Thus only being female enabled me to experience the everyday life beyond the public sphere.

At the same time being female put certain restrictions on my behaviour in the public sphere. While my identity as *Bideshini* meant that I could cross some of the borders that would have been relevant for a Bangladeshi female researcher, I still could not move as freely as a male researcher (similar experiences are for example documented by Forsslund 1995). While I participated in *adda* rounds and spent time at tea stalls, overdoing this in public would have caused rumours in the communities, where I experienced that gossiping was a favourite pastime. Thus I was restricted in participation, especially in the late evening rounds of *adda* that developed at the tea stalls when most men had come home from work and spent leisure time discussing recent developments and politics. I only caught a glimpse of these discussions a few times while sitting at my landlord's shop as it was getting late. Sitting at the neighbouring shops was accepted at all times, indeed other women were also doing so, but going to the next tea stall and mixing with the male customers would have crossed the line and would have earmarked me with a negative identity.

Furthermore being addressed as *Apa* by men underlined my unmarried status. If I had been married I most probably would have become *bhabi*, sister-in-law, at least for the men. As in Bangladesh it is very unusual for a woman at the end of her twenties to be unmarried this was a very common question I faced. Once more, my identity as a *Bideshini* resulted in acceptance of my explanation of how the concept of marriage differs in Western European society and culture. However, it evoked curiosity and an identity of 'being different' that as an unmarried woman I walked through the study settlements very extensively, always trying to establish new contacts with both men and women.

Contestations – Becoming 'political'

When I started to live in one of the study settlements as part of the participant observations I was aware of the necessity to reflect on my role and to discuss the implications of my presence continuously with my landlords, in order to adhere to the ethic principle of 'minimising harms' (Israel, Hay 2006). However, my experience in the end only demonstrated how limited the ability to fully understand the complex network of a community really is. As I – and my Bangladeshi field assistant – were convinced that my landlords would openly communicate if my presence would lead to any problems, I was all the more surprised when they told me in April 2010 how at the time that I moved in, in March 2009, my presence had become a matter of internal politics.

In brief the following had happened. At the beginning of March 2009, in broad daylight, some AL (Awami League, one of the two main political parties which has been in power at the national level since January 2009) supporters had broken down a few tin sheets that my landlords had stored at the back for a future extension of their house. This presumably had been done because my landlord was a member of the opposition party BNP (Bangladesh Nationalist Party, see more details on both parties in Chapter 5.1.1), and the household was known to support this party. Thus the act of vandalism was interpreted as claim making and power demonstration by the now ruling AL followers, who also expressed their anger about my landlords who had done fairly well before the AL government, i.e. they had acquired their own shop and house and had recently extended the compound onto the lake, making them landlords with tenants instead of owner-occupiers only. My landlords did not take the act of vandalism lightly, but went to the police to file a General Diary[15] (GD) against the offenders. This was not perceived well by the local committee who were thus looking for a chance to have the GD removed. Accordingly, when they realised that a *Bideshini* was living at my landlord's place they started threatening him by telling him that he should not allow me to stay for they were not able to guarantee my safety. Unfortunately, my landlords did not tell me about these contestations at that time. They had decided that they wanted me to stay and thus tried to ensure my safety with their own capabilities. After a while, when there were no doubts any more about my research intentions and it was clear that my presence was not harming anyone, these contestations ceased and I was accepted as *Bideshini* and as *Apa*.

In retrospective, and after having conducted research in the community on the negotiations of spatial claims in everyday life, the above story does not surprise me any longer. But when I set out to conduct this research, while being aware of issues of power and positionality, I would never have imagined such incidents and the implications of 'embedded research'. However, the above 'story' did not in the first place happen because of my personality but because my presence provided an excellent opportunity to put pressure on the family I stayed with to achieve the aims of a few political actors – to maintain a white vest with the police. Had I not stayed in the field at that time, they would have found another means to exert pressure, which was later on realised by exploiting the family's most vulnerable point – the problematic marriage of their eldest daughter[16].

15 General Diary is an expression that has made its way from English to Bengali and refers to a notification at the police station, thus of less consequence than filing a case would be.

16 To solve the GD matter, the local committee members finally applied a different strategy: in another conflict of my landlords with their neighbours they threatened to include my landlords' daughter in a GD against my landlords – which was a very effective way of blackmailing them, given the fact that this daughter had just been 'returned' by her husband and negotiations were underway about the conditions under which he would 'take her back'. A GD

Religion

What I felt mattered least was my religious background in the study settlements predominantly inhabited by a Muslim population. Of course I was asked about my religion and there was one interviewee with whom I spent a lot of time doing *adda* who was very interested in talking about the different religions and his knowledge of the bible in particular. Given the highly tolerant Bengali-Islam culture and hospitality, I was included naturally into religious celebrations. Thus I witnessed two *milads*, thanksgiving prayers, held by two families in each of my study settlements. Furthermore, I had the opportunity to join one person from Manikpara and his friends to visit the *Bisho Ijtema* – the annual congregation of the *Tabligh* movement held in Bangladesh since 1979 and presumably the second largest congregation of Muslims worldwide after the *Haj*. While my experience indicates religion being a very minor aspect of my positionality and identity, I am aware that I might not have understood hidden meanings attached.

4.4.3 Synopsis of reflections on research ethics and positionality

Above I have outlined some of the challenges that I experienced while being in the field. While conducting my research I set out to be "context-sensitive, honest and 'up-front'" about my own interests and how these affected the relationships I built with the communities (Brydon 2006: 28). The examples above further indicate how difficult it was to fully understand the complex power relations that social research, and furthermore intercultural research, brings about. Additionally, when conducting research in development studies and sympathising with the researched, especially when the research reveals the exploitation of the urban poor and the denial of elementary rights, it becomes a challenge to "tread a thin line between research and political action (advocacy) and the methodological imperative of academic objectivity" (Brydon 2006: 37). While especially from critical theory and feminist researchers there have been demands that research should contribute to the political struggles of oppressed groups (Aktinson, Hammersley 1994: 253), this at the same time raises questions on how to position oneself towards the more powerful elite groups (Israel, Hay 2006: 110).

In showing the ambiguity of being a researcher and a participant observer and representing the voice of interviewees adequately I have decided to not follow what has been framed a 'formal style' for research reports that encourages distance (Butler 2001: 265p.). Instead I decided to actively demonstrate my personal involvement by giving myself a voice through the use of the personal pronoun 'I' where appropriate, and by giving interviewees a voice through the use of direct quotes (Butler 2001: 265pp., 272; see the preliminary remarks for information on how interviews are quoted in the text).

against her would have ruined any chance of successful negotiations and therefore my landlords were prepared to do everything to avoid the GD.

5 SETTING THE CONTEXT

The context this research is set in is discussed in the following sub-chapters. It seems particularly relevant to discuss some key issues of Bangladesh's political and socio-cultural legacy that set the background for the empirical study (5.1). Furthermore, a brief overview of the urbanisation of Dhaka megacity is provided, with reference to the main characteristics and critical notions concerning its urban development (5.2). In the sub-chapters 5.3 and 5.4 the two study settlements Manikpara and Nasimgaon are introduced, covering important milestones of their development history and current urban form, public spaces and economic livelihood activities.

5.1 BANGLADESH – POLITICAL AND SOCIO-CULTURAL LEGACY

The results of this research on the spatiality of livelihoods and the negotiations of access to public space can only be understood by considering the broader set-up in which they are located. Accordingly, this sub-chapter discusses the specific characteristics of Bangladesh's political system, the implications of Islam being the religion of the majority, the nature and implications of gender relations as well as the continuing importance of traditional (rural) institutions and power structures.

5.1.1 Political system and effects on local governance

The political tradition of Bangladesh goes back to the viceregal system established during British colonial times and continued during the Pakistan period between 1947 and 1971 (Jahan 2004b). The viceregal system was characterised by influential and dominant power holders, and by a sharp divide between administration and citizens. This 'political heritage' did not disappear during the short period of parliamentary democracy after independence in 1971, the 15 years of civilian and military autocracy and the return to parliamentary democracy since 1991. Instead, there continues to be widespread concern about the state of democracy, with reference to the existence of an "illiberal democracy" (Andaleeb, Irwin 2007; Jahan 2007); both authors refer to Zakaria (2003) in their use of the term "illiberal democracy"). This indicates a lack of political legitimacy despite electoral democracy, where the government fails to serve its citizens and the legitimacy problem arises due to excessive corruption and confrontational politics (Andaleeb, Irwin 2007: 101–105). Political leaders in Bangladesh, regardless of the political system in place and the party in power, have resorted to a style of leadership that can be characterised as a patron-client relation (van Schendel 2009: 215) and that builds on the misuse of state power for partisan and personal gains (Jahan

2007). While political supporters are rewarded, the political opposition is repressed.

This resulted in continuous confrontational politics between the two main political parties, Awami League (AL)[1] and Bangladesh Nationalist Party (BNP)[2], instead of them working jointly and in dialogue with each other to serve the citizens of the country (Jahan 2004a; Andaleeb, Irwin 2007: 104; Jahan 2007). Although various visions were developed, including a series of Poverty Reduction Strategy Papers, the welfare orientation of the state must at least be seriously questioned. Past governments have failed to produce a national consensus for a long-term vision supported and implemented by all parties, regardless of which party is in power. Instead, the ruling party's activities tend to concentrate on producing short term results within the five year electoral cycles.

These confrontational politics embrace all political spheres: they are ever-present in parliament where the opposition frequently boycotts parliament sessions, but also move beyond, resulting in a "politics of street agitation and violence" that increases pressure on the already "fragile democracy" (Jahan 2007). These politics of street agitation are what is experienced in everyday life, where especially the parties' front organisations (see Figure 8), for example the powerful student wings, compete with each other for resources. For example, this includes extorting money from businessmen, claiming a share from public contracts or occupying and managing the university halls of residence (Siddiqui et al. 2010: 324). The dominance is often with the sub-organisations of the ruling party, however, the opposition party and its sub-organisations are eager to "unnerve the government in power through agitation politics" (Siddiqui et al. 2010: 324). It became thus very common for the opposition to call for regional or national *hortals*, general strikes, which normally lead to a standstill of public life. The confrontations

1 The AL (translating as 'People's League') has its roots in 1949 in East Pakistan, when its founder broke with the Pakistani Muslim League to form the Awami Muslim League. In 1955 it was renamed Awami League to underline its secular character, building on a linguistic identity rather than religion, that it upholds till today. Six years after the 1975 assassination of the first president of Bangladesh, Sheikh Mujibur Rahman, his daughter, Sheikh Hasina Wazed, became the AL's president and continues to hold this position today (van Schendel 2009). Since parliamentary democracy was restored in 1991, the AL has ruled from 1996–2001 and again since the 2008 elections in a twelve-party alliance, after it achieved a landslide victory winning 230 of the 300 parliamentary seats (262 for the alliance), following a two-year period of caretaker government. (Website ICG)

2 The BNP (*Bangladesh Jatiyotabadi Dol* – Bangladesh National Party) was founded with an ideology of national conservatism by Bangladesh's first military ruler General Ziaur Rahman (van Schendel 2009: 199–201). After his assassination in 1981, his daughter Khaleda Zia took the leadership of the party, which she has held until today. The BNP under Khaleda Zia has been in power twice since parliamentary democracy was restored, in 1991–1996 and in 2001–2006. When their last legislature period ended in 2006, a caretaker government was established to hold the next parliamentary elections. However, the caretaker government was considered partisan to the BNP and thus the opposition AL boycotted the elections scheduled in January 2007 (Jahan 2007). In the December 2008 elections held under the new caretaker government formed after January 2007, the BNP won only 29 of the 300 parliamentary seats and consequently went into opposition (Website ICG).

between the two main parties escalated last when the elections scheduled in January 2007 were boycotted by the AL due to suspicion of fraud by the then ruling BNP. Thus at the beginning of this research, until December 2008, Bangladesh was governed by a military-backed caretaker government. While the caretaker government set out to fight corruption, its lack of legitimacy and human rights violations were also criticised strongly (e.g. Naik 2007). Since January 2009, the country has again been ruled by an elected government consisting of an AL-led twelve-party alliance.

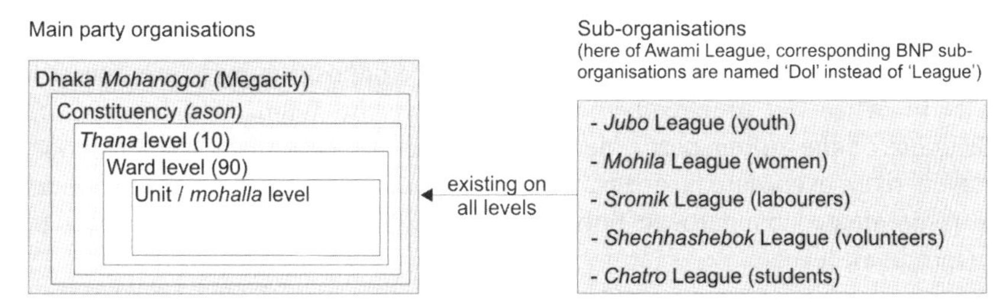

Figure 8: Party and party sub-organisations on different (spatial) levels in Dhaka

Corruption is widely practised both within political parties as well as in the government's administrative institutions. The initiative of the caretaker government in 2007 and 2008 achieved little change. The Corruption Perceptions Index 2010 measured by Transparency International (Website TI) was 2.4 (out of 10) and Bangladesh thus ranked 134th among 178 countries. The rate of households falling victim to corruption, i.e. payments of bribes and unauthorised money, and harassment in 2010 while accessing services was highest in the judiciary (88%), followed by law-enforcing agencies such as the police and the Rapid Action Battalion (RAB)[3] (80%), but it was also significant in the areas of land administration (71%) and local government (44%) (Transparency International Bangladesh 2010). While these figures are alarming, van Schendel (2009: 253) points out the long history of corruption as a pattern of behaviour in the region. This, for sure, does not excuse such practices, but it helps to understand how this system is deeply rooted in society, reflected also in the practice of *chadabaji* (extortion). This is carried out on a neighbourhood level by *mastans* and *chadabajs*, Bengali terms referring to local petty criminals and extortionists, who can at the same time be accepted neighbourhood leaders (for more details on the roles of *mastans, chadabajs* and similar characters see Chapter 9.1.3). Often, *mastans* are the "local tools" of national politicians and bureaucrats who enforce *hortals* and generate party funds; this leads van Schendel to refer to a '*mastanocracy*' (van Schendel 2009: 253; Siddiqui et al. 2010: 327–328). The *mastans* and similar local leaders are thus actively 'produced' by the political parties.

3 The anti-crime and anti-terror force RAB was introduced in 2004 following a time of deteriorating law and order under the 2nd BNP government (Jahan 2004a). RAB continues to exist today and is highly criticised by human rights organisations for its involvement in torture and extra-judicial killings during crossfires (Human Rights Watch 2011: 282).

Corruption and political agitation are also ever-present at the lowest tier of local government, the Ward level in Dhaka City Corporation (DCC)[4]. Ward Commissioners have only very limited resources and responsibilities (Siddiqui 2004: 375, 390) and thus their performance very much depends on their own initiative and commitment (Banks 2008: 362). Furthermore, the Ward Commissioners are highly dependent on the support of the political party structures at the Ward level, as the current situation demonstrates. The 90 Ward Commissioners are normally elected every five years and the last tenure period ended on 14th May 2007. While at that time the caretaker government was in power, since the take-over of the new political government the Commissioner elections of Dhaka City Corporation keep being delayed. Many Ward Commissioners from the opposition party are thus still officially in office but practically powerless, while the Ward-level leaders and organisations of the ruling party, despite their lack of legitimacy via elections, have taken over everyday business. This again demonstrates how the political sphere is dominated by partisan politics, undermining democratic principles of governance at all levels.

5.1.2 Islam in politics and culture

Today about 90% of Bangladesh's population are Muslims, following sixty years of migration history when minority religious groups, especially the Hindu population, today only making up 9% of the population, left the country for India. While Bangladesh was founded in 1971 as a secular state, prohibiting religious political parties, this soon changed after the assassination of Sheikh Mujibur Rahman (Lorch 2008). Thereafter, the Islamist political party Jamaat-e-Islami (JI), whose leaders had been involved on the side of West Pakistan during the Liberation War, resumed its activities in 1979 and Islam was made the state religion in 1988. In the years of parliamentary democracy, Islamist parties have often been decisive for establishing political majorities, and the BNP has included the JI in its alliances (Lorch 2008: 12). In the 2008 parliamentary elections, however, the JI and other Islamist parties could not win a substantial amount of votes, although they remain present in the BNP opposition alliance and their student organisation is involved in many campus activities all over Bangladesh. Despite a continuous presence of Islamist political parties, Bangladesh builds on a largely secular political tradition and thus presents an example for the compatibility of Islam and democracy (Lorch 2008: 5–6).

The development of political Islam in Bangladesh is also reflected in everyday life culture. While in the 1970s a liberal vision of Bangladesh dominated,

4 The DCC covers an area of 145 km² and in 2004 had an estimated six million inhabitants (Islam 2005). The megacity of Dhaka, however, has for a long time far exceeded this administrative boundary (see also Chapter 5.2.1). To improve service delivery to its citizens, on 29th November 2011 it was decided to split DCC into two parts, Dhaka North and Dhaka South.

expressed among other things in the clothing of women who wore *sari*[5] but hardly ever *borka*[6], later on the *borka* style was revived (van Schendel 2009: 254–255). The increased wearing of *borka* indicates how women are considered responsible for maintaining 'purity', and the wearing of *borka* in public spaces can be understood as a "compromise between the need for women to take part in society and the desire to keep them in seclusion" (Rozario 2006: 378; see also Chapter 5.1.3 for a more detailed discussion of religion, society and women's movement in public spaces). A growing influence of Islam in society is witnessed with the spread of social religious institutions, for example the private *madrasas* are often financially and ideologically supported by Arab Islamic organisations, that have gained importance following the weakness of the Bangladesh welfare state (Lorch 2008: 13).

At the same time, notions of a mystic Islam, rooted in the tradition of Sufism, are still widespread in Bangladesh, and one of its expressions is the culture of *mazars*, shrines of Muslim saints, and charismatic spiritual guides, known as *pir*. This is often critically viewed by mainstream society for it is said to violate the "true interpretation of Islam" (Siddiqui et al. 2010), but such representations fall short of accepting the diverse and syncretic Islam characteristic to the region (Gardner 2000: 230). When considering religious identities of the Muslim and the Hindu population, it is worth noting that "there has always been strong cultural resistance in Bangladesh to such bipolar categorisation, not only with regard to social stereotyping but also at the most basic religious level" (van Schendel 2009: 37). For example, the *Baul* community relates their spirituality to Sufi, Tantric and Vaisnava traditions (van Schendel 2009: 37). However, such 'hybrid' identities do not go uncontested. For example, recently, a group of Lalon[7] devotees in Rajbari was assaulted by members of a local mosque, who forced them into a rite to make them 'real Muslims', shave off their beards and cut their long hair (The Daily Star 2011). Such incidents indicate an increasing conservatism and Islamisation that challenges the traditionally diverse and multiple socio-cultural and religious identities and frontiers.

5 *Sari* is the traditional clothing of women in most of Bangladesh (except for some ethnic minorities), which consists of a tight blouse leaving the belly uncovered, an underskirt referred to as a petticoat and a long piece of cloth, the *sari*, which is draped around the body, commonly in 'Bengali' style.

6 In Bangladesh *borka* usually refers to the combination of a long coat that is worn by women on top of other clothing with a veil covering the head. This can mean covering the head tightly, with a headscarf (*hijab*) ensuring that no hair is shown, or covering the face completely and only revealing the eyes. Some women in Dhaka also wear only the coat without covering the head with a veil or covering their head only loosely (I have for example seen this with many female garment workers). The style chosen depends on the purpose and location, a person's perceptions and the norms and expectations of the (local) society.

7 Lalon Shah or Lalon Fokir is one of the most famous *Bauls*, who lived in the 19th century and is widely appreciated among Bengali Muslims for his spiritual poetry and songs. Lalon Shah in one of his songs referred to his multilayered identity: "People ask if Lalon Fakir is a Hindu or a Mussalman. Lalon says he himself doesn't know who he is."

The various public holidays celebrated in Bangladesh also indicate the complex and multi-layered identities prevalent in society. Expressions of liberal nationalism, attracting huge crowds, are the celebrations commemorating the martyrs on International Mother Language Day on 21st February (*Ekushe*), marking the day in 1952 when the Pakistani armed forces fired at students demonstrating in favour of keeping Bengali the official language, and the Victory Day on 16th December, marking the end of the Liberation War in 1971. Both holidays change public spaces considerably. The first one produces an atmosphere of shared mourning, where especially at a memorial on the Dhaka University campus, the *Shohid Minar*, flowers are laid down, but also at the many replicas of the memorial within each neighbourhood. The Victory Day is then joyfully celebrated and both the campus and the urban neighbourhoods are decorated with national flags. The Bengali cultural identity is celebrated on the first day of the Bengali New Year on 14th April (*Pohela Boishakh* – the first day of the *Boishakh* month) when especially the urban population, regardless of income, dresses in red and white and millions gather in Ramna Park and on the Dhaka University campus, but also at smaller neighbourhood fairs. The main Muslim celebrations are the two *Eids* (see Chapter 4.2.5). Public life all over Bangladesh changes considerably during the fasting month *Ramadan*, when during the daytime all food stalls are hidden behind cloth curtains to protect those not fasting, and only open up for *iftar*, the breaking of fasting at sun-set. During the two *Eids,* the *Eid* prayers are held on large open fields, referred to as *Eid Gah Math*, which are spread all over the city and are used for various other religious and non-religious purposes throughout the rest of the year. Furthermore, Buddhist, Hindu and Christian public holidays also exist for all citizens, but their overall public visibility is much less than the liberal-national, cultural or Muslim celebrations.

5.1.3 Female positions in gender relations

In the following I will discuss the position of women in society in Bangladesh with specific focus on their economic participation, religious and social norms and women's mobility in public spaces. The sub-chapter will conclude with a discussion of changing practices versus resilient gender ideologies.

In the past decades women's labour force participation has increased tremendously, especially that of young women aged 20–24 years. This is particularly due to the growth of the ready-made garments sector in urban areas. In 2003 the national women's employment rate stood at 26% (World Bank 2008: 55pp.). While 26% can be considered low, this is a considerable rate for a country in South Asia, where women's employment rates tend to be among the lowest worldwide. The World Bank study considered all strata of society, i.e. also included those households where women do not have to contribute to the household income out of economic necessity. Accordingly, the Urban Livelihoods Study carried out in 1997 among slum dwellers in Dhaka-Mohammadpur slum areas arrived at different results arising from the necessity of many low-income households to have more than one income earner. Among the women interviewed, 40% were in paid employment, while another 27% had worked in the past (Salway et al. 2005: 326). With regard to the predominant gender relations, however, both studies derive similar conclusions. Women's employment in many ways challenges traditional norms and gender relations in Bangladesh, but at the same time these norms show a high level of persistence and resilience and women's employment is far from overcoming them.

Socio-culturally women's employment is regarded as secondary and inferior to men's employment (Salway et al. 2005: 339), and women as additional income earners are an indicator of a household's poverty and the inability of a husband to provide for his family, and thus translate into a loss of family prestige (Salway et al. 2005: 342). In contrast to this traditional, low-income and middle-class perspective are a considerable number of women with an academic education who are well respected for their work. But for women of low (formal) education levels employment options are more limited than male ones, as women are excluded from several manual occupations. Brick-breaking, often performed in public spaces and/or on construction sites, is a surprising exception. Women are restricted from many street trading occupations, including those requiring high mobility (Salway et al. 2005: 340).

Very important for women's position in society is the social norm of *porda* (also spelt *purdah*) that refers to the seclusion of women in traditional society. Maintaining *porda* includes the control of women's mobility and their exclusion from public, male space (Salway et al. 2005: 320). The concept is tied to notions of family honour, purity and prestige. What is understood as *porda*, however, differs both geographically and socially, and "purdah cannot be understood as a fixed or homogenous set of beliefs and practices, but instead as multidimensional" (Gardner 1994: 2). The contestations and multiple interpretations of *porda* are underlined by the World Bank Study (2008: 77) that failed to identify strong links between *porda* and women's welfare and outcomes for gender equality. Similarly, Gardener analysed that women's relative access to resources, especially property, was determined by the economic position of a household rather than her maintenance of *porda* (Gardner 1994: 16). Furthermore, age is a decisive factor for women's necessity to adhere to *porda* rules. While young and especially unmarried women are under close surveillance by their families and (potential) in-laws, elder women, beyond the menopause, are able to move more freely, change behavioural patterns and as elders are also respected by male society and its institutions (Gardner 1994: 15; similarly Bose 1998 for Kolkata, India and Tarlo 1996 for Gujarat, India, see Chapter 2.2.2). The contradictions arising from such traditional notions where women are "expected to maintain the purity and modesty required to safeguard their menfolk's honor" (Rozario 2001): 162) and their increasing participation in the job market, in politics and in education is obvious. Furthermore, through the NGO sector that often focused on women, for example in micro-finance schemes, women have become more powerful in society, however, as discussed in Chapter 2.2.2 this does not necessarily lead to increased political agency. On the other hand, this challenge to traditional gender relations and traditional male power-holders resulted in anti-NGO campaigns by both Islamists and traditional elites (Naher 2010; Rozario 2006). Women's reactions as recorded by Naher (2010): 323) were especially "everyday forms of resistance of the weak" in the form of gossip and jokes that captured the hypocrisy of the powerful.

While participation in employment outside the home, especially working in garment factories and as domestic helpers, has increased women's mobility, this new mobility tends to still be controlled by husbands, indicating the "persistence of a gender ideology that asserts male authority and gives a husband the *right* to control his wife's movements beyond the home" (Salway et al. 2005: 328, original emphasis). Community-level restrictions furthermore continue to prevent women from visiting certain places such as tea stalls or road-side eateries (Salway et al. 2005: 328). In a society where marriage is central to retaining women's honour, physical security and social recognition are also closely tied to family and 'being married' to reduce potential harassments (Salway et al. 2005: 343). As already indicated above, the *borka* and other styles of veiling are increasingly visible in public spaces, and this can be interpreted in manifold ways: as a strategic choice for increased personal protection and security in a male environment, as instrumental for enabling socially acceptable freedom and movement, but also as a matter of identity emanating from a commitment to an overtly religious lifestyle (Rozario 2006: 376).

The above summary of women's position in society, although far from complete, indicates the multilayered nature of gender relations in Bangladesh. An increasing participation in the labour market and participation in NGO activities suggest that women challenge traditional gender relations. Representations of women as the "powerless victims of patriarchy" are challenged at the same time by powerful female characters, for example goddesses or other superhuman protectors and literary characters, but also by an array of influential women in politics, education, advocacy and the arts (van Schendel 2009: 34). But other events point out how resilient the gender ideologies, especially the social norm of *porda*, still are. The events that followed several rape incidents by a group of male students at Jahangirnagar [Jahangirnogor] University in 1998 resulted in the rape victims being blamed by the media for moving around the campus after sunset, and thus not maintaining their honour and purity, while the rapists were not taken to court (Rozario 2001). Here, *porda* implied that

> "[i]f a woman loses her purity/virginity, the issue of utmost importance to the vast majority of the Bengali people is that she is a fallen woman, who has dishonoured herself, her family, and the community. It is of little or no concern that she was violated against her wish" (Rozario 2001: 163).

Any analysis of gendered access to public space thus has to consider this setting of persistent gender ideologies and simultaneously changing gender relations.

5.1.4 Traditional institutions and power structures

Besides the institutions of statutory government (see Chapter 5.1.1 on local government and Chapter 5.2.2 on the planning authority's performance), especially rural Bangladesh is to date characterised by strong traditional institutions and power structures. To some degree these rural institutions and power structures also play an important role in urban areas. This is especially the case in those urban areas which are characterised by a relative absence of (legal) statutory power, i.e. most of the low-income settlements without (initial) planning approval, where rural-urban migrants have settled in recent decades and brought with them some of the rural traditions of community organisation. But this is also the case in some middle-income areas, where development takes place mostly independently of state policies, as well as in the settlements of Old Dhaka, where for a long time similar structures have existed and indeed continue to exist even in the newer residential areas adjacent to the traditional Old Dhaka areas. Given these rural-urban continuities it seems useful to draw attention to some of the institutions prevalent in rural Bangladesh, serving to explain some of the power structures common in parts of urban society. As the further analysis in this research will show, these traditional institutions of society do exercise considerable influence in both study settlements, and the inherent logics are reflected in the ways these communities are organised and function.

In rural Bangladesh patron-client relationships have been dominant in organising society for a long time, and as indicated above this is also very much the way the political system is organised. The underlying hierarchies between families are constructed considering symbolic power, especially (patrilineal) kinship relations and social position, and economic power, although such hierarchies are always contested and dynamic (Gardner 2000: 128–151). Such a tradition of hierarchies and patron-client relations is also seen as the reason why, despite development projects targeting the poor, it has often been observed that those most marginalised were not reached, while instead powerful individuals enjoyed the benefits[8]. Consequently, (rural) communities are characterised by differentiated power structures, which explains geographies of inclusion and exclusion and is also reflected in the membership of traditional institutions.

Three important institutions based in social norms and cultural practices and differentiated by Lewis and Hossain (2007: 282–284) are the *shalish*, the *shomaj* and the *gusthi*. The *shalish*[9] is a local conflict resolution system that is officially recognised by the legal system, but does not have legal authority in criminal cases

8 Prominent studies in this regard are the BRAC (Bangladesh Rural Advancement Committee – a large NGO in Bangladesh involved especially in microfinance both in Bangladesh and abroad) study 'The Net' of 1983 about relief distribution to landless households during a drought in 1979 (see the discussion of Lewis, Hossain 2007) and the book of a previous employee of the German Federal Ministry of Development Cooperation 'Tödliche Hilfe' ('Lethal Help') on failed development projects in Bangladesh (Erler 1985).

9 Alternatively, this is referred to as *bichar* (the process of doing justice), while the committee giving the verdict is referred to as *ponchayet.*

and marriage disputes (Lewis, Hossain 2008: 54). Many local disputes are nonetheless taken to the *shalish*, whose members are local leaders and elites (*matbors*), rather than consulting the formal judiciary which is especially inaccessible to the poor due to its high levels of corruption (see Chapter 5.1.1) and lengthy procedures. However, the *shalish* is also criticised for only slowly allowing women's participation, often restricted only to attendance, and for its elite domination that reinforces traditional power structures and gender ideologies (World Bank 2008: 85–86; Lewis, Hossain 2008: 56).

The *shomaj* refers to the local residential community and is based on the religious congregation and local neighbourhood, but the term can also generally be translated as society (Lewis, Hossain 2007: 289). The *shomaj* builds on the power of locally respected 'elite' leaders who engage in patron-client relationships with their supporters (Lewis, Hossain 2008: 72). These leaders are often, but not necessarily, members of the local *shalish*. Finally, the *gusthi* refers to the patrilineal kinship group, and often this results in patron-client relations between poor and well-off households belonging to the group, as the poorer households seek benefits from membership in the *gushti* while the wealthier households have a social obligation to share with their poorer kin.

These different social institutions, all based on patron-client relationships and locally respected leaders, are commonly interlocked with each other. Furthermore, these traditional structures are overlaid by alliances with the political parties and their leaders (Lewis, Hossain 2008: 73). In addition, both local leaders and politicians often build their influence on the mosque and religious practices, as Lewis and Hossain describe, based on Bode:

> "At election time, candidates favoured by the committee are invited to address the *jama'at* (muslim congregation) and even tour the village during their election campaigns, being careful to maintain a non-political façade by presenting themselves as 'patrons of the congregations' ready to donate materials for mosque repairs and improvements (Bode 2002)." (Lewis, Hossain 2008: 73p.)

Another religious tradition reaching urban areas is the *pirbad*, the acceptance of advice from spiritual leaders (Gardner 2000). Religious leaders like *pirs*, *moulanas* and *imams* exercise considerable power in local communities and are frequently linked to political processes (Lewis, Hossain 2007: 290). Furthermore, the practice of *porda* as a concept of maintaining the respect and honour of a family can also be considered a religious institution securing traditional power structures (see Chapter 5.1.3; Gardner 1994).

The *gusthi* does not seem to be as widespread in urban areas as the concepts of *shalish* and *shomaj*, but nonetheless it points to how deeply engraved patron-client relationships are in Bangladesh's society. While these structures prevail, Lewis and Hossain (2007: 293) also point to a diversification of strategies of local elites who previously maintained power mainly through agricultural dependency relations: "These [strategies] include an active local business association, the building of widespread 'party political' networks and more recently the phenomenon of "NGO-ing"" (Lewis, Hossain 2007: 293). In urban areas especially party political networks have become the main access strategy to traditional institutions and local leadership (Siddiqui et al. 2010: 324, 328; Banks et al. 2011: 14). Despite traditional institutions dividing citizens along constructions of power, it has to be noted how the boundaries continue to be dynamic and fluid, and such less rigid power structures tend to allow a "higher degree of bargaining and negotiation" and "more social mobility", opening up "room for manoeuvre" for the marginalised (Lewis, Hossain 2007: 295).

5.2 DHAKA – SCARCITY OF SPACE AND DIFFERENTIATED CITIZENSHIP

The exact time when the city of Dhaka started to develop is unknown. There is evidence that a new commercial town developed in proximity to an old Buddhist and Hindu temple, the Dhakeshwari Temple (Khatun 2009: 675p.), which today is the oldest Hindu temple of the city, built in the 12th century. In pre-Mughal days the city was a small town of craftsmen with localities that today still bear Hindu names such as Lakshmisbazar and Kumartuli (Chowdhury, Faruqui 2009: 58). By the late 13th or early 14th century Islam had reached the region and early mosque inscriptions tell of a pre-Mughal Muslim past (Chowdhury, Faruqui 2009: 58). In 1610, Islam Khan Chisti established Dhaka as the capital of the *Subah* (province of the Mughal Empire in South Asia) (Karim 2009: 34). The city expanded rapidly until the beginning of the 18th century (see Figure A-1). Under the influence of the British East India Company, the built-up area of Dhaka expanded only slightly and the population even declined as Kolkata had become the new capital of East India. Only with the separation of India and Pakistan in 1947 did Dhaka start to urbanise rapidly. After the Liberation War against Pakistan in 1971, Dhaka became the capital of the new nation Bangladesh and today continues to be its primate urban centre.

5.2.1 Urbanisation and the urban poor's struggle for space

Today Dhaka is one of the world's largest megacities with a population of 12.6 million estimated for 2005 and 20 million inhabitants projected for 2014 (Islam 2005: 15; see also Table 4)[10]. The high average density of 30,000 inhabitants per

10 Despite the key figures I provide in Table 4, a cautionary note seems necessary. The population estimates for the megacity of Dhaka relate to the 2001 census carried out by the Bangladesh Bureau of Statistics (BBS). The 2007 census results for Dhaka were not available at the time of writing (i.e. in 2011). Furthermore, for this research the availability of exact population data is not essential, for it would not impact on the findings on the socio-spatial conditions of everyday life. It seems advisable to handle the available population estimates with care, for they are based on ten year old census data and few facts are known about the real development that has taken place in this decade. The administrative boundaries of DSMA, commonly referred to as Dhaka megacity, include vast areas that can hardly be characterised

km² in Dhaka Metropolitan Area (DMA) (CUS et al. 2006: 40), the core urban area (including the low-density urban fringe in the East), indicates that space is a scarce resource in Dhaka. In comparison, the average residential density in the urban core of Shanghai is 26,000 persons per km², compared to the London urban core with 4,800 persons per km² (Burdett, Sudjic 2007: 198–199). Although the exorbitant growth rates of up to 10% per annum belong to the decade following independence, the city is still growing rapidly with an estimated 300,000 to 400,000 new rural-urban migrants every year (Bangladesh Bureau of Statistics (BBS), in World Bank 2007: xiii) and additional natural population growth.

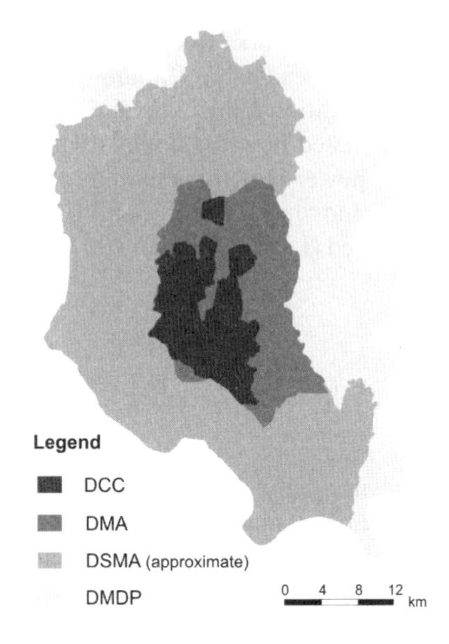

Figure 9: Dhaka megacity administrative boundaries

Land is in high demand, both by the urban poor and the middle to high-income groups. Land availability for the expansion of the city is highly limited as Dhaka is surrounded by rivers and their natural retention areas. Hence the current development in the urban fringe area is severely increasing the vulnerability of the city to natural disasters, especially flooding. The real estate sector is eager to develop both prime locations and urban fringe areas into office complexes, condominiums and shopping malls. This is accompanied by a rise in land values in central neighbourhoods (Islam et al. 2007: 33), which are the preferred location of poor migrants, as proximity between living and working places is of vital importance for socio-economically poor groups, enabling them to circumvent opportunity costs (Payne 2002: 152). As a result, rural-urban migrants and urban poor struggle to secure their share of city space, and these developments have triggered off unauthorised housing on newly created 'spaces' along and over water bodies and swamps using bamboo posts to support fragile platforms (Hafiz 2007: 55). These houses are highly unsafe, several of these structures have already collapsed, they lack sufficient services and are prone to disasters (Hafiz 2007: 64), but they enable the urban poor to occupy a space within the city.

as urban, indicating how administrative boundaries and urban frontier definitions may vary between countries and thus limit systems that rank megacities by size (see for example UN-Habitat 2008: 6). Furthermore, population figures are often manipulated to serve the interests of politicians and/or donor agencies (about reliability and comparability of populations statistics see Satterthwaite 2007).

Furthermore, the income ranges provided for 2005 also need to be considered with care, as at the time of this research, and even during my first field visits in 2007 these figures did not match reality. For example, a household with an income of 5,000–10,000 Tk could at the time of this research not be considered to be of lower middle income.

Area (2005): (see Figure 9)	145 km^2 (Dhaka City Corporation, DCC) 306 km^2 (Dhaka Metropolitan Area, DMA) 1,353 km^2 (Dhaka Statistical Metropolitan Area, DSMA, commonly referred to as Dhaka megacity)
Population:	5.9 million (DCC 2004) 9.14 million (DMA 2005, estimated) 12.6 million (DSMA 2005, estimated)
Population growth: (per annum)	1981: 8.1% (DSMA) 1991: 6.5% (DSMA) 2001: 4.5% (DSMA)
Population density:	29,857 persons/km^2 (DMA, 2005) 220,246 persons/km^2 in slum areas (DMA, 2005)
Population living in slums (see footnote 11):	3.42 million / 37.4% of the DMA population (2005)
Household income:	Monthly household income 2005 in Tk (DCC, estimated, 100 Tk equalled 1.21 Euro on 01.01.2005) Hardcore Poor (< 2,500) 20% Moderate Poor (2,501–5,000) 15% Lower Middle (5,001–10,000) 25% Middle Middle (10,001–25,000) 20% Upper Middle (25,001–50,000) 13% Lower Upper (50,001–100,000) 5% Upper Upper (> 100,000) 2%

Table 4: Key figures for Dhaka megacity (Sources: Islam 2005; CUS et al. 2006)

In 2005, 37% of Dhaka's inhabitants (DMA) were living in 5,000 'slum settlements'[11], ranging from clusters of only a few households to areas with up to 100,000 inhabitants (CUS et al. 2006: 20; see Figure A-1). These areas only claim 5% of the urban area, and thus densities are extremely high, with an average of 220,000 inhabitants per km^2 and maximum densities of up to 500,000 inhabitants per km^2 (CUS et al. 2006: 40). The living spaces of the urban poor are further

11 In the CUS-survey, slums were defined as settlements with a minimum of ten households and characterised by at least four of the following five criteria: predominantly poor housing, high population density and room crowding, poor environmental services, low socio-economic status and lack of tenure security (CUS et al. 2006: 14; see complete definition in Chapter 4.1.1). This definition is similar to the definition applied by UN-Habitat (2003: 8).
While overall the survey gives an overview of the locations of 'slums', a note of care is also necessary. Some settlements designated as slum areas are in fact characterised by a diversity of living conditions and in practice cannot be entirely designated as 'slums'. While some population lives in slum-like conditions, for example in Manikpara, and there are many slum pockets, these areas also have many clusters of middle-income housing not fitting the slum definition. Probably due to the regular appearance of slum clusters within them, these areas have been completely identified as slums – which, however, does not adequately represent their urban structures. Thus the clusters of big slums appearing on the map (Figure A-1) especially in the south-west and the east have to be viewed with care.

threatened by eviction, which unfortunately has a long history in Bangladesh, despite existing legal commitments (Hackenbroch et al. 2008). The state authorities frequently evict slum dwellers from public-owned land with little advance notification and without provision of adequate resettlement or rehabilitation schemes. The main reasons for the slum evictions of recent years were development of land for road construction, reclamation of drainage channels, housing projects, beautification of the city or removal of illegal encroachments (Hackenbroch et al. 2008). Space in the city of Dhaka is thus clearly contested with the urban poor being actually 'squeezed' within the city, always under the threat of displacement.

As illustrated in Figure A-1, the 'slum' areas as identified by the CUS are not located randomly in the urban agglomeration but follow some of the main development axes, both economically and geographically. A cluster can thus be found along the western flood protection embankment, where the leather industry provides employment, while at the same time land especially outside of the embankment is available. Other clusters can be found where the city is expanding towards the urban fringe; however, these areas can only be termed low-income with care, as most of them are rather developments by the lower middle class with scattered low-income pockets (see also footnote 11). The two study settlements of this research constitute more consolidated developments, located at economic centres of the city – Manikpara with its proximity to the trading hubs of Old Dhaka and Nasimgaon with its proximity to the service-oriented high-income areas and to the industrial areas including garment factories.

5.2.2 'Informal settlements' and notions of citizenship

The 'slums' identified by the CUS survey are often referred to as 'informal settlements', distinct from planned or formal settlements (see discussion of terms in Textbox 2, Chapter 2.3.2). Informal urbanisation practices do not, however, solely prevail in the low-income settlements without planning approval but form an 'organising logic' (Roy, AlSayyad 2004a) of most of Dhaka's neighbourhoods. For example, many up-market housing and land development projects, both by private real estate developers and government agencies, have been developed in violation of the Dhaka Metropolitan Development Plan (DMDP) guidelines. Of the 17 housing and land development projects found to be located in flood flow zones, harming the urban environment and thus illegal according to the DMDP guidelines, a Review Committee formed by the Ministry of Housing and Public Works found that two projects were being implemented by the development control authority RAJUK[12] itself (The Daily Star 2010b). This example indicates how in

12 RAJUK stands for *Rajdhani Unnayon Kortripokkho*, which translates into Capital Development Authority. It emerged in 1987 as a new governmental institution with a view to improving the urban environment through urban planning, land development and building control in the city and its periphery (Website RAJUK). It is responsible for developing the three-tier DMDP, consisting of the Structure Plan, the Urban Area Plan (both approved in 1997) and

Dhaka – similar to many other cities of the world – informal practices in middle and high-income neighbourhoods are often tolerated by statutory authorities, resembling what Yiftachel (2009) refers to as 'whitening' of 'gray space'. In contrast, such 'whitening' practice is largely absent for low-income settlements without planning approval and these at best remain as 'gray spaces' or are 'blackened' and consequently, due to their insecure land tenure arrangements, are exposed to the threat of forced eviction. These practices can be understood to deepen unjust spatialities, as analysed in Chapter 2.4 especially with reference to Soja (2010).

Underlying this is a differentiated notion of citizenship that renders part of the population as welcome urban dwellers, while the majority of the urban poor are viewed only as temporary urban dwellers, whose urban status is reduced to contributing labour to the industrial sector and providing services to middle and high-income households. This notion is also reflected in the neglect of urban poverty by state policies and action programmes (Banks et al. 2011). Despite urban poverty having become a severe problem that is not likely to diminish in the coming years and can rather be expected to increase in severity given the expected impacts of climate change on cities like Dhaka (Banks et al. 2011), there is still a rural bias in politics and society in Bangladesh (van Schendel 2009: 202; Banks et al. 2011: 9–10). A recent update of the publication 'Social Formation in Dhaka, 1985–2005' (Siddiqui et al. 2010: 328) identifies Dhaka as "basically a city of poor and uneducated rural people". The urban poor tend to be criminalised and "[e]lite perceptions remain focused on rural areas as the rightful home to the poor" (Banks et al. 2011: 23). This differentiation of citizenship also becomes obvious from the statement of a RAJUK planner, who expressed his opinion about the planning system's responsibility to cater for slum dwellers: "The informal, poor people do not own land, so how can I do planning for them?" (RAJUK planner, 02.03.2008). The question about for whom the city is planned carries with it the question of the degree of spatial justice achieved by planning decisions. The responsiveness of planning in catering for the full strata of its inhabitants is further hampered by the complicated governance structure of Dhaka. As many as 19 ministries and 40 government organisations are involved in the planning and development of Dhaka with practically no coordination between them (Islam et al. 2003; Siddiqui et al. 2000).

While middle and high-income settlements, as discussed above, can be subject to informal practices despite their secure land tenure, statutory authorities are far from absent in the low-income settlements without planning approval. The semi-autonomous utility authorities Dhaka Water Supply and Sewerage Authority, Dhaka Electricity Distribution Company and Gas Distribution Company, extend their services to these settlements with the consent of state authorities (see for example Hossain 2011). This service extension does not, however, follow the formal price system. Instead, the charges for provision in low-income areas, organised via middlemen, far exceed the charges of those areas 'formally' supplied

five Detailed Area Plans (DAPs, approved only in 2010) and for controlling the implementation of the plans.

– thus the urban poor pay most for the basic utilities water and electricity. Besides the state's involvement in low-income areas, a variety of actors mainly operating outside of the statutory sphere are filling the gaps of state power and the rules of the game within a particular settlement (this is further elaborated in Chapter 1 with reference to the study settlements).

5.2.3 A critique of planners' perceptions of the city

While Dhaka's problems are manifold, some of them discussed above, the dominant view of planners and architects on future development is characterised by rather technical solutions, including undoubtedly important infrastructure improvements and new projects, the planning of satellite towns at the urban fringe and the relocation of industries to the urban fringe. While touching on important issues for Dhaka's future, the urban development plans, including the DMDP consisting of the Structure Plan, Urban Area Plan and the DAPs (see footnote 12), seldom follow an integrated approach and fall short of recognising the specific requirements emerging from the socio-spatial fabric of the city.

The debate that evolved around the 'Proposal for a Housing Development Programme in Dhaka City' (Islam, Shafi 2008) also exemplifies the rather technocratic approach applied. On the one hand the proposal is progressive in recognising the need to accommodate the urban poor in the city, a notion that has often been neglected (see Chapter 5.2.2). However, the approach to housing the urban poor, namely to provide multiple-storey houses on government land, fails to recognise and integrate the existing structures and mechanisms that enable the urban poor to access housing. Instead of building on the potentials these 'informal' housing markets might offer and integrating the poor into the process of planning, the proposal returns to solutions that have failed to help the poor in the past (for example the Bhashantek resettlement project that in the end only benefited middle-income groups). Thus it has also received criticism for its production-oriented view:

> "The proposal has been a classic example of the native 'self's' portrayal of what the 'other' should have. This portrayal, soaked in unmasked parochialism, originates from the narrow, uncritical positivist urban studies a premise that delivers commissioned surveys more comfortably than original studies" (Ghafur 2008).

The discussion of the housing proposal for Dhaka points at the general necessity for urban planning to become more responsive to the needs of citizens. As citizens of Dhaka are diverse, socio-economically but equally so concerning other characteristics such as gender and religion, planning responses that do not take note of diversity cannot serve the city in the long run and cannot create spatial justice (see Chapter 2.4).

5.2.4 City representations: hopeless, fragmented, smooth or ordinary?

In representing Dhaka, a tendency can be found among authors to create a picture of doom, like the one Davis creates of a 'planet of slums' (Davis 2007). Thus, for example, a publication titled 'Dhaka – A City of Dirt, Darkness and Deprivation' (Rahman 2003) immediately triggers images of hopelessness, despair, chaos and apathy, terms often used in describing the city (see for example Eckardt 2009: 113). While Dhaka is surely 'bursting at the seams', suffers from an environmentally critical situation, and widespread and severe urban poverty, I call into question the helpfulness of such one-sided accounts, for they do not recognise the potentials that lie in the bustling megacity, and fail to recognise local perceptions that may tell a different narrative, one of the possibilities that urban life in Dhaka offers to many. A conceptualisation of Dhaka as a 'hopeless city' at best reflects an elite perception of the megacity, recreating existing stigmas and justifying the dominant politics of exclusion of the urban poor from the city space. I thus understand Dhaka as both a space of possibilities and opportunities and as a space where urbanisation has reached a highly critical level. Bertuzzo (2009: 89pp.) also finds that in the narrative of its inhabitants the city means an opportunity and the availability of infrastructure, and at the same time inequality, exclusion and injustice. Thus Dhaka also seems to be a city of contrasting elements, but a closer look reveals how closely entangled its elements are.

Although the urban poor face difficulties in securing their share of city space, Dhaka is not a highly segregated city, compared to other cities worldwide. The image of stark contrast that is widely used in publications on Dhaka – that of Banani [pronounce Bonani] Lake with the largest '*bosti*' Korail and its CI-sheet huts in the foreground and the up-market model-towns of Gulshan, Banani and Baridhara in the background – is hardly representative of the city's urban fabric. In almost all other parts of the city one finds a dense and mixed pattern of industries and residential areas of different income groups, with rather smaller clusters of low-income areas (see Figure A-1 for distribution of 'slums' in DMA according to CUS et al. 2006). In contrast to what the prominent image suggests, Dhaka is less segregated and everyday life in a rather fragmented Dhaka contrasts many of the divisions that exist in housing structures. At different moments the urban space is frequented by different population groups, ranging from the ever-present street vendors and rickshaw pullers in all parts of the city to the movement of garment workers to their workplaces and the leisure seekers visiting the few green spaces for morning or late afternoon walks. Accordingly, Bertuzzo and Nest (2008b), referring to a reflection on space by Deleuze and Guattari ('1440: The smooth and the striated', Deleuze, Guattari 1987 [1980]) which discusses space as a continuous transition between striated, planned space and smooth space (Bertuzzo, Nest 2008a), conclude how 'smooth and striated' "not only coexist in Dhaka, but they are generating the constant tension and transition that make the city vibrate under huge debates, contrasts, and struggles" (Bertuzzo, Nest 2008b: 28). A trend for segregation is, however, visible and lingering, for example in the up-market model town developments of the real estate sector that create secluded city spaces or gated communities often at the cost of low-income communities who are moved out to make space or because they cannot afford the increasing land prices in the city.

In a last note I would like to refer to Dhaka as an ‘ordinary city’, a term coined by Robinson (2006) in her attempt to move beyond developmentalist conceptions towards a post-colonial framework for thinking about the world’s cities. In the approaches that rank cities as global cities or world cities in relation to advanced producer services (for example Friedmann 1986; Sassen 2005; Taylor 2004), Dhaka does not appear on the list for it is not an economic, financial centre of the world economy. Nonetheless, Dhaka is affected by global dynamics, for example by being a production centre of the global garment industry, but also by the growing demand for land (for agriculture and real estate) and resulting speculation.

Expressions of these global processes on the city scale are the export processing zones of the ready-made garments industry, but also the development of new model towns and shopping malls along the urban fringes on previously agricultural land, especially paddy fields within the river flood zones. Within the city, these processes affect also the organisation of neighbourhoods and lead to the displacement of especially low-income areas. As the example of the locations of the plastic industry shows, such global processes also affect the micro level. When in 2009 the oil price and subsequently the price of new plastic pellets declined, exports from China became more attractive than the plastic recycling process. Almost immediately the plastic recycling industry reacted by lowering production, which considerably affected everyday life spatial practices and threatened the existence of many small businesses. But the city’s residents are not without agency, and although sometimes politically fuelled, the regular protests of garment workers for improved working conditions and minimum wages underline this.

The production of Dhaka’s city space is thus embedded both in local and global processes. Global city rankings, emanating from a Euro-American tradition of urban theory building, tend to put the megacities of the Global South off the map, seeing them as “‘big but powerless’ entities” (Roy 2009a: 821). Criticising this bias in urban theory in her article on ‘The 21st-Century Metropolis: New Geographies of Theory’, Roy (2009a) calls for a ‘worlding’ of cities disrupting the standard geographies of core and periphery and recognises “that while distinctive and alternative modernities are produced in multiple urban sites, such experiences can speak to and inform one’s analysis of other places” (Roy 2009a: 828). Following these lines of argument I consider Dhaka an ‘ordinary city’ and believe that this research will not only help to understand the spatiality of everyday life and urban informality in the Global South, but will also inform a wider debate on the modes of the production of space and the preconditions of planning at this time of globalisation and urban restructuring. The two study settlements that are at the focus of this research will serve as examples to discuss the current modes of the production of space in Dhaka. As already outlined above (see Chapter 4.1.1) they are not representative for the whole city, which was also not the aim of this qualitative research. However, the findings from these two rather different settlements will contribute to a discussion of similar issues taking the whole city into account.

5.3 MANIKPARA – A DIVERSE AND HIGHLY DYNAMIC URBAN QUARTER

A first visit to Manikpara left me with a picture full of contrasts that persists until today. Walking through one of the narrow lanes in between multi-storey houses requires constantly giving way to workers carrying enormous bags with plastic materials on their heads. Everywhere are workshops and factories of the plastic recycling industry, from small-scale businesses where two persons are sorting used plastic materials, to large and automated factories. The activities of crushing plastic waste, and burning and smelting it, contribute to considerable air pollution in the area, and the working conditions in many of the enterprises are harsh. But then there are other, surprising places, like a garden that was established on the flood protection embankment by a community initiative. Especially in the late afternoon hours of the summer season in April and May, residents of all socio-economic backgrounds take a rest in the shade of the trees and enjoy the swift breeze coming from the riverside. Mobile vendors come to the area, selling typical snacks (e.g. puffed rice with chillies and spices), cucumbers and roasted nuts. Beyond the dull picture of the plastic recycling industry thus emerges a picture of a dynamic and diverse everyday life shaped by multiple realities and the underlying contradictions and fragmentations that so often characterise Dhaka's vivid urban quarters. In the following sub-chapters, I attempt to present the key features of Manikpara's urban fabric, society and everyday life that are of central relevance to this research.

5.3.1 Development and living conditions

Manikpara is located close to the historic area of Dhaka under Mughal rule. At the time of the Mughal Empire today's Manikpara was part of the Buriganga River; later on it became a low-lying marshland that until the 1970s was mainly used for agriculture. Housing development started from the 1970s with rural-urban migrants arriving mainly from areas south of the Dhaka District. The first settlers constructed houses on bamboo platforms above the water or swamp held by bamboo pillars. At that time boats and narrow bridges constructed from bamboo poles were used for moving from one place to another. A pond was filled gradually and the Eid Gah Math, still existing today, emerged as a central landmark from where development continued. Because the flood protection embankment did not yet exist, large parts of what today is Manikpara were still part of the river.

Since the 1980s the appearance of Manikpara has totally changed, since the flood protection embankment was built and the area became dry land. It started to develop as one of the centres of the plastic recycling industry in Dhaka and Bangladesh (see Textbox 3). Improved road communication, partly the result of urban upgrading schemes from the 1980s as well as community initiatives for road widening and paving, allowed the industry to thrive because now heavy machines could be transported to the workshops. The thriving economy resulted in a highly dynamic urban transformation process in this mixed residential and industrial area. The land values reported by two key informants confirm this: in the beginning, when the area was still a marshland, land was available for 400–500 Tk per *katha*[13]. About 10 years ago land values were estimated at 150–200,000 Tk per *katha*, and in 2010 5–6 million Tk per *katha* (750–900 Euro/m²) were paid for land beside accessible roads and 3–4 million Tk per *katha* (450–600 Euro/m²) in the less accessible areas.

Textbox 3: The recycling industry in Dhaka

The recycling industry in Manikpara and the surrounding areas comprises of a number of different activities in a value chain. Small waste dealer shops (*bhangari dokan* in Bengali, i.e. shop for discarded materials) and larger wholesalers buy plastic that has been collected all over Bangladesh in large quantities. This plastic is then sorted by type at both the *bhangari dokans* and the wholesalers (both are referred to as *mohajon* locally, which refers to the owners of these businesses). Normally three to five persons are employed by the sorting shops that can be found all over Manikpara. The next step of plastic recycling are the shredding enterprises where the materials are cut into small pieces using machines. Often the *mohajons* sub-contract the plastic shredding enterprises and afterwards their own workers are involved in washing and drying the shredded plastic pieces. This is mostly carried out along the embankment slopes, where the plastic pieces are first washed in the river and then dried in the sun. But a few businessmen have also specialised in this washing and drying process and offer it as a service to the *mohajons*. After washing and drying the plastic pieces are taken to the pelletising enterprises where plastic granulate is produced. This granulate is then sold either directly to the factories that produce products from the recycled materials or to retail shops for plastic granulate. While generally young men tend to work in the plastic recycling industry, the share of women among the employees in pre-processing activities, i.e. sorting, shredding, washing and drying, and pelletising, stood at 20% in 2008 (Staffeld, Kulke 2011: 216–217).

Today, Manikpara is a consolidated, if not saturated, multi-storey inner-city neighbourhood. The history of development can be read from the existing housing structures. Around the Eid Gah Math and the accessible roads branching off it, multi-storey *paka* structures have developed (*paka* is the Bengali term for more permanent housing structures normally made of bricks). Along the narrow lanes in the west and along the Embankment Road – until April 2009 unpaved and full of dust and noise – the houses are predominantly made from CI-sheets, some of them with up to three storeys, or *semi-paka* (brick structures with CI-sheet roofs; often the ground floor is *paka* while the upper storeys are made of CI-sheets). Both housing types are organised into single rooms occupied by one family, while there are shared bathrooms and kitchens. Only some of the apartment blocks comprise of self-contained flats. However, the socio-economic status of a household tends

13 The Bengali term *katha* is a common measurement of land sizes; one *katha* equals 66.9 m².

to be reflected in their housing choice. The low-income households predominantly occupy the CI-sheet houses in which they rent a room from the landlords. The middle-income groups rent rooms or are landowners of the *paka* houses, where despite often similar room cluster organisation they do not have to share kitchen and bathroom facilities with as many other households as in the CI-sheet houses.

When looking at Manikpara from the Embankment Road, the 'frontier' of transformation is clearly visible, in the background the multi-storey blocks and in the foreground the CI-sheet houses (see Photo A-1 for a panorama of the urban landscape). In the years of my research I have witnessed the dynamic and continuous re-development of single storey CI-sheet houses transformed into multi-storeys and of CI-sheet houses transformed into *paka* buildings. This process is continuing and making it difficult for low-income households to remain in the area, because the house rents in the *paka* buildings are often double the amount (2,000–2,500 Tk in the *paka* buildings, 1,000–1,500 Tk in the CI-sheet houses, both including service charges).

Today, Manikpara is characterised by a high building density with many three to four-storey buildings and a few higher ones, an infrastructure that was designed for a considerably lower density, and very few public spaces. The densities calculated by the CUS survey for the area characterised as 'slum' (approximately 6 ha) was 265,000 persons/km^2, which is quite high but could well match the reality and the figures of the 2001 BBS census. The male-female ratio in the BBS survey is interesting. About 60% of inhabitants were male, which indicates that at least in 2001 there was still an in-migration of men searching for work in the factories while leaving their families in the villages.

Local identity

Interestingly, the CUS in their Slum Mapping of 2005 characterised almost one complete *mohalla* (sub-unit below the administrative Ward level) of Manikpara, and thus most of the study area, as a 'slum'. Despite the changes described above, the living conditions especially in the multi-storey CI-sheet houses are still characterised by poor quality of housing, largely without ventilation, and room crowding. Water, electricity and gas are available but supply is erratic especially in summer. Thus some households, especially the socio-economically poor, are still living in conditions similar to those associated with 'slum'. However, as outlined above, many of the 'local middle class' – which I use to refer to the households of comparatively higher income commonly involved in plastic recycling as businessmen – do not live in more spacious conditions, except for being able to choose a *paka* house often including a narrow balcony. However, no one would today consider them as 'slum dwellers' and a closer look at the local perceptions reveals how technocratic terms and definitions made by urban planners can deviate from the perceived reality.

In the local understanding people do not consider Manikpara a *bosti*. Only a small area north of my study area was locally identified and referred to as a *bosti*.

Their own area they referred to as industrial or residential, while the *bosti* was considered 'another place'. The following quote from a resident about the development history of Manikpara even reveals that despite slum-like conditions, Manikpara was not perceived as a *bosti* in the past – although this was the 'label' it was given for many years by the media and donor-financed upgrading schemes:

> "There was a *bosti*-like settlement. But the houses were not located densely like in a *bosti*. As an example, I have two *katha* of land [appr. 134 m²]. I have constructed two row houses keeping an open space in the middle. That was not like a *bosti*. [...] In this location, all of the houses were *kaccha*[14]. There were ditches, the river at this location. A lot of big boats used to arrive here. The boats were full of straw, clay materials. That was such a beautiful scenario to us." (Jahangir, 19.04.2010)

This self-representation, here expressed by a local businessman, together with the reality, where many of the households belong to the local middle class, reveals enough for the area not to be considered as a 'slum'. The mismatch with the classification of the CUS may be attributed to the fact that in the five years since the survey in 2005 the area has transformed considerably.

Especially among the businessmen of Manikpara who have been residing in the area for many years or were born there, Manikpara is home, *bari*[15], and during interviews they expressed a strong sense of belonging. These roots become evident from the following statement by a businessman referring to a conflict about the exact location of a road, because this determined Ward boundaries:

> "Now if the road was provided at the back side then this area would become Ward A. Then we would have become inhabitants of Ward A. [...] We claim to live in Ward B. Our birth and death – everything is here. On this matter we did grouping, confrontation, demonstration – many things [of resistance] against the MP." (Mesbah Uddin, 31.03.2010)

Furthermore, the local businessmen referred to Manikpara as part of Old Dhaka, although the area is a rather new development. This Old Dhaka identity carries with it a specific set of social values and norms expressed in many aspects of everyday life which will be taken up again in Chapter 7.1 and 7.2.2.

The Old Dhaka connotation of Manikpara also becomes evident from its early organisational set-up. Some of the persons who were involved in the land selling process from the 1970s were connected to the Nawab family (pronounced Nobab)[16]. The Nawab family were *jomidars*[17] and one of the most influential fam-

14 In the context of housing, *kaccha* refers to houses made from nondurable materials (bamboo or polythene). Here I differentiate *kaccha* from CI-sheet houses, although it sometimes is also used for these.

15 *Bari* refers to the village home and the place someone originates from. It thus carries a sense of belonging and ownership, although an increasing number of people are landless. In the city it is furthermore used to denote ownership of a house, while the rented house would be *basha*.

16 Nawab was a title from the Mughal Empire and later on became a high title for Muslim nobles. In Dhaka this referred to one (politically) highly influential family of Old Dhaka and the main *jomidars* (see next footnote), whose influence declined during the East Pakistan times (read more in Rashid 2009).

17 The term *jomidar* (often spelt *zamidar* or *zamindar*) originally referred to local lords settling in rural areas during the Mughal Empire. In the British period the *jomidars* became *de facto*

ilies in Old Dhaka in the 19th century. The names of these land brokers also reflect their respected status in society for they all have acquired religious titles, i.e. Haji and Hafiz. Especially one of them was mentioned by several interviewees as having contributed to the development of the locality considerably by building a mosque and a school and donating the Eid Gah Math. Till today the *bishisto murobbi*, notable elders, are highly respected in local society. Furthermore, development initiatives by groups of businessmen are also common. Additionally the Ward Commissioner and local MP are influential elected politicians, while the political parties' sub-organisations are also powerful (for details see analysis in Chapter 9.1.3).

Representation in official plans

In the DAP of RAJUK (see footnote 12), Manikpara is identified as a location of insufficiently serviced low-income areas (Website RAJUK-DAP). An indication of prevalent prejudices against the urban poor is the mentioning of crime and anti-social activities in connection with low-income areas and in consequence the suggestion to relocate 'squatter settlements' to Bhashantek (the failed resettlement scheme) and similar areas (Website RAJUK-DAP).

Manikpara's existing land use is characterised as mixed use while the proposed land use is residential along the Embankment Road and mixed residential-commercial use in the internal area (Website RAJUK-DAP). The plan further suggests urban upgrading (housing and environmental) and renewal schemes for the area. Interestingly, the plan specifically mentions the system of local, traditional institutions and underlines their usefulness for development initiatives:

> "Old Dhaka has local Sardar [*shalish/ponchayet*] system (local Matbors/Leaders) and it is a longtime [sic] patronage system of old [sic] Dhaka society to control old [sic] Dhaka's social harmony and culture and social integration. This can be an effective media via which initiation of urban renewal program can be done" (Website RAJUK-DAP).

However, this notion fails to recognise and discuss how such institutions may as well result in exclusionary practices, as the term 'patronage' already suggests. Generally, the problems of narrow roads, low accessibility, mixed land uses and over-densification are recognised, however, the proposal remains vague and ambivalent: on the one hand it is proposed to preserve the distinct urban fabric of Old Dhaka, on the other hand vertical expansion is suggested to make way for open spaces, utilities and roads.

landowners and collected taxes payable to the government. Although the system was changed in 1950 and land ownership restricted to a maximum of 13 ha, the *jomidar* families and their intermediaries remained influential. (van Schendel 2009)

5.3.2 Public spaces

Public spaces (i.e. roads, footpaths and squares) within the built-up area of the study settlement, excluding the space beyond the Embankment Road, account for only 12% of the surface area. There are only two larger public spaces, the Eid Gah Math and the extension ground for a school. The Eid Gah Math was donated by one of the land brokers and religious leaders of the early days. Some years ago it was fenced by the local mosque and *madrasa* committee which is in charge of the Math, and today it is normally not accessible to the public in everyday life but only during the *Eid* prayers. The school ground was a swamp and waste deposit site in 2007 and was filled gradually, until in 2010 it was covered with sand and made available for children to play on (see Photo A-25). Except for these two areas the public space inside the built-up area is limited to the network of roads and footpaths and road junctions. Most of the roads are not or hardly accessible for cars, and transport is thus limited to rickshaws and pushcarts. A number of lanes are not even accessible to rickshaws and accordingly goods are transported by carriers. Nonetheless, the roads and footpaths as well as street junctions are occupied by street vendors, shops and production units extending their business premises onto these public spaces.

This scarcity of public spaces in the internal settlement area is, however, not the complete picture. Adjacent to the settlement there are public spaces of differing quality. During the course of this research these also turned out to be highly dynamic. For example, a waste deposit site existing in 2008 was transformed into a place for rickshaw garages in 2009. At the focus of this research, however, are the large open space of the flood protection embankment and the community garden.

The public space at the centre of this research is the flood protection embankment which was constructed from the 1980s until the end of the 1990s to protect the city from flooding caused by the surrounding rivers in the west and south, especially the Buriganga River. On top of the embankment dyke is a busy main road, Embankment Road. The road is separated from the slopes and the river by an almost two metre high wall that is perforated only at the *ghats*, Bengali for boat or ferry terminals, where stairs lead down towards the water. At three *ghats* small boats take up to ten passengers to the other side of the river, and sometimes cargo vessels load or unload materials (see Photo A-2). Of the other two *ghats* included in the study area one is not in use by passenger boats or cargo vessels, while at the other one cargo vessels unload their goods from time to time but passenger boats do not depart from there. The operation of the *ghats* in Manikpara (Manikghat is used as an umbrella name for the *ghats* of the area) is contracted on an annual basis to a leaseholder through a tender process by the statutory Bangladesh Inland Water Transport Authority (BIWTA).

Besides the *ghats* the slopes are used for the activity of washing and drying plastic waste (see Photo A-15 and Chapter 6.1.1). Additional uses on the slopes include two green spaces developed privately (one of them 'Tariq's garden' which is especially discussed in Chapters 8.1.3 and 9), a temporary mosque (until August 2010) and a number of small shops.

On the top of the embankment slope and along Embankment Road the community garden, locally known as Bagan (Bengali for garden), was developed. This space was previously used as a dump site for industrial plastic waste and was highly polluted. One of the local factory owners took the initiative to clean the place and later on a garden was planted, financed and managed by a group of businessmen from the locality. A large part of the Bagan is accessible for everyone, while a smaller area is fenced off and not open to the public at all times. During the time of this research I witnessed considerable changes, ranging from construction of a monkey cage to the replacement of the wooden fences demarcating the garden towards Embankment Road by brick walls and entrance gates.

Above I have given an overview of Manikpara's public spaces which are scarce inside the built-up area. In order to later on be able to discuss spatial patterns of everyday life, it is also worth taking a look at semi-private spaces inside the housing compounds. In the low-income houses, outdoor spaces are mostly reduced to the corridor from which the family rooms are accessible. This corridor is rather narrow in most houses and only opened up towards the cooking area, which, however, is not always outdoors. In the apartment blocks, it is also common to share an area for cooking and doing laundry, while some of the rooms also have access to a private balcony. Overall, many houses do not have spacious courtyards open to the sky, while semi-private areas for cooking and laundry as well as access to the rooms are common in all buildings.

5.3.3 Economic livelihood activities

The economic livelihood activities of the inhabitants of Manikpara are primarily based on the plastic recycling process with its various production steps. Furthermore, the plastic recycling industry depends on the service sector and other industries, such as the transport sector with bag-carriers, pushcart and rickshaw-van pullers, and small-scale metal industries which produce casting moulds. The factories and workshops are mostly operated by local businessmen who then employ members from low-income households of Manikpara and the surrounding areas for sorting and crushing plastic and for operating the machines. Women are mostly involved in sorting, crushing or assembling the final products, while men are involved in all activities of the recycling process. Children are also involved in the industry, and according to one of the teachers of the local primary school it is common especially for boys to drop-out early from schooling in order to join the workshops and factories to sort and pack.

A normal working day in the plastic recycling industries lasts for 12 hours, commonly from 8am to 8pm including a short lunch break. On Friday the factories are closed and the other workshops tend to open only for half a day or remain closed as well. Wage levels differ with experience, but especially according to gender. For the same working hours, women earn considerably less: while the average wage level of male plastic sorters was 800 Tk/week (7.50 Euro) in November 2007, women earned only 500 Tk/week (4.70 Euro) for the same work (explorative household interviews December 2007). Similarly, a study of 2008 on the plastic recycling industry in the area of Lalbag found that on average women's salaries were 35% less than men's (Staffeld, Kulke 2011: 216–217). In plastic pre-processing (including sorting, shredding and pelletising), the average monthly salary (of men and women) was 2,931 Tk (Staffeld, Kulke 2011: 216–217).

Besides the livelihood activities centred on the plastic recycling industry and supportive businesses, a range of other activities was found. To characterise the area further, the most relevant are presented here. Especially the internal roads of the settlement are occupied by grocery shops, fruit and vegetable sellers, pharmacies, barber shops, restaurants and numerous tea stalls, just to name the most common. Especially tea stalls, drinks sellers and *pitha* (cake or bread commonly made from rice flour) shops occupy public spaces along the roadside to sell their goods. Mobile vending of snacks is especially common in the Bagan and around the Eid Gah Math. Some women also work in the garment industries located close by. Men are also involved in rickshaw pulling, renting the rickshaws for 50–100 Tk per day from nearby garages.

5.4 NASIMGAON – A VIVID YET PARTLY 'INVISIBLE' URBAN NEIGHBOURHOOD

When I spent my first night in Nasimgaon, I experienced the neighbourhood as the most tranquil place in Dhaka at night. The surrounding city with its almost never-ending traffic jams suddenly seemed miles away, while I observed the late evening hustle and bustle in the Bazar area. I sat in my landlords' shop, which from 8am to 11 or 12 at night is open to the road. Customers pass by on their way home from the garment factories and pick up and pay for their orders; the night guards drop by to collect the monthly contribution; and at 11pm money is collected from those shops who want an additional hour of generator power. The bench at the street corner tea shop is occupied by a group of men who enjoy their tea and cigarettes while being involved in *adda* (Bengali for a lively discussion among friends, see Chapter 4.2.2). After 12 midnight everything becomes quiet; the tranquillity of the night is only interrupted by the occasional barking of dogs. My image of Nasimgaon as an energetic lively settlement, and at the same time as a settlement that in some ways is far from the surrounding neighbourhoods of Dhaka, remained throughout my research. The sense of familiarity and community identity which I experienced was also expressed during many of the interviews that I conducted here.

5.4.1 Development and living conditions

Nasimgaon is surrounded by neighbourhoods that were planned and designed for middle and high-income households. When development of the model towns started, the area of Nasimgaon was still an open space and in parts a swamp area with a few scattered houses mainly used for agriculture. By the mid 1990s a rapid development process had started, fuelled further by extensive forced evictions in other areas of Dhaka in the late 1990s. At that time local leaders claimed the public land and erected *kaccha* and CI-sheet houses for renting out rooms to tenants[18].

Today Nasimgaon is densely built-up with mostly CI-sheet housing structures organised as housing compounds with a shared semi-private courtyard in the centre and private rooms opening towards this courtyard. In 2005, the CUS calculated a density of 220,000 persons per km^2. While there are hardly any options for construction of new housing compounds in the internal settlement area, a continuous restructuring of existing housing compounds can be observed. This includes, for example, the reconstruction of roof structures from flat roofs to gabled roofs which are favourable during rains, or the transformation of CI-sheet structures to *semi-paka*. But transformations also serve to accommodate an increasing population. From 2009 onwards two-storey houses of CI-sheets started to be developed. These investments were a consequence of the increased security felt after the elections in 2009 which triggered housing transformations. Later on, however, the security situation changed (see below), and thus only investments with an almost immediate return remained attractive.

Extension of the settlement continues along the external boundaries by the lake (see Figure 10). Gradually the lake space is being claimed, often by first building houses on bamboo supported platforms and later on filling the land below the platform. This mode of urban extension is a prominent example of how the urban claims of the poor are criminalised – for example when they block natural drainage channels and waterways, an activity often used to justify forced evictions. On the other hand similar claims and encroachments by high-income groups used to go on unabated. It is only recently, following the launch of the 'Save rivers, save Dhaka' campaign by the local media in May 2009, that these 'up-market encroachments' have been actively contested by government agencies and public opinion (The Daily Star 2009b).

18 In 2010 the rent for a CI-sheet room ranged between 500 (rare) and 1,200 Tk per month, depending on location (the accessible bazar areas were more expensive), quality of the house (especially floor and roof) and size of the room. Additionally, electricity costs stood at 100 to 120 Tk per device, i.e. if a household had a light bulb and a ventilator (most common), this cost 200 to 240 Tk per month; a TV counted as a third device. Water supply, delivered once a day by connection to the main pipe, cost between 100 and 120 Tk per household per month.

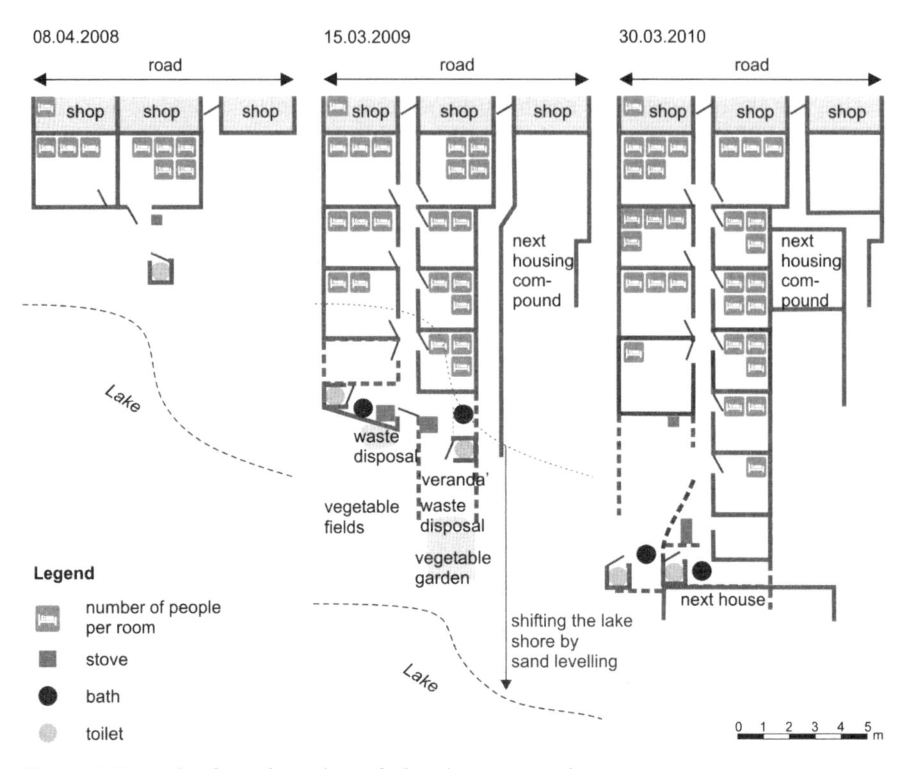

Figure 10: Example of transformations of a housing compound, based on own field observations 2008–2010

Security of tenure

During recent years, the area's location adjacent to some of Dhaka's high-income areas designated it a prime site for condominium development in the eyes of planners and real estate developers. Thus, despite the settlement's long existence, the economic and symbolic value (given the high visibility of this prime site) of the area result in recurring threats of evictions. Following a severe threat of forced eviction in November 2008[19], a human rights NGO filed a writ petition at the High Court and achieved a stay order for Nasimgaon residents. For some time this, combined with the presence of a stable political government after two years of uncertainty under the caretaker government, increased feelings of security among the inhabitants. Manifestations of this included various improvements of infrastructure, e.g. roads, pavements and drainage channels, and of buildings, e.g.

19 In November 2008 a notice was issued by government bodies to Nasimgaon residents "to vacate the possession of the bosti land immediately", threatening to carry out forced eviction if they would not adhere to the directive (The Supreme Court of Bangladesh, Writ Petition of 2008).

the transformation of CI-sheet houses and shops into *semi-paka* and the construction of the first two-storey buildings.

But soon land use was contested again when the new government developed new plans for the locality. While nothing has been decided yet, and my experience was that many of the ordinary residents were not informed at all about the ongoing discussions, in March 2011 new rumours started to spread. This time, it was heard that the land had been sold by the government to private companies who would soon ask the residents to leave in return for payment of compensation. The residents expressed their fear that they would not receive adequate compensation, rather the local political leaders would take the bulk of compensation money leaving a marginal amount for distribution among the ordinary. While I could not confirm these rumours told to me by some of my interviewees on the phone, the incident nonetheless shows the common perceptions of political leaders as using their power to appropriate resources for themselves. Furthermore, it indicates the government's attitude towards the urban poor, characterised by the production of invisibility and visibility on demand. Politically the area is visible, for it is a vote bank for every MP and Ward Commissioner, but when it comes to urban development the area is reduced to being a hindrance for development and the claims to civic citizenship remain unheard and thus 'invisible'.

Representations in official plans: unmapping?

The production of visibility and invisibility by demand is striking in Nasimgaon's representation in official documents. From a technocratic planner's perspective, Nasimgaon is a temporary land use and thus would not 'need' to be recognised as an 'official part of the city'. This perspective fails to consider the urban poor as citizens who have the same right to the city as all other inhabitants, but it is nonetheless the dominant perspective. In the relevant DAP section Nasimgaon is identified as a *bosti* that lacks basic utility services and connection to roads. The plan continues that this development "jeopardized the possibility of further extension [of settlements] on the other side of the lake; which will be required in future to complement the development on the other side of the lake" (Website RAJUK-DAP). A further interpretation of this comment is not made but it points to Nasimgaon's existence limiting the extension of other residential areas – this argument carries with it a differentiated notion of citizenship and fails to recognise the inhabitants of Nasimgaon as having equal rights to a share of urban space. In its recommendations for action, the urban poor are then deemed invisible and it is proposed that "spontaneously developed areas" be declared as government reserve land for developing upper class residential quarters and infrastructure (Website RAJUK-DAP). The land use proposal then suggests planning of an alternative resettlement scheme for Nasimgaon in the eastern fringe area, on government reserve land. This shows the low importance attached to giving the urban poor a right to the city: these eastern fringe lands where Nasimgaon's inhabitants could resettle are at the same time the lands to be brought into the government reserve to develop future high-income residential areas. This seriously questions whether any long-term perspective is sought for housing the urban poor and indicates a strategy of 'unmapping of space' where settlements of the poor as current land uses are largely neglected and thus the urban poor are invisible on official mappings.

5.4.2 Public space

The public space in Nasimgaon amounts to 11% of the total settlement area. This is even less than in Manikpara, but the availability of semi-private outdoor spaces (see definition in Chapter 2.1.1) is much higher than in the other study settlement, with every housing compound sharing a courtyard between the residing households (see below). Public space in Nasimgaon mainly comprises of roads and footpaths. The internal main roads are accessible by rickshaws and this is the most common means of transport used by shop owners for stocking up. However, many of these roads generally accessible to rickshaws also constitute the bazar areas and are thus lined with shops and at most times of the day are crowded with people.

Apart from the road network, there are a few public spaces used by vendors, small scale producers, for social and religious activities and for two graveyards. Two public spaces are at the focus of this research: the largest open space inside the settlement, the Nasimgaon Eid Gah Math (in the following chapters also referred to as Nasimgaon Math or the Math), and a small open market place, referred to as Khalabazar. The development of both spaces is a long story of negotiations over space and power struggles and is continuing to date, however, here only a brief overview is given, while the details are discussed in the subsequent chapters of empirical analysis.

The Nasimgaon Math provides space for both economic activities, including storage of goods, production and vending, and social or religious activities, like *adda*, children's games or the *Eid* prayers. In 1996 the first *Eid* prayers took place and the committee of the nearby mosque took over the management. According to one religious leader, at this time a 'social consensus' to keep the Math open for security reasons, e.g. fire accidents, was established (Rabiul, 23.03.2010). Furthermore, the Math was kept free from buildings due to a fence erected in the BNP period 1996–2001. However, the fence gradually deteriorated and thus during the following BNP government period the barrier preventing occupation of the Math was low. Consequently, a dump site next to the Math was filled and the space was rented out to rickshaw garages. What saved Nasimgaon Math from further encroachments was a serious fire incident in 2004. This consolidated awareness of the need for an open space in order to evacuate people and operate fire fighting vehicles. However, during the course of this research Nasimgaon Math underwent several changes and, due to encroachments, was also reduced in size.

Khalabazar was a small open market place when I first explored it during a field visit in 2007. For some years mobile and semi-mobile vendors sold their products (fruit, vegetables, spices, snacks, fish and meat) there, while others were engaged in small-scale cloth production, restaurants and tea stalls. During the course of this research the market place has changed considerably: while until 2008 it was a sandy open field from which the vendors sold their goods using *chouki*, wooden beds, as platforms (see Photo A-3), in 2009, after the election of the AL government, *paka* platforms, i.e. concrete structures, were introduced by the political management committee. This was followed by further improvements like more stable CI-sheet roofs and brick walls between some of the shops (see Photo A-4). In June 2010 one part of the market space was transformed into housing units, reducing the size of Khalabazar considerably (for a detailed discussion of the highly contested changes see Chapters 8.2 and 9 on the negotiations of access to space).

While public spaces are rare inside Nasimgaon, most of the settlement's housing compounds include outdoor semi-private spaces, i.e. the corridors and courtyards of the housing compounds, normally separated from the public street spaces by a CI-sheet door. The individual family rooms branch off from these semi-private spaces, which are thus shared by all members of a housing compound. Commonly, this semi-private space contains an area for cooking, a detached area for bathing and access to the toilet (see Photo A-5 and Photo A-6). Considerable differences can be found in terms of quality of space: while some courtyards are very small, barely leaving enough space for cooking activities, other courtyards resemble the rural *uthan*, the traditional courtyard within a homestead, in terms of their spaciousness and greenery.

5.4.3 Economic livelihood activities

Many of the livelihood activities of the inhabitants of Nasimgaon are closely linked to the surrounding high-income neighbourhoods. Many residents earn their living by providing services to these areas, for example men primarily work as rickshaw pullers or three-wheeler drivers, then as day labourers or in the garment factories. Most of the women work in garment factories, others work as housemaids. This matches my experience on those days when I entered Nasimgaon at 10pm and walked along with a constant stream of mainly female garment workers towards Nasimgaon bazar area. A considerable number of women do not carry out any income-generating activities but stay at home. The second focus of livelihood activities is on operating small businesses, either within the slum such as carpentries, barber shops, tea stalls, grocery shops, tailor shops or market stalls mainly serving the local market and demand, or as mobile vendors within Nasimgaon and the surrounding model towns.

This brief overview shows that the income generating activities of the Nasimgaon residents are much more varied and diversified than those of Manikpara. They are not based on one industry but centred on the provision of various goods and services both to the surrounding high-income areas and to the local neighbourhood.

The household incomes differ considerably. While the majority of the residents are socio-economically poor with household incomes below 5,000 Tk or between 5,000 and 10,000 Tk, there are some households who represent a 'local middle class'. These are especially the early settlers who are involved in local committees and who own housing compounds with several rooms which they rent out to tenants. Furthermore, many of the shops within the area have existed for a long time and especially where they are owner-occupied the monthly incomes exceed 5,000 Tk or even 10,000 Tk. Although socio-economically Nasimgaon is not as diverse as Manikpara, it is nonetheless more diverse than a first observation reveals.

PART II – THE SPATIALITY OF LIVELIHOODS

The two chapters of Part II of this research present the first part of the empirical analysis. Both chapters aim to answer the research questions concerning the spatiality of livelihoods. Chapter 6 mainly describes how public spaces are used in everyday life as well as in extra-everyday life, referring to special events that change the use of public space considerably. These uses of public spaces then provide the input for Chapter 7 which analyses the spatial practices against the background of the embeddedness of the study settlements into the urban structure of Dhaka and against the background of relevant norms, especially with regard to gender relations determining the use of public space by men and women. The negotiation of access to space will be discussed later, in Part III, the second part of the empirical analysis.

Main characters appearing in Part II

This overview serves to introduce the ‘main characters’ appearing in the stories and analyses of this part of the empirical study. It serves as an orientation for the reader, while the complete list of interviews can be found on page 377.

Manikpara

Fahima	works in a plastic recycling shop, age ~40, married, 3 children
Gulshana	works in a plastic recycling shop, age ~35, married, 4 children
Jahangir	previous leaseholder of the Ghat, now operates a plastic recycling shop on embankment slopes, age ~45, married, 3 children
Mashrufa	does not work outside or regularly, occasionally sorts plastic for her husband’s business from home, middle-class household, age ~30, wife of Mesbah Uddin, 2 children
Mesbah Uddin	plastics businessman of the area, well connected and respected in the locality, age 46, husband of Mashrufa, 2 children
Momena	together with her husband Rokib operates a plastic washing and drying business on embankment slopes, takes work on commission from *mohajons* (owners of plastic businesses), age ~30, married, 2 children
Rokib	Momena’s husband, see above, age ~55, married, 2 children
Roxana	changes her jobs (mostly in factories) almost every month, her mother works as a marriage broker, age ~25, unmarried
Tariq	young plastics businessman, together with his friends established the small garden at Dokkin Ghat, age ~25, married, 1 child

Nasimgaon

Aklima	operates a semi-mobile cold drinks business on Nasimgaon Math together with her husband, age ~30, married, 2 children
Dipa	in 2009 operated a vegetable stall or tailoring unit (depending on weather conditions) at Khalabazar, had to leave the place in June 2009, struggled for a long time to settle at a different location, but eventually in 2010, with NGO support, started to successfully operate a tailoring shop and in November 2010 even joined AL politics in the neighbourhood, age ~37, divorced, 1 child
Firoza	does not currently work, plans to work in garments factories again, age ~22, recently married
Ishrat	does not currently work, previously worked as a housemaid, at the moment her mother provides the household income by begging, age ~45, husband left, 2 children
Meher	operates a shop together with her husband (owners) and rents out five rooms to tenants, age ~37, married, 4 children
Mokbul Hossain	operates a construction materials shop (shop owner) and uses Nasimgaon Math for production activities, age ~45, married, 2 children
Rabeya	13-year-old girl, supervises the *karom* board business of her father
Rehana	operates a rice shop at Khalabazar (renting), age ~40, married, 2 children
Rohima	operates a fruit shop and tea stall (renting two units) at Khalabazar together with her husband Shahin), age ~25, married, 3 children
Taslima	operates a *pitha* stall on Nasimgaon Math, age ~40, married, 4 children

6 USING PUBLIC SPACE IN EVERYDAY AND EXTRA-EVERYDAY LIFE

The spatiality of livelihoods firstly relates to the way individuals make use of public space in their everyday life (sub-chapter 6.1). But at specific times the everyday life routines change, especially around religious and cultural events. These extra-everyday life activities will be explored in sub-chapter 6.2. Sub-chapter 6.3 will move from activities to discuss the quality of public spaces within the study settlements. Finally, the discussion of the three sub-chapters will be summarised in sub-chapter 6.4.

6.1 EVERYDAY LIFE SPATIAL PRACTICES

The everyday life has been elemental in Lefebvre's writings, centring around a series of publications on the critique of the everyday and quotidian. In his understanding, everyday life constitutes the core of the spatial practices of a society, and in 'Everyday Life in the Modern World' he states:

> "If we wish to define everyday life we must first define the society where it is lived, where the quotidian and modernity take root; we must define its changes and perspectives, distinguishing from an assortment of apparently insignificant phenomena those that are essential and co-ordinating them. The quotidian is not only a concept but one that may be used as a guide-line for an understanding of 'society'; this is done by inserting the quotidian into the general: state, technics and technicalities, culture (or what is left of it). This seems the best way of tackling the problem, and the most rational procedure for understanding society and defining it in depth." (Lefebvre 1971 [1968]: 28p.)

Accordingly, in this chapter I seek to analyse the manifold activities which, by claiming public spaces in various ways, produce the everyday life spatial practices in Manikpara and Nasimgaon. These activities will be discussed below, as they form the basis and starting point of this research; only by understanding these spatial practices was it possible to go further into the underlying negotiation processes. The presentation of everyday life in this sub-chapter, as well as in the subsequent sub-chapters, is based on the analysis of solicited photography and related interviews as well as semi-standardised and participant observations (see Chapter 4.2). Accordingly, the following categories of everyday spatial practices will be discussed: economic activities, free and leisure time activities, reproductive activities and religious and spiritual activities. Although included in the above categories, night-time activities will be explored separately.

6.1.1 Economic activities

The economic activities carried out in public space are numerous and accordingly three sub-categories are discussed here with reference to the activities and locations: vending activities; production activities and storage of goods and materials; and (briefly) urban agriculture.

Public space used for vending activities

Public space in both study settlements is commonly used for vending activities, although to varying degrees in terms of the type and intensity of activities and the involvement of men and women. The vending activities – as well as the production activities – are classified into five categories based on the degree of mobility and the structure of the unit (see Textbox 4).

Textbox 4: Categories of vending and production activities

Based on the degree of mobility and the temporariness of the spatial claims made, vending and production activities are classified in the following. The distinction is made with reference to both the potential mobility of activities and the physical location and permanence of the structure used for the activities. A similar distinction has been used by Etzold (2011) in his research on street food vending in public spaces in Dhaka.

Mobile: This refers to vendors/producers who due to the small equipment they carry are able to follow an explicitly mobile strategy by constantly moving around selling their goods from different locations. This can be done by carrying all the goods, for example in a vendor's tray. However, while mobility is easily possible, there can be mobile vendors who chose only one specific location to sell their goods, for example by sitting on the ground with goods laid out in front of them.

Semi-mobile: This refers to vendors/producers selling from a device which can be moved to a different location. For example this includes selling from a rickshaw van which can be moved at specific times but is not constantly in motion. However, the choice of the vendor/producer may also be to remain in one specific location.

Semi-permanent: This refers to vendors/producers who are bound to selling from a specific place. Although it would generally be possible to move their equipment to another location, it would not be comfortable to do so. For example, this includes vendors/producers selling from *chouki* (low wooden beds) or setting up stoves which are generally not moved to a different location because the equipment is too heavy or bulky to be moved around continuously.

Permanent: This refers to a vendor/producer selling from a structure that is fixed to the ground, but not of a very stable nature. This could, for example, be a makeshift polythene roof or CI-sheets on bamboo pillars. It also refers to having some equipment at the stall, for examples stoves or benches, which could not easily be removed and taken to an alternative location. These vendors/producers are thus relatively immobile.

Stable structure: This refers to a vendor/producer selling from a stable structure, i.e. a stall made from CI-sheets or even concrete, which cannot easily be de-constructed and moved any more. The existence of such units in originally public spaces points at a relatively irreversible transformation of a public space into built-up space.

Mobile vending, especially of snacks, occurs in both Nasimgaon and Manikpara. In Manikpara, mobile vendors mostly come into the area, and especially to Bagan, in the afternoon and early evening hours. Generally, these times are also the most crowded times of the day in both Manikpara and Nasimgaon, as people come to spend their free time and children come to play and buy snacks. In Nasimgaon, Wednesday afternoon is the busiest time for mobile vendors – as Wednesdays are also the main bazar days due to the fact that many garments factories close on time (5pm) and thus everyone goes grocery shopping.

In Khalabazar, a group of beggars regularly came to sit in the south of the bazar and sell small amounts of rice which they had collected from households and shops (see Photo A-7). Among these rice sellers were also women. One of the women, Ramisa, did not sell rice obtained through begging. Instead, she got the rice she sold sitting on the ground from her daughter and son-in-law who operated a tea stall and grocery shop, and sold rice. Although these rice vendors were able to just pack up and move to a different location, following a mobile strategy, Ramisa always sold her rice from the same location and thus did not make use of her vending activity's potential mobility. Women are normally not mobile vendors, as the constant moving around this implies is not accepted in society (see Chapter 5.1.3). The beggars selling rice were accepted in 2009, but in 2010 the developments at the bazar meant that they were reduced in numbers (see also Chapters 8.2 and 9).

Similar to mobile vending, during the afternoon and evening hours the Bagan and Nasimgaon Math are frequented by many **semi-mobile** vendors who bring along their goods, especially selling snacks, on pushcarts. Some vendors, however, also use semi-mobile devices and equipment but remain in one location permanently.

Aklima and her husband sold drinks made of water with a bit of sugar and lemon juice and ice from a pushcart. They operated two of these pushcarts. While Aklima stayed in the same position on Nasimgaon Math throughout the day, her husband went to different places in the surrounding settlements in the afternoon. The strategies of many vendors at Khalabazar can also be classified as semi-mobile. For example, a woman sold chicken scraps (mainly innards) from a large bowl, after she had boiled them at home. She normally sat at one of the semi-permanent or permanent vendors' places, but her style of vending allowed her to change position in the bazar, depending on where she got permission to stay – and thus she needed to constantly negotiate access to space.

In irregular intervals, canvassers came to Nasimgaon and especially to the Math selling different kinds of homeopathic or 'spiritual' medicine. The canvassers often came by car or other vehicles, due to their heavy and bulky equipment, including loudspeakers. I consider them semi-mobile, as they could not as easily move around as mobile vendors, but they also did not always use the same location for their vending activities. These canvassers praised their goods from one place by announcements or singing, or walked the streets with their megaphones. Rohima, at that time operating a fruit shop together with her husband, took a picture of one canvasser (see Photo A-8) and explained:

> "This is in the Nasimgaon Math. A person is selling some medicine made from tree leaves and grass and I liked it. [...] I like the medicine and also buying the medicine from him which helped me to cure some problem. He has the leaves and will make medicine if anyone asks. [...] He is mainly selling medicine for gastric pain or pain from high blood pressure. He also gives *tabij* [an amulet containing written verses of the Quran]. The medicines are cheap, ranging from 100–120 Tk to 60 Tk depending on the amount needed, and so I like them. This man is coming here once a week for one or two hours. If I have any problem I go to see him." (26.04.2010, Rohima)

Due to their entertaining character such vending strategies draw a large crowd in public space and thus considerably shape these places at specific times. The gathering crowd is of mixed gender and also includes many children.

On Nasimgaon Math, the most common strategy for vending is **semi-permanent**. This involves the selling of goods from *choukis* or other wooden platforms which are not permanently connected to the ground (see Photo A-9). Such *choukis* or benches are also used as extensions (e.g. to present more materials) by many of the stable shops surrounding the Math and other bazar areas in Nasimgaon and thus claim additional space. Selling from *choukis* was also the common style of vending on Khalabazar (see Photo A-3) until the upcoming AL leaders considerably changed the physical structure and access rules to vending units. Until then, the *chouki*-vending allowed the bazar area to be used for other programmes as the *choukis* could be removed easily. In Manikpara such semi-permanent vending activities are less common and vendors resort to more mobile strategies or sell their goods from permanent/stable structures.

Permanent vending structures are found both in Nasimgaon and Manikpara (see Photo A-10). Especially the *pitha* stalls often are permanent in structure, being less easy to remove. Taslima's *pitha* stall in Nasimgaon, for example, consisted of three stoves and a polythene roof held by bamboo pillars and, occasionally, depending on the contestations she faced, a wooden bench for customers. In a transition period, the stalls at Khalabazar could also be considered permanent, although here the distinction to stable is difficult, as the *paka* platforms put up by the AL leaders in 2009 present a structure that despite being open towards the sky can be considered stable, as they cannot be removed as easily as the *pitha* stalls.

In Manikpara, most of the vending activities take place in **stable** shops built along the main roads and footpaths and the bazar area in the adjacent settlement. The tea stalls that have been constructed along the road from Bagan to the bazar made of CI-sheets (see Photo A-11) can also be considered stable as they cannot easily be dismantled. However, they are different from the stable (*paka*) structures in the surrounding areas which are integrated into buildings.

In Nasimgaon, there are a lot of stable shops in the main bazar areas. These considerably shape the spatial practices along the roads and public open spaces. The events observed from 2009 to 2010 indicated that semi-permanent vending structures were slowly but continuously transformed into stable structures (see Photo A-12), especially in the case of Khalabazar. Here many of the vendors, who before 2009 used to sell from *choukis* on the sandy open field, later sold from stable structures (see Figure 18 in Chapter 8.2 and Chapter 1 for the analysis of these transformations).

One of the vendors at Khalabazar, who is also politically active, made an interesting comment about the quality of open space for economic activities as compared to the transformation process that led to the open space being completely covered by stable structures:

> "But there is no such place to stand [i.e. where people could gather for example after fire accidents] in Nasimgaon. Open space is very necessary here. The *bosti* is a densely populated area. So in-between there is a necessity of open spaces. [...] If the shops had not been built, then a normal *bazar-ghat* could take place, then there would be no problem. Then people will do their business as well as coming to see if anyone else has any kind of problem." (Afsana, 23.03.2010)

The term *bazar-ghat* is used as a phrase in Bengali and refers to a place with the characteristics typically found at a bazar and at a *ghat*. A bazar is perceived as a public space of selling and buying products, a place of public gathering and a place of endless gossiping at tea stalls. A *ghat* is also perceived as a crowded public space, a meeting place of different people and a gossiping place. The common characteristics are the crowd, the gathering, the meeting opportunities and the gossiping – generally indicating a place where something is happening. Thus Afsana's comment seems to suggest that a specific quality is assigned to an open space where economic activities take place in the style of a *bazar-ghat*, and indicates a preference for a space which represents an atmosphere of solidarity among vendors, as opposed to stable shops in the ownership of others.

With regard to gender relations, the two study settlements differ considerably. The vending activities in Nasimgaon, except for those based on actual mobility, are carried out by both men and women. If the business requires goods to be bought from the wholesale bazar outside of Nasimgaon (often Kawranbazar, the largest wholesale market in Dhaka), this is done by the men in case of joint businesses. But those women who operate their own businesses go to the wholesale bazar by themselves. In contrast, in Manikpara only the *pitha* stalls are commonly operated by women, while all other vending activities, independent of the mobility, are carried out by men. This is supported by the rather negative societal perception of women working outdoors (see discussion in Chapter 7.1.1).

Seasonality has an impact both on the profit of vendors as well as on the accessibility of public spaces and thus also shapes the spatial practices. While for example *pitha* selling is most profitable in winter, Taslima closed her *pitha* stall during the rainy season due to lack of customers. On the other hand, Aklima, selling the iced water on Eid Gah Math, profited most in her business on hot days while winter and cloudy days limited her business. When she was still operating her shop in Khalabazar (see Chapter 8.2), Dipa decided whether to sell vegetables or put out her sewing machine based on the weather. At that time the *paka* platforms on Khalabazar had not yet been properly roofed and thus on rainy days she decided to sell vegetables and left her sewing machine at home. Generally, during the rainy season mobile, semi-mobile and semi-permanent vending are reduced due to the weather conditions. The more permanent and stable vending structures, in contrast, do not have to close down because the rains do not affect their shops. On rainy days the movement of people and thus customers is generally reduced, making selling less profitable.

In terms of the use of public space in everyday life, economic vending activities considerably influence the spatial practice. The mobile and semi-mobile vendors appearing in the study settlements especially in the afternoons and early evenings lead to a gathering of people in open spaces, producing a spatial practice resembling what Afsana referred to as *bazar-ghat*. The transformation and consolidation of vending units in public spaces from semi-permanent to stable structures, as happened in Nasimgaon, changes this spatial practice considerably. While in Nasimgaon public spaces are considerably shaped by vending activities, in Manikpara most of the activities taking place in public space focus around the plastic recycling as a production activity.

Public space used for production activities and storage

Public space in both study settlements is used for manifold production activities and as a place to store goods. In Manikpara this is especially related to the plastic recycling industries and its side-industries, which are the focus below. In Nasimgaon the production activities are not equally dominated by one specific activity, and thus a diversity of activities is discussed.

The plastic recycling activities extend considerably into public space. Many of the sorting shops open up towards the road spaces, and sorting of different plastic materials takes place outside of the shops (see Photo A-13). Thus during weekdays, the public space is considerably reduced by these activities. Furthermore, the roads are occupied by manual carriers and pushcart operators who deliver or pick up plastic goods to and from workshops and factories. Thus everyday life in Manikpara is dominated by the processes of the plastic recycling value chain (see Textbox 3, Chapter 5.3.1). In contrast, on Fridays most of the factories and workshops are closed, which has a direct impact on the accessibility of the area. The difference to weekdays becomes evident in Photo A-14. A similar effect followed the decrease of global prices for plastic in 2009, when in consequence Manikpara remained quieter than usual and many sorting shops were closed for a considerable time.

The sorting of plastic outdoors is conducted both by men and women, although rarely in mixed teams (see Chapter 5.3.3 regarding working conditions and salary). It is generally accepted that women from low-income households work in the plastic recycling industries and in public space out of economic necessity. During the focus group with working women, Gulshana and Salma underlined that they did not face problems of teasing while working outdoors. However, they complained about the working conditions in public space:

> Gulshana: "I feel bad. I work amidst dust. I never did this type of work before. I feel bad between the people, vehicles and dust. I can't have peace. I get diseases in my body. I needed medicine of 350 Tk after coming here. I never did these works before. I have sorrow in my

fate. That's why I have to do these things now. I need more medicines. All my income is spent on medicine."

Salma: "I feel the same way. When the sound of the vehicles enters into my ear I feel bad. It seems like the sound enters into my head. There is too much dust here. I am working as I got into trouble. So I have to work even if it is difficult." (Manikpara FG1, 02./09.04.2010)

The mentioning of the necessity to carry out this hard work and the mention of peace, sorrow and fate indicates that although they have adapted to the work in plastic recycling, they would prefer a different life or workplace. This is also expressed in their preference for a different lifestyle (see Chapters 6.1.4, 7.2.2). Middle class women in Manikpara do not consider working outside acceptable at all because of the need to maintain their status (see Chapters 6.1.4, 7.1.1), and are thus only involved in home-based plastic sorting.

The embankment slopes in Manikpara are used for livelihood activities from the early morning. One section is used by several stable plastic sorting shops throughout the day. But the most dominant use of the slopes is plastic washing and drying. This requires both closeness to water – which depending on availability and affordability could be substituted through tap water – and availability of a large open space – which is impossible to realise inside the built-up area due to the high density and low percentage of open spaces in the settlement. Thus from early morning until afternoon, various parts of the embankment slopes are occupied by businesses involved in this activity. At 7am the plastic drying businesses are already laying out cloths to dry the plastic pieces and the washing of plastic has started, both on weekdays and Fridays. Until the afternoon, plastic drying is the dominant activity conducted in front of the Bagan and between some of the *ghats*. Rokib and Momena operated a plastic drying business on commission from plastic businesses and factories. Almost every day they received new materials and together with their employees washed and dried them before delivering them back to the factories. For women in Manikpara this kind of work, involving direct exposure to sun and heat throughout the day, is uncommon. For Momena this was only possible because she was 'protected' by the respected status of her and Rokib's relatives in Manikpara (see further discussion in 7.1.1).

During the rainy season, the above patterns of plastic drying on the embankment slopes change. While the sorting shops are able to continue their activities, plastic drying becomes difficult, both because of the rain and the high water table of the Buriganga. From April onwards the drying process occasionally gets disrupted by rains and from July/August until September, the water table height causes the plastic drying businesses to periodically abandon the embankment slopes (see Photo A-15 and Photo A-16). Three strategies are then employed:

- The factories washing and drying their materials occasionally make use of the rooftops within the neighbourhood, although this was described as unfavourable because the heavy plastic bags and cloths for drying have to be carried to the rooftops and water for washing is not as easily available as by the riverside.

- The businesses drying plastic on commission and some of the businessmen prefer to shift activities to an alternative location, where the embankment is higher. Thus the plastic is transported by boat and washed and dried there. This alternative strategy entails different rules of access (see Chapter 9.2.1).
- Some *bhangari dokans* involved in both sorting and drying halt the drying process during the rainy season and store the sorted and crushed materials for the three months until drying can resume on the Manikpara riverside.

In Nasimgaon the Eid Gah Math is used for a lot of production and storage activities. Most of these activities can be understood as semi-permanent, as they take place without any fixed structures. One exception was a blacksmith who constructed a concrete water basin to cool down his products in public space. The other activities such as the preparation of firewood (see Photo A-17) or the production related to construction depend on the time of the day and the demand for such goods. Some goods stored on the Math during the day are moved inside the stable shops at night. During the daytime, a considerable number of rickshaws are parked on the Math and the rickshaw pullers also engage in repairing these (see Photo A-18). The Nasimgaon Math thus serves as an important extension of the business premises for production-related activities. At the same time this reduces the space available for other activities, and over time the multiplicity of use interests has recurrently led to contestations about access to the Eid Gah Math (see Chapter 8.3).

In Manikpara, everyday life in public spaces is dominated by the plastic recycling industry. The scarcity of public space in the area and the space requirements of the industry put considerable pressure on the existing public space. In Nasimgaon, production activities mainly centre on internal consumption and thus take place on a less dominant scale. The remaining public spaces, especially the Eid Gah Math, are, however, used intensively for storage and production activities.

Public space used for urban agriculture

Urban agriculture can hardly compete with other land uses in the densely built-up city of Dhaka. Accordingly, only niche (public) spaces continue to be used agriculturally and the trend is for urban agriculture to become a pastime instead of an economically viable activity.

This is especially the case in Manikpara, where the high density and the high land values (see Chapter 5.3.1) eliminate agriculture as an economic livelihood activity. The embankment slopes, which cannot be used for building purposes, are the only space that can be used for such activities, although this is not statutorily permitted. Thus Tariq and his friends in their garden at Dokkin Ghat cultivated fruits and vegetables and reared chickens as a pastime. Here, an image of countryside scenery is produced within the urban by residents who identify themselves primarily with the city and their neighbourhood. While the garden changes the appearance of physical space in Manikpara, it does not contest the dominant mode of the production of space. Rather, as will be discussed in Chapter 1, the garden would not exist if its establishers did not possess dominating power themselves.

In Nasimgaon, on the other hand, agriculture is part of the economic livelihood activities of a small number of residents. While mainly taking place at the fringes of the settlement area, it nonetheless is not in contrast to the settlement's urban fabric, which resembles a rural setting despite its high density. Previously, agricultural activities used to be common all over the area and three interviewees mentioned cultivating vegetables and rearing ducks and chickens to support their families. With increasing density, however, the in-between spaces for agricultural activities disappeared, leaving only the fringe areas and two places where cows are kept indoors. At the fringes of Nasimgaon, the agricultural activities are highly organised and under the jurisdiction of local leaders. The pressure to capitalise land, however, has resulted in the transformation of many vegetable fields into housing compounds. The considerable amount of pictures taken of these fields by participants in the solicited photography indicates the value attached to such open spaces.

In Manikpara, agricultural activities are not able to compete with the high land values and densification processes within the settlement and thus are reduced to pastime activities in 'niche spaces'. The developments in Nasimgaon point in a similar direction, although at an earlier stage. The great pressure for the further extension of the settlement makes urban agriculture a disappearing economic livelihood activity, as it is too space-intensive to co-exist with high urban densities and land values.

6.1.2 Free and leisure time activities in public spaces

Public spaces play an important role in both Manikpara and Nasimgaon for spending free or leisure time. However, there are considerable differences between the two settlements mainly due to differences in urban fabric and economic livelihoods, but also with regard to norms and specifically gender relations. In both settlements, men and children spend much of their free and leisure time outdoors, while women tend to spend such time at home.

Men's free time: spending time with friends and relatives outdoors

In Manikpara, it is common for men to move around with their friends, especially if they are of a younger age and from slightly better-off households. Especially an unmarried participant of solicited photography, working in a plastic sorting shop, moved around a lot together with his friends. The places they used to go to include the Dhaka University Campus, parts of Old Dhaka as well as the other side of the Buriganga River where they could find more rural settings. These locations are partly a reaction to the dense urban fabric of Manikpara where, for example, open spaces to play football are not available. Taking boat rides on the Buriganga is also a favourite pastime (see Photo A-19). For several times during my fieldwork presence, launch[1] picnics were organised by the local community. A launch was chartered by a group of people and, in return for a contribution to cover costs, everyone was able to join the day-long picnic tour traversing the rivers of Bangladesh.

Generally, men who were already married and of older age indicated that there was no longer time for friends because of their family responsibilities, and thus their movement patterns concentrated more on the neighbourhood:

> "There is now no time for friends. Friends are busy with their wives. I'm also busy with my wife and my children. Still, during the holiday or in leisure time we have some *adda* at the laundry/saloon or roof after 8/9 o'clock in the night and laugh and have some fun. Sometimes one friend spends 20 Tk for me and again sometimes I spend. [...] I don't go anywhere. I can't go even though I wish to go. I don't have that mentality now. Previously I led a bachelor life and went wherever my friends asked me to go. But now, even if I want I can't go, I have to do everything like to work, to maintain my family and to do shopping. " (Mesbah Uddin, 21.05.2009)

Thus married men spend more time within the settlement, meeting their friends at tea stalls for *adda* (see below), at the Bagan, or at specific shops and on the roof tops of 'famous' houses. Tariq together with his friends even constructed a small garden on the embankment slopes at Dokkin Ghat, and they spent much of their leisure time taking care of and extending the garden (for further discussion of the creation of this space see Chapter 8.1.3). In general, the participants of solicited photography all referred to the beautiful green space of the Bagan and took pictures of it, and it seems that it was perceived as a green oasis and a valuable space within the otherwise densely built-up settlement. This common image of men (and children) spending their free time in public space was also indicated by Gulshana during the focus group with working women. She referred to the value of open spaces and the deeply rooted understanding that public space was predominantly male space:

> "If there is open space, children can play. Men can sit. It looks good when people move. Children are playing in the narrow lanes here because there is no open space. There are accidents. If there is a field, they can play in the field. (Gulshana, 02./09.04.2010)

In Nasimgaon, only three men mentioned spending their leisure time outside of the settlement. Abdul, a rickshaw puller, whose family was in the village, took pictures of places which he liked to visit mostly by himself to enjoy the environment (see for example Photo A-20). Assad on the contrary, who was still young and had only recently become a father, spent much of his free time with friends moving around in the Cantonment, where he also took many pictures. His flexible working hours as a contract carrier in garment factories allowed him to spend such times with friends. Hamidul, a boatman, visited places in the surrounding settlements together with his friends or children. All the other men I interviewed

1 The term 'launch' refers to the large ferries traversing the rivers of Bangladesh and connecting different regions (as opposed to the motorised trawlers used along the Buriganga River within Dhaka city).

in Nasimgaon did not visit the surrounding neighbourhoods during their everyday life routines, but some indicated visiting places in their rural homes or *mazars*, the shrines of Muslim saints, in other regions of Bangladesh. The men spending leisure time within Nasimgaon mainly mentioned the tea stalls where they would take tea and engage in *adda* (see below)[2].

Women's free time: visiting friends, relatives

While men in Manikpara spend much of their free time outdoors, women tend to spend such free time at their own home or at the homes of friends and relatives. The women interviewed in Manikpara all reported that if they had free time they would visit friends and relatives, but they would not remain in any outdoor places. The women working in the plastic recycling industry did not even have time for such visits considering their 12-hour working day and their domestic obligations. The results of the solicited photography also reflect that women of Manikpara hardly spend any free time outdoors. Of the four female solicited photographers, only Roxana took pictures of the Bagan. This she did because at that time she was unemployed and she wanted to show me the beauty of the place (see Photo A-21). On this specific day she went to the Bagan with her brother, but she said that she also went alone whenever she was unemployed. Later on, however, she said that she was no longer allowed to visit these places on her own, which may be attributed to the fact that she was yet unmarried. The other women only took pictures of their family and workplace or on the way to their workplace, indicating that they do not spend their free time outdoors. Women in Manikpara thus spend their free time in the private spaces of the home and with neighbours, friends and relatives. There, if time and workload allows, they get involved in gossiping and *adda* just as their male counterparts do in public.

In Nasimgaon women more frequently talked about spending some free and leisure time outside their housing compounds. Those who had young children sometimes moved around with their kids to let them play in the open spaces and towards the lakeside (see Photo A-22). Furthermore, many women took pictures of places they enjoyed visiting, for example on the way to the bazar they passed the vegetable fields at the lakeside and took pictures because they liked the place. However, they would not simply go there and spend time sitting 'aimlessly' which is only acceptable behaviour for males. Dipa spending free time at a *mazar* in Mirpur (another part of Dhaka, west of the Cantonment) is rather unusual (see further details in Chapter 6.1.4). While some women left Nasimgaon for economic

2 The comparatively better-off households in Nasimgaon, especially those of political leaders, showed a different behavioural pattern towards spending leisure time outside of Nasimgaon, however, I did not conduct these interviews with them. However, Mokbul Hossain, among the better-off in Nasimgaon, did not leave the neighbourhood unless for business related activities. Furthermore, the shopkeepers near to my housing compound also did not display such mobility patterns.

activities, she was the only woman who went not only to the near neighbourhoods, but to many other places in Dhaka and beyond in her free time.

Adda and gossiping

The Bengali *adda* is one of the favourite pastimes of both men and women in Bangladesh (see Chapter 4.2.2). In both Manikpara and Nasimgaon, *adda* rounds take place at tea stalls and other outdoor spaces. In Manikpara it is very uncommon for a woman to join these rounds and only Roxana, a young female participant in solicited photography, did so. I met her the first time while taking tea at a tea stall. Her behaviour in entering a male-dominated public sphere and thus crossing gender norms immediately caused people to talk negatively about her, and the first times I met her and visited her house I was warned by others not to mix with her and her family. In Nasimgaon there is a higher acceptance of women joining these *adda* rounds, although it is also not very common and women rather spend their *adda* times at people's homes. Dipa frequently did *adda* with a group of men at the banyan tree of Khalabazar. She enjoyed sitting at that place very much, but at some point her position in the market was contested and she started to avoid meeting certain people out of a belief that they were wishing her bad. Her trespassing of gender norms, see also Chapter 6.1.4, may have contributed to her exclusion and finally dislocation from the place where she used to live, sell her products and spend her free time. Mostly, then, women have their *adda* rounds at home, either their own or that of friends and neighbours.

The importance of *adda* and the tea stall as a place for *adda* is underlined by the following incident. Ripon, a plastics businessman, was also involved in the community police, thus his socio-economic status was comparatively high and he was respected in the community. He decided to open a tea stall at one *ghat*:

> "I used to run a tea stall. My leisure time was not being passed. That's why I thought to run a tea stall. My leisure time would pass as well as I would earn something. I invested in the tea stall from my pocket, the money I got as house rent. I brought goods for the stall but people ate on credit." (Ripon, 13.04.2010)

Beside the tea stalls, other shops also serve as gathering points for *adda* rounds, and this seems to depend more on the person operating the shop than on the kind of shop operated. For example, *adda* rounds meet in the rickshaw garages, at pharmacies, at snack shops, at a blacksmith's or at cloth shops. Furthermore there are some locations in the open spaces, close to trees, or along the lakeside, where such discussion rounds gather.

While religious uses of public space will be discussed below (Chapter 6.1.4, 0), two *adda*-related activities should be mentioned here. In Nasimgaon, Mokbul Hossain, a male participant of solicited photography, spent much of his free time discussing religious matters with his *pir baba*, a religious and spiritual guide of Sufi Islam, while sitting in front of his shop facing the Nasimgaon Math. In Manikpara, women regularly join *Tabligh*[3] meetings at other women's houses where they read *Hadis*, the quotes of the Prophet Mohammed, and *Talim*, Islamic advice. These *Tabligh* meetings were mentioned both by working women and the women mostly staying at home and doing the domestic work (see also further discussion in Chapter 6.1.4).

Watching TV / cinema

An important pastime, especially in Nasimgaon, is watching TV or *cinema*[4]. Throughout the day in Nasimgaon there are gatherings at the tea stalls, especially where a DVD recorder is available or cricket matches are broadcast. Some tea stalls indeed start to look like little public cinema halls with high volume sound (for example making my interviews in a nearby housing compound impossible) and a gathering of a concentrated, male-dominated crowd (Photo A-24). Thus, tea stall owners are eager to buy a TV and DVD recorder to increase profits. Women mostly do not join these public cinemas, although in a few cases I found women doing so, for example from the opposite side of the road, separating the crowd into a male one in the first row and a female one in the second row. Most of the time, however, women watch *cinema* at home, in the evenings. In my housing compound there was only one room with a TV which was occupied by the women of the compound almost every evening at the time when the local TV station broadcasts a specific Bangla movie. At that time this room became a female space, and only occasionally a few young men watched from the threshold (Photo A-23).

In the working environment of Manikpara watching TV at tea stalls is not as prevalent, which is accounted for by the fact that most male inhabitants are involved in plastic recycling throughout the day and do not have time to watch TV. The better-off households mostly have TVs at home, switched on for most of the day. But for neither male nor female groups was the *cinema* as important as for the inhabitants of Nasimgaon, which can be attributed to both the working hours and the specific social structures that cause men to prefer other activities, especially extensive *adda*.

Playing football and other games

Children and young men also use public space to play football or cricket. In Manikpara, however, the urban fabric and high density does not leave space for such

3 The *Tabligh* or *Tabligh Jamaat* is an orthodox Islamic movement founded in India in the 1920s. The annual congregation *Bisho Ijtema* held once a year in Tongi, north of Dhaka, is presumed to be the second largest Muslim congregation after the *Haj* to Mecca.

4 *Cinema* refers to watching a movie either at the public tea stalls or in private rooms. I found that a considerable number of families own a TV at home.

activities, and only the site of the planned extension of a school was used by young children for football and cricket, after it had been sand-filled in 2009 (Photo A-25). Given the lack of space to play games, Hortem and his friends visited a playing field on the Dhaka University campus to play football and cricket.

In Nasimgaon, given the density of the built-up area, the only available space to play football or cricket is the Eid Gah Math. Traditionally, every locality has a *math* for children to play on, both in rural and urban areas. On Nasimgaon Math the local youths also sometimes organised football tournaments. At the times of these tournaments, the space became completely male-dominated and there were no women in the watching crowd. Especially during religious or Bengali cultural events, but also during other times of the year, there were playing devices on the Eid Gah Math, i.e. a *nagordola* (big wheel) and a merry-go-round (Photo A-27). These were never permanent, but whenever they were on the Eid Gah Math then they were highly frequented by children, although a fare had to be paid per ride to the commercial operator. Apart from these activities, the schools sometimes hold programmes with games for the children (see Chapter 6.2.4). The use of the Eid Gah Math also triggers land use conflicts, as will be discussed in Chapters 8 and 1. In addition to the Eid Gah Math, young children use the roads, footpaths, open spaces at the fringe and the *ghats* for their games, including swimming.

A favourite pastime among young men in Nasimgaon is the playing of *karom*[5] on boards operated commercially in a few public spaces – this activity does not require a large public space (Photo A-28). Women do not play these boards, thus these public spaces are basically male spaces[6].

Concluding remarks

The above analysis of everyday life free and leisure time activities indicates that men generally spend their free time outside of the home and in public spaces. In Manikpara men tend to travel more frequently outside of the settlement for leisure time purposes than in Nasimgaon. Women in Manikpara tend to spend their free time at home, while women in Nasimgaon also spend time within the settlement, although often in the vicinity of their houses.

5 *Karom* is a kind of billiard played with the fingers on a wooden board. It is common in South Asia.

6 Rabeya, a 13-year-old girl in charge of collecting the money from the players of her father's *karom* boards, told how one day a foreign woman came to play the boards, trespassing common behavioural patterns: "But the Bengali women do not play as many men stare at them. Some days ago one female foreigner came wearing a T-Shirt and they [men] were staring at her." (Rabeya, 05.04.2010)

6.1.3 Reproductive activities

The reproductive activities include shopping for the household, especially grocery shopping but also other consumables, as well as activities commonly carried out inside the housing compound, such as cooking, fetching water, bathing and doing laundry.

Shopping for the household

The bazar plays an important role in all interviewees' everyday lives. Shopping is conducted daily, but who goes shopping varies considerably. Shopping, especially the daily grocery shopping for the family, is mostly conducted by men in Manikpara and by women in Nasimgaon. The reasons for this difference will be explained below.

In Manikpara, going to the bazar is closely tied to maintaining a family's respect and honour. While some women from low-income households occasionally go to the bazar, this is not done by the middle-income households. Also women of households whose income is low but whose family acquired a respected status in the community, for example via relatives, do not go to the bazar. Among those groups it is considered dishonourable if a woman goes to the bazar to shop. Mesbah Uddin, a well-known plastics businessman, upon being asked whether his wife went grocery shopping, exclaimed:

> "No no, why would I send her to the bazar? If I would send her to the bazar would her honour/prestige remain [*oke bazare pathale man-ijjot thakbe?*]? It is Old Dhaka and people of this area will not like it. This work is not prestigious for a woman of Old Dhaka who is a house owner; then my brother is a secretary of the BNP Chatro Dol of this area's *Thana*[7]. As well I am a leader of the workers; I am a *dada* [brother] of Manikpara and everybody, that means the elder and younger, all call me *dada*. Once upon a time, foxes and dogs used to live in Manikpara. All the killers, thieves and criminals used to live here. We removed them after coming here. That's why everybody calls me *dada*." (Mesbah Uddin, 21.05.2009)

Both due to the political activities of his brother (non-kin) as well as due to his own commitment to the community, Mesbah Uddin's family was respected in the local society. If his wife did the daily grocery shopping and would be seen at the bazar, Mesbah Uddin's family would lose this social respect and status (*man-ijjot* or *ijjot*). Within this middle-income group, only a woman's position in terms of life cycle can provide for exceptions. Tariq's mother-in-law, for example, did the grocery shopping for the family, and Subina, a widow aged about 40 with grown-up children, went to the bazar herself – both of them were in a later position in the life cycle.

7 A *thana* is the area of jurisdiction of one police station, normally consisting of several Wards. The *thana*, however, is not part of the urban local government system. Nonetheless it is the next level of party organisation (see Figure 8 in Chapter 5.1.1).

In contrast, in Nasimgaon all interviewees indicated that the women of the household do the shopping. For example, Nasrin's statement, when asked whether her husband went to the bazar, demonstrates that this is considered the 'natural way' of splitting household tasks: "Women do the women's shopping themselves" (Nasrin, 11.05.2009). Like other women of Nasimgaon, she furthermore enjoyed going to the bazar and accordingly took many pictures inside the bazar and on the way to the bazar, underlining that this routine was important for her and that she felt comfortable (Photo A-29). The difference between low-income and middle-class households observed in Manikpara also does not exist as distinctly in Nasimgaon. The wife of Mokbul Hossain, who can be considered as being among the better-off within Nasimgaon, also did the household's grocery shopping herself. Female Nasimgaon residents expressed the importance of going to the bazar by taking several pictures of shops where they bought their grocery items, both for the household and for their businesses (see Photo A-29). Such pictures, however, were not taken by anyone in Manikpara.

Domestic activities

Carrying out domestic activities, especially cooking, in public spaces is not common and is perceived as an indicator of poverty. Exceptions are bathing and doing laundry, which can take place at public water tanks, lakes or the river, similar to the pond in rural areas. However, this depends on gender and the perceived environment of the neighbourhood.

In Nasimgaon, Meher, a female participant in solicited photography and a shop owner together with her husband, took a picture of a woman cooking food along the street and described the picture as follows:

> "She is cooking in a dirty place. After preparing bread and cooking rice she keeps these along the street side. She also allows her children to eat there, in the street. [...] She is very poor and living in great misery" (Meher, 08.04.2008)

The fact that the photographer found this incident worth a picture indicates that it is not common and contravenes the social norms. During my participant observations I never found a similar case within the two study settlements.

Within Nasimgaon, men often take their baths at the several water tanks, and while doing so they are exposed to the public. This is mainly done by men who live alone, while their families may be in the village, because for one person it is cheaper to get the water from the tank than to get water supplied to the house. Women who do not have water supply at their houses do not take their baths at these water tanks but rather buy water from the tank and take their baths at home with considerably more privacy. Among the interviewees only Dipa, a self-supporting woman, collected water from the tank, while Aklima, a vendor selling cold drinks in public space, did so in times of erratic supply. At such times she also went to the nearby lake for her bath, and from her quote the adherence to gender separation and privacy becomes obvious:

"Again, I have to take a bath in the lake if water is not available. In the lake, a small space has been made kind of *paka* with some broken bricks. [...] [Women and men take their bath] in the same space. When we find there are no men then we take the bath. When men finish bathing and leave, then women go." (Aklima, 18.04.2010)

The Nasimgaon image resembles a rural lifestyle. In rural Bangladesh, men and women often take their baths in the pond or river. Thus the Nasimgaon image of bathing at the water tank resembles a rural lifestyle. On the contrary, Manikpara is officially connected to the water lines of the Dhaka Water and Sewerage Authority, pointing at its urban integration. Although water supply is erratic especially in summer time, there is no 'culture' of water tank bathing. I saw a few incidences of men taking a bath in the Buriganga and a woman doing her laundry in the river, but considering the general perception of the environment of the river – polluted – these seemed to be exceptions and expressions of poverty. However, with the change of the season, washing laundry and bathing in the Buriganga become more accepted (see also Photo A-15 and Photo A-16 which depict a higher number of domestic activities on the embankment slopes than observed outside of the rainy season). Roxana indicated that both men and women took a bath in the river in the rainy season and her family, which could be considered socio-economically poor, also did the laundry in the river. The general acceptance of 'using the river water' during the rainy season was also indicated by Hortem, who with his friends enjoyed swimming in the Buriganga during the rainy season (see Photo A-19).

6.1.4 Religious and spiritual activities in everyday life

Religion is also part of spatial practices in everyday life, while extra-everyday religious festivals and functions are discussed below (Chapter 0). The everyday practices discussed here primarily refer to Muslim practices. Although some Hindu families live in both Manikpara and Nasimgaon, I have not extended the research accordingly; expressions of Hinduism were hardly visible in public spaces in everyday life[8].

8 Only one Hindu place was mentioned by the interviewees of the middle-income women's focus group in Manikpara. This was the Hindu crematorium which was considered a no-go place: "Yes, there is [a place where people do not want to go], there is a graveyard on that side. Women never go to that side. The Hindu dead bodies are burned on that side, those are called *shoshan*. We do not even go to the places which are just near to that place. None of the women does. Is that clear?" (Mashrufa, 03.04.2010). This short comment of course cannot represent the perception of all residents, nor does it tell much about the relationship between Hindu and Muslim populations in Manikpara, as avoiding the crematorium could be interpreted as simply avoiding the place of an unfamiliar custom of dealing with dead bodies. However, the addition of Hasna could be understood as a reservation against other spatial practices and points at an understanding of space shaped by Muslim dominance: "We go to good places; we do not go to those places. [...] We want to go to Islamic places." (Hasna, 03.04.2010). Furthermore, Rabeya's account of the Muslim graveyard in Nasimgaon as a beautiful place where "people come and take pictures [...] with their mobiles" (05.04.2010) indicates that a similar avoidance of Muslim graveyards does not exist.

Religious place: mosques and mazars

In Bangladesh mosques are commonly only visited by men. Thus praying at the mosque is included only in men's activity patterns and most of the men visit the mosque for the Friday noon prayers. Of the interviewees of solicited photography, only Rokib went to the mosque regularly every day for his prayers. The mosque to him also seemed to be an important place to meet with his relatives and the *mohajons*. The importance and value of the mosque as a religious place in the associations of the rickshaw puller Abdul is made very apparent in a picture he took of a mosque in a high-income area:

> "I like this place a lot. I am also going there sometimes for prayer. I mainly like the religious programmes over there. Also one day one of my kids was sick and I brought some water from a religious person from there and because of that my kid recovered. Since then the place is a part of me and I am proud of it. That's why I took the picture of this place." (Abdul, 26.04.2010)

In Nasimgaon, apart from the mosque there are *mazars*, often with a Sufi connotation. Especially for Mokbul Hossain, the local *mazar* and religious talks with his *pir* were important. While women commonly do not visit religious places in their everyday life, Dipa used to visit a *mazar* in Mirpur once a week[9]. She was the only woman I met who moved around the city in everyday life for religious purposes. During the solicited photography, she took six pictures of different *mazars* and religious places and five pictures while travelling to these places[10]. At the Mirpur *mazar* she used to work every Thursday night from 8pm to 5am serving tea and biscuits to the guests of the caretaker of the *mazar* and cleaning the compound. She stopped this work, where she earned a considerable amount of money, because of the gossiping that started due to her absence at night:

> "They gave me 1,000 Tk for three months at first, and then they gave 5,000 Tk for four months as salary[11]. [...] But some women started to tell many bad talks and then I left the job. Now I don't go there. I left the job about three months ago. I did the job over there for 15 years." (Dipa, 25.04.2009)

9 Other interviewees also mentioned visiting *mazars*, but these were special occasions and not so much embedded in everyday life practices, they are thus discussed in Chapter 0.

10 It seems important to add some more specific aspects of Dipa's background. She was born to a Hindu family and married off by her uncle to a Muslim man at the age of ten or eleven. Her in-laws treated her badly and she thus left the family and came to Dhaka, where she was brought to Nasimgaon by someone who saw her on the street. After some time in Nasimgaon she was forcefully married to another man. She finally divorced, but her husband had taken all her possessions, so she had to start all over again. Despite her sufferings, she always appeared to have a very strong belief in Islam. Only once, after she had been dislocated from Khalabazar (see Chapters 8.2 and 1), she expressed her frustration about having become a Muslim: "If I would have known that people in this religion are so bad, like the way they treated me in the past, took my shop, took my land, my house; if I had known that before, I wouldn't have changed my religion and become a Muslim." (Dipa, 17.06.2009). Afterwards, however, she again showed a deeply devout Islamic belief.

11 The amount she claimed to have earned surprised me, as it seems rather high for work of nine hours once a week, especially compared to other occupations.

The example shows that Dipa's way of expressing her religious belief by working at a *mazar* was at the same time contesting gender norms and was not considered decent by the local society.

Porda influencing women's movement in public space

Instead of Dipa's publicly expressed religion, women are expected to express their religious belief differently, especially by maintaining *porda* (seclusion, see Chapter 5.1.3), prestige and honour (see further discussion in chapters 7.1.1, 7.1.2, 7.2.4). *Porda* especially was maintained by women of the middle class in Manikpara. When discussing working inside or outside of their home, they referred to the need to maintain their prestige and honour:

> "If the work can be done inside the house, I do it. But I do not go outside. It is a matter of shame [*shorom*] for us. We have never worked outside. Now if we go outside, we will lose our prestige [*man-ijjot*]. That is why we accept hardship, but still remain inside the house." (Subina, 03.04.2010)

The Bengali word *shorom* here can be understood as a feeling of shame and shyness that is expected to be expressed by women. Subina referred to the fact that women of social status like hers are not used to working outside. *Man-ijjot* refers to her and her family's feeling of prestige, honour and self-pride (social reputation), which she seeks to uphold by behaving accordingly. While the working women of Manikpara and the women in Nasimgaon, most of them whom work, could not afford a *porda*-oriented lifestyle, their dream-lifestyles and the advantages they saw in rural life reflected the same norms of maintaining honour and prestige. In Nasimgaon, Firoza expressed this during the focus group discussion with working women:

> "I want to live inside home, to cook, to call the name of Allah, to take my prayers five times a day. [laughs, presumably because this 'lifestyle' will remain a dream]. I don't want to work." (Firoza, 24.04.2010)

Similarly, Meher, a shop owner together with her husband, indicated how she disliked having to move around outside and work because of poverty, while if she had a choice she would prefer to stay at home and conduct a more *porda*-oriented lifystyle:

> "Which one is better for me [village or city life]? I am a woman. I will stay at home, I will cook, I will take care of my children – that should be it. Now, here [in the city] we are poor people. I have to work in the bazar to feed my children. These things are not good. If we had properties, I could just sit and eat. Now that is not going to happen." (Meher, 30.03.2010)

The working women in Manikpara expressed a similar wish to conduct a *porda*-oriented lifestyle during the focus group discussion. They also understood village life as enabling them to lead such a lifestyle:

> "We can do the prayers and fast properly [in the village]. We can maintain *porda*. We do not need to meet men [referring to men who are not family members]. We can take care of our

> house properly. We have a pond, a garden in our village. We can do everything, there is no problem. We can work conveniently. That's why we like the village more." (Salma, 02./09.04.2010)

The discussion indicates how the religious norm of keeping *porda* influences and determines women's everyday life practices. Not only for those who are able to adhere to the rules of *porda*, but also for those who are not able to do so comprehensively, but whose wishes and desires for a dream-everyday life are nonetheless guided by *porda.*

A private female Muslim space: the Tabligh

The above analysis of men and the mosque, but also of the rules of *porda*, has shown that the public space/sphere does not offer female religious spaces. As already indicated in Chapter 6.1.2, the *Tabligh* movement presents a private female Muslim space in Manikpara. The women from both middle-income and low-income households referred to these congregations held rotationally in different private spaces of members. The moral authority this movement has over female socio-spatial practices becomes apparent from the following comment by a female resident of a low-income household:

> "This time I went to *Tabligh*. That's why he [my husband] asks me to move carefully [in a decent way by using *porda*]. He does not like me to walk within a lot of people. He does not let me go. When there are very few people in the shower, he tells me to have a shower then. He does not let me go in the middle of gatherings. Allah will give us sin if we walk without *porda*." (Salma, 02./09.04.2010)

The moral authority established and re-produced by joining these meetings is also intended, and Mashrufa's statement underlines the missionary objective:

> "They [conveners of the *Tabligh*] will convince the persons who do not pray yet to take prayers. They will tell you to remember the name of Allah all the time or to read out *Kalema* [religious words]. [...] After listening to these, you will get the feeling of having the kindness of Allah with you in your heart. You will realise that listening to these will increase your knowledge, your intelligence. Many of us do not know anything, even not the *Alhamdu Sura* [the first surah of the Quran]. If she goes to the *Tabligh*, she will get to learn that, she will get to learn many prayers." (Mashrufa, 03.04.2010)

Religious and spiritual beliefs shaping the everyday

Apart from the religious norms influencing everyday life activities, there are also mobile activities carried out in public space with religious connotations: especially in Nasimgaon, 'mobile' preachers regularly teach about religion using megaphones. Often the preaching is at the same time an economic livelihood activity, in that the preaching person sells specific medicines (see the canvasser in Chapter 6.1.1, Photo A-8). These faith-healers sell treatments like *jhor fuk* (reciting of verses from the Quran and blowing them), blessed water and *tabij* (amulets).

Their powers of healing are widely believed in, and so is the belief in the works of supernatural beings and spirits, especially in rural areas but also in some urban areas (Miaji 2010: 96p.) In Nasimgaon, the belief in the effectiveness of faith healing is widely spread, for the protection and curing of sick children and even as a method of female contraception. Especially Dipa repeatedly suspected that her sicknesses were related to evil-wishers, as the following story demonstrates:

> "Once I was suffering from stomach pain. I could not even say any word. At that time a boy of ten years entered into my room and addressed me as 'mother'. I did not recognise the boy. The boy told me 'Mother, are you severely suffering from stomach pain? Some people want you to die. They put three amulets within your house to kill you. They don't like your betterment [that she became a shop owner after she had been among the most deprived in 2009]'. I didn't believe him at first. I didn't see him before. Then the boy told me 'Search. You will find one amulet in front of the door under the bricks. You will find one in your shelf and another one on the west wall of the room.' Believe it or not, I found three amulets at those places and suddenly I felt better. I didn't feel any pain in my stomach." (Dipa, field note, 12.11.2010)

This is only one of many examples where Dipa linked her well-being with spiritual powers and evil wishers. Similarly, I came across a fear of ghosts among the women in Nasimgaon. While most of the women only feared ghosts in the village and rural areas, where it is a common belief (see Chapter 7.2.2), Rehana, a rice vendor, mentioned ghosts as a reason to restrict her movements at night:

> "Late at night, I am afraid of ghosts. But I am not afraid of people. I am just a bit afraid of the ghosts. Otherwise I don't have any problem with anyone. I can go anywhere, I don't have any problem." (Rehana, 23.04.2010)

Rehana's fear of ghosts leading to a control of her movements at night indicates how in Nasimgaon the social norms are shaped by rural traditions. Although the *tabij* is very common all over Bangladesh, especially to protect new-born babies, I did not come across a similarly deep-rooted belief in religious healing and supernatural powers in Manikpara. The *tabij* and similar amulets were also common, but no one talked about other methods of religious healing or fear of ghosts.

Concluding remarks

The everyday life practices of both men and women indicate a more dominant influence of religion on everyday life in Manikpara compared to Nasimgaon. This is on the one hand reflected in the adherence to religious norms, especially the *porda* of middle-class women in Manikpara and the wish expressed by the lower-income women to maintain a similar lifestyle. Another indicator of this is the importance of the orthodox *Tabligh* movement in Manikpara (also in extra-everyday life, see Chapter 0). In Nasimgaon the religious everyday life practices indicate a more multidimensional and syncretic understanding of Islam and spirituality mixed with folk beliefs. The analysis of religious practices in extra-everyday life (see Chapter 0) supports this observation of religion in Manikpara being a more dominant factor in producing spatial practices.

6.1.5 Night time activities

At night time[12] there are several restricted areas, both for women and for men. In both settlements there is a community police/night guard system to supervise the area at night and report any incidents to the *thana* police. Furthermore, the frequent electricity cuts in the summer months change the activities taking place at night considerably, as residents leave their houses to get fresh air (further discussion in Chapter 7.3).

The embankment in Manikpara is not a socially accepted place at night, and residents warned about going there as it was frequented by 'unsocial youths' who are said to consume cannabis and wine. The *adda* group of respected members of the local community held at one *ghat* until late at night, however, is an exception (see Chapter 7.3). The only place along the embankment which used to be frequented by a larger public until late in the evening and during power cuts even late at night (see Chapter 7.3) is the Bagan. From the evening hours onwards, there are vendors selling snacks like roasted nuts and *chanachur*. A reason for the acceptability of the Bagan during the early night hours may be its non-seclusion. While the embankment slopes are not visible from Embankment Road due to the wall, the Bagan is only separated by a fence[13].

For women, moving around at night is hardly acceptable and dependent on their position in terms of life-cycle and status in society. Mesbah Uddin referred to the matters of status and respect when he exclaimed about the embankment at night: "Women go. But isn't there difference between women!" (Mesbah Uddin, 21.05.2009), indicating that there are 'different types of women' who apply such social norms in different ways, while he did not appreciate deviant behaviour. Similarly, Roxana was immediately teased by local boys about whether she had been at Ramna Park (a place with a bad reputation at night for drugs and prostitution) when she came back from the factory late at night.

In Nasimgaon it is more common for both men and women to stay outdoors even until late at night. In the bazar areas, most shops are kept open until 11 or 12 at night[14]. Until that time the streets remain busy, as it is also the time when many

12 In Bengali, the evening (*shondha*) ranges from 5pm to 8pm, and the night (*rat*) ranges from 8pm to 3am. This delineation, however, does not suffice to differentiate 'night time' with reference to the activity patterns, which is rather defined by the end of working hours when everyone returns home. With night time I thus refer to the late night hours when the activity patterns considerably change and being outdoors becomes more regulated according to what is deemed socially acceptable behaviour. In Manikpara, this night time starts from 9/10 pm, after the plastic sorting shops have closed at 8pm and workers have returned home. In Nasimgaon, this night time starts from 11/12 at night when the shops of the bazar areas mostly close down and the garment workers have returned home after an 8am to 10pm (including two hours overtime which is common in many factories) duty.

13 In April 2010, however, a wall was constructed between the Bagan and Embankment Road. This decreased visibility of activities may have resulted in shifting use patterns, however, this was not investigated within this research.

14 In case of electricity cuts, power from a generator is offered to the shops in the bazar area until 12 midnight.

garment workers return from their duty. After midnight the area is very calm and quiet and there is no sign of the surrounding city. The last group of people returning to Nasimgaon are the garment workers working the night shift until 3am. As long as the shops are open, women do not perceive any issues with staying outdoors, although they avoid certain narrow lanes where young men meet and consume cannabis and wine, and prefer to stay close to the house later at night. The night guards confirmed that the security of the area allowed women to move around rather freely even at night hours. Thus many women did not express fears about night time movements, except for Rehana who feared ghosts (see Chapter 6.1.4).

Night time activities in public space are restricted depending on user groups. Here gender norms become much more explicit in the use of public space, especially so in Manikpara. The further discussion in Chapter 7.1 will point at some of the underlying concepts of gender relations leading to these restrictions for women.

6.2 EXTRA-EVERYDAY LIFE SPATIAL PRACTICES

In 'Rythmanalysis' Lefebvre (2004 [1992]) differentiates the extra-everyday from the everyday. He concludes how extra-everyday events can be seen as an extension of everyday life and understood as a continuation and re-enforcement of everyday life spatial practices:

> "With these places are we in the everyday or the extra-everyday? Well, the one doesn't prevent the other and the pseudo-fête emerges only apparently from the everyday. The former prolongs the latter by other means, with a perfected organisation that reunites *everything* – advertising, culture, arts, games, propaganda, rules of work, urban life… And the police keep vigil, watch over." (Lefebvre 2004 [1992]: 36, original emphasis)

In this sub-chapter, I thus seek to explore how the spatial practice in Manikpara and Nasimgaon changes (or continues) in extra-everyday life. The differentiation into everyday and extra-everyday activities emerged during the empirical fieldwork. Subsequently, I particularly analyse the changes and continuations during religious holidays and functions (I here include the *Ramadan*, although it could also be understood as part of everyday life during this one-month period), Bengali cultural events, political activities and Bangladesh national celebrations and educational activities.

6.2.1 Religious holidays and functions

Religious holidays and religious functions[15] considerably change public space. The spatial practices shaped by these extra-everyday events are discussed in the following. The extra-everyday religious practices solely refer to Muslim festivals and programmes, as do the everyday practices, and I did not visit the settlements during the Durga Puja, the largest Hindu festival in Bangladesh, or any other event.

Changing spatial practices during Ramadan

During the Islamic fasting month of *Ramadan*, the religious norm of fasting produces a specific spatial practice that changes the appearance of public space considerably. During the daytime, all tea and food stalls that throughout the year are just open towards the public sphere are covered by cloth curtains. The curtains give privacy to those who do not fast while the public sphere is maintained according to religious norms (see Photo A-31 and Photo A-32). After *iftar*, the time of the fast-breaking meal at sunset, both Nasimgaon and Manikpara become quiet and calm for about half an hour. Afterwards life on the streets continues again after everyone has taken *iftar* at home.

The increased consumption during the Eid holidays following Ramadan usually accelerates the economic activities of plastic drying in *Ramadan* considerably, and thus street sides are largely occupied by plastic recycling activities. In both 2009 and 2010 the *Ramadan* was parallel to the rainy season, and thus for the plastic drying businesses a high workload was coupled with the need to move activities to the alternative location. Momena, drying plastic on commission on the slopes between two *ghats*, decided not to adhere to the rules of fasting in *Ramadan*, due to the heavy workload while being exposed to the heat plus the family obligations of running the household. Most of the sorting shops and factories change their working hours, so that after *iftar* most workshops remain closed.

Especially in Nasimgaon, during *Ramadan* (and the *Eid* festivals) a number of temporary stalls are set up. The most prominent new vending units are stalls selling *Eid* greeting cards and toys. These range from semi-mobile to permanent stalls (see Photo A-33), which are removed after the holidays. Furthermore, many vendors and shop owners use already existing or new extensions to their shops (commonly benches or *chouki* set out in the public space in front of stable shops) to sell *iftar* items (see Photo A-34). Accordingly, the opening hours of some businesses also change. Zakir, operating a restaurant at Nasimgaon Eid Gah Math,

15 As discussed in Chapter 4.2.5, I was not able to attend the two *Eids* in the study settlements, but these periods were covered by my field assistants, and by the internal observers as part of the solicited photography (see Chapter 4.2.3). I only witnessed some *milad* ceremonies at the home of interviewees and one *Mussolmani* celebrated in public space. Regarding other religious functions, I was only able to attend some preparations but not the events themselves.

kept the shop open during the whole night so that people could buy food, while outside of *Ramadan* they kept the restaurant open until 12 or 1am for the rickshaw pullers who came back from work. In the week before the *Eid*, some interviewees reported a lower business turnover, because people started to go home to their villages. For Mokbul Hossain, his shop became less important in *Ramadan* and he spent much of his time at the *mazar*, taking a rest from business and concentrating on his religious life. While the shops that sold items in demand for *iftar*, or that adapted to the changing food consumption patterns in *Ramadan*, generally made additional profits, other businesses did not do well, for example the *pitha* stalls or the selling of cold drinks.

Changing spatial practices during the Eid days

In both settlements, one public place is specifically designated for the prayers at the Muslim *Eid* festivals, thus the name Eid Gah Math (see also Chapters 5.3.2 and 5.4.2). The funeral prayers also take place on these Eid Gah Maths. At the time of the *Eid* prayers (but also the funeral prayers), only men are present in those spaces (see Photo A-35). Apart from the prayers, the appearance of public space and the spatial practice change considerably during both *Eid* days.

On the *Eid-ul-Fitr* day in Manikpara there are a *nagordola* and a merry-go-round as well as different food vendors in the Bagan. In the evening there is reciting of the Quran and a musical programme. The Eid Gah Math is surrounded by shops selling dolls and different foods after the morning prayer. For the lease operator Foyez, the *Eid* days meant less income as no boats were unloading goods and most other economic activities taking place at the *ghats* and the embankment slopes came to a halt. Thus for the *Eid* days he only paid a reduced lease fee to the leaseholder (see Chapter 8.1 for the leasehold arrangements).

The goods on sale also change towards the *Eid* day, when special food items and ingredients, cloth, shoes, jewellery and toys are especially in demand. In Nasimgaon additional shops selling snacks are set up by the established shop-keepers with the permission of the Bazar Committee. On the night before *Eid-ul-Fitr*, Rohima and Shahin kept the shop open for the whole night and the next day. For the *Eid* days, they changed the stock of their shop. On *Eid-ul-Adha*, for example, they started to sell toys in their shop. From Rohima's description of the pictures she took at that time, the importance of the *Eid* day and the practices of this extra-everyday event become very obvious:

> "Everybody was having fun during *Eid* day. There were *fuchka* shops, *chotpoti* shops, *shorbat* shops [food and drink]. Children were so happy. [...] We were selling. I liked it so much. I was having fun. Many children came and bought from me. I have taken this picture. People were buying those things from me. We were selling and they were buying. This was so much interesting. [...] We kept toys on display on the day before *Eid* so that people could see those and wished to buy them on *Eid* day. How would people know that there are a lot of toys in this shop if we did not keep those on display?" (Rohima, 18.01.2010)

Other religious functions in public space

Especially in Manikpara, religious functions are celebrated regularly. The previous leaseholder of Manikghat and plastics businessman Jahangir regularly organised religious functions at his shop on the embankment slopes close to one *ghat* in Manikpara (see Photo A-36). This included organising, normally in July, an annual *orosh* – a yearly meeting of the followers of a Muslim saint – with reciting of the Quran, donations, prayers and *Baul* songs. During these events, food is distributed to the participants, and while he was still the leaseholder Jahangir also gave out food and *lunghi* (the traditional loincloth of men) to the boatmen at the *ghat*. For the days of the function in July 2010, the shop area was decorated with garlands, cloth banners and lights, and a stage and loudspeakers were installed. The pictures taken during the function in the evening only show men and children, while women were absent. A similar function had been held at the Bagan just a few days before Jahangir's function. This function was held for the whole day, and the pictures of the afternoon indicate that a few women also participated, although men and children formed the main part of the crowd of participants.

In Nasimgaon public space is also used for religious functions. I witnessed the celebration of a *Mussolmani,* the celebration of a boy's circumcision, on an open space at the fringe (see Photo A-37). The family of the boy had invited their relatives and neighbours and cooked food for everyone. According to my landlords, *Mussolmanis* are commonly celebrated, but only those families who are among the better-off organised big celebrations in public spaces. Other religious functions of *orosh* were organised by the *mazars* in Nasimgaon, but mainly took place within these compounds. Repeatedly, a night of *Baul* songs and prayers was organised by the political committees or the mosque committee, and a stage was setup on the Math, for example in February 2009 and November 2010.

In February 2009 the Eid Gah Math was occupied by a large tent for four days. The tent had been put up by the devotees of a *mazar* located in Faridpur District. According to the organisers, this tent was put up every year to provide a shelter for donations for the ceremony taking place at the *mazar* in February, especially for the goats, buffalos and cows. The tent in Nasimgaon was erected to collect the donations from the surrounding areas for joint transport to Faridpur. Inside the tent, people cared for the animals by giving them water and food. Here, Nasimgaon Eid Gah Math served as a provider of a storage space probably in the absence of a large choice of public spaces available for such purposes.

As already mentioned above, many interviewees talked about visiting *mazars*. Except for Dipa's case (see Chapter 6.1.4), this was an extra-everyday activity performed on special occasions rather than regularly. For example, Hamena's sister decided to go to a *mazar* in Sylhet after the *Eid-ul-Adha* 2009 as she had vowed to do so if she became a mother. My landlord together with other shopkeepers of the area went to a *mazar* in June 2010 to attend an *orosh*. Together they had bought a goat for 3,000 Tk as a donation to the *orosh*. In Manikpara, Jahangir also travelled to *mazars* and had built one of his own in his home district. Several interviewees, both men and women, especially from Nasimgaon, expressed their wish to visit specific *mazars* in Bangladesh, indicating the syncretic Islam practised in Bangladesh and in Nasimgaon specifically.

The *Bisho Ijtema* as the main congregation of the *Tabligh* movement is highly important in Manikpara. In 2010 I joined a group of men for the final prayers at Tongi in the north of Dhaka. Due to the crowd, it was impossible to reach the main praying ground, and thus we sat down along the river banks during the time of the prayer. The importance of this event in Manikpara is manifested by the large number of people joining the prayers. Every year the community organises a launch which takes both men and women (mainly of the middle class) to the *Ijtema* grounds where they stay for the full three days. Jahangir, with whom I went to the *Ijtema*, normally spent the whole week on the grounds and helped to organise accommodation and food for the pilgrims. His wife normally joined the launch journey, but in 2010 she was pregnant and thus did not join. The leasehold operator Foyez reported that on the day of the *Ijtema,* Manikpara became considerably quieter, which also affected his *ghat*-business as fewer boats came to unload goods. In Nasimgaon I also found people going for the final prayers, but in Manikpara's middle class it seemed to have greater importance, as also the women's practice of *Tabligh* (see Chapter 6.1.4) suggests.

Concluding remarks

The above accounts show a rich diversity of religious festivals and functions held in public spaces. At the same time, they show the importance of having adequate public spaces to celebrate religious festivals. According to my observations, religious events tend to have a more pronounced expression in public space in Manikpara, where more functions are held regularly. One explanation may be the presence of more middle-class households, but also the more orthodox view of Islam as influenced by the *Tabligh* movement. In general, religious expressions of extra-everyday life tend to be male-dominated and the spatial practice on these days seems to re-produce the dominant patterns of gender relations and social norms.

6.2.2 Bengali cultural celebrations

The cultural celebrations discussed here are the holiday of *Pohela Boishakh* and the holding of Bengali fairs (*mela*). Although there are other events[16], for example the celebration of springtime, these do not much spread to Nasimgaon and Mani-

16 Although I did not witness any such event, marriages in Nasimgaon are occasionally celebrated on the Eid Gah Math and on a public space at the fringe, according to interviewees. The political leaders I interviewed took this as one reason why a community centre should be built on the Eid Gah Math (see Chapter 0). In Manikpara, I witnessed one *Gay Holud*, the event when the family say goodbye to either bride or groom by feeding her/him and painting her/his face with yellow turmeric paste. This event took place in the school compound of the primary school, and thus not in a public space.

kpara. The other cultural programmes tend to be religiously connoted, as in the case of spiritual *Baul* songs.

Pohela Boishakh – Bengali New Year

In Manikpara, the Bengali New Year of *Pohela Boishakh* on 14th April (see Chapter 5.1.2) is widely celebrated. Usually a programme is held in the Bagan including the presentation of songs and special foods (see Photo A-38). On *Pohela Boishakh* women of the middle class, who normally do not leave the house, also show up at the Bagan to enjoy the programme. Both children and adults take part in the celebrations at the Bagan and at other places – for example private celebrations are hosted on rooftops. However, the division of society into a working class and a middle class of businessmen becomes most obvious on this day. While the middle class enjoy the holiday, many of the workers continue to work in the plastic recycling businesses as usual. The day also serves to invite family and supporters for food. Rokib and Momena usually invited the employees of their plastic drying business and the *mohajons* for lunch, and also Mesbah Uddin and Jahangir held similar programmes, together with other family members and friends or by themselves.

In Nasimgaon the extra-everyday event of *Pohela Boishakh* is less visible and many inhabitants continue with their everyday life activities, e.g. those operating shops and stalls within Nasimgaon and the rickshaw pullers. Before *Pohela Boishakh* many of the (stable) cloth and tailoring shops start to advertise and sell special *Pohela Boishakh* designs, either red and white garments or garments with special prints. While many residents wear new cloth of *Pohela Boishakh* designs, the day is mainly celebrated by children, youths and those whose workplaces are closed, e.g. the garment workers. Thus the changes on the Nasimgaon Math are not very apparent, only visible due to larger gatherings at the vendors' stalls (see Photo A-39). It is mainly more crowded with children and the vendors state that they normally experience a higher business turnover as especially children and youths buy snacks and drinks. The observation that *Pohela Boishakh* in Nasimgaon is mostly celebrated by children and youths is also reflected in Rabeya's (a 13-year-old girl) enthusiasm for the celebration. Together with ten of her friends, she had organised a cooperative to which each of them contributed 10 Tk per day before *Pohela Boishakh*. They had made a list of expenditures for *Pohela Boishakh* and needed to save 2,000 Tk in order to cover their expenses on the day. Rabeya's desires associated with *Pohela Boishakh* become obvious from the importance they attached to dressing up and looking beautiful on the day of *Pohela Boishakh*. Similarly, I found many parents buying/sewing *Pohela Boishakh* dresses for their children, while they themselves mostly did not dress up and the day rather remained like any other day for them.

Despite my general observation of life in Nasimgaon seeming more everyday than extra-everyday on *Pohela Boishakh*, in 2010 a stage was put up on the one open space at the fringe. The crowd gathering in front of the stage to listen to the songs and cultural programme was dominated by children, youths and young men and young women, many of them dressed up in the dresses of the day. The composition of this crowd supports the general observation that for many inhabitants everyday life goes on almost as usual and only the younger generations celebrate.

Bengali fairs (mela)

Fairs (*mela*) are very popular in Bangladesh, both in rural and urban areas. Often they last for a month and besides stalls selling food, decorations, cloth and jewellery they are often accompanied by cultural programmes of songs and theatre and attractions for children such as the *nagordola* and merry-go-rounds. In Nasimgaon a winter *mela* was held from 24th December 2009 until 7th January 2010 on the Eid Gah Math. In Manikpara I did not come across a similar event.

The *mela* in Nasimgaon had been organised by the local Committee of the Freedom Fighters of the 1971 Liberation War to support the disabled and poor freedom fighters with the revenues from the *mela*. The organising committee had to pay 120,000 Tk per day generated from the *mela* to the police, party leaders and freedom fighters. The *mela* was opened by the AL Ward Commissioner candidate and his supporters. The *mela* changed the appearance of the Eid Gah Math and the spatial practices considerably. A large tent was put up temporarily (see Photo A-40) to host puppet shows, concerts and dramas for those who bought tickets beforehand. Another stage was set up at the north-eastern corner of the Math where gambling and raffles took place. Especially for children, the *nagordola* and merry-go-rounds set up on the Math provided attractions (see Photo A-41).

Furthermore, a number of food shops of permanent structures were put up surrounding the Math, but removed after the *mela*. Mokbul Hossain, one of the interviewees of solicited photography, also decided to put up a shop selling bethel and cigarettes for the duration of the *mela*, in addition to his usual business. He put up a shop with four bamboo pillars fixed to the ground and a polythene roof, and removed this structure soon after the end of the *mela*. He had counted on a large crowd visiting the *mela* and thus a profitable business, and made a profit of 200-300 Tk per day. The number of mobile vendors also increased considerably. Aklima, who sold drinks on the Math, also indicated that these events were profitable for her, while at the same time this meant she had to pay more to those keeping watch:

> "When there is any *mela* then the selling goes well and they know it. When there is any programme of school then also the selling goes well. On those days I have more income and have to give more money to the night-guard." (Aklima, 18.04.2010)

The fair was planned to be continued from 9th January 2010. The mosque committee, however, protested against the continuation of the fair. On 8th January after the noon prayer, the committee and their followers convened a demonstration against the fair and demanded that the gambling and drama show be banned as they were not in accordance with their religious views. A member of the mosque committee explained:

> "There was a definite duration to run the *mela*. They wanted to carry out the *mela* for some more days after the deadline. Then the mosque committee and the social leaders stopped them. [...] The mosque committee stopped them because there were some unsocial activities done by the people like gambling. That's why we tried to stop the *mela*." (Rabiul, 23.03.2010)

A previous *mela* which the AL Chatro League (the student sub-organisation of the AL) of the Ward wanted to organise in May/June 2009 was not held because of the protest against it by the inhabitants of the adjacent settlement. The reason for the protest was especially the activity of gambling, which was perceived as a disturbance and not socially accepted:

> "[...] the people [of that settlement] are opposing the *mela*. The reason is that there will be some gambling going on in the *mela*. The people [of that settlement] want to save their children from that gambling. Also before the Nasimgaon leaders were planning to have a *mela* but the people [of that settlement] successfully protested it." (Mokbul Hossain, 24.05.2009)

This account of the *mela* in Nasimgaon indicates that it is an extra-everyday spatial practice contested by different forces for different reasons, while at the same time it presents a chance for additional income to the residents and the attending crowd indicates that it is enjoyed by many.

Concluding remarks

Pohela Boishakh does not change the spatial practices as distinctly as the religious holidays and functions discussed above. This is significantly different in other areas of the city, for example on the Dhaka University campus where *Pohela Boishakh* is celebrated most intensely. The spatial practice in Manikpara points at socio-economic status as an explanation of the lower visibility of *Pohela Boishakh*. While middle-class households can afford to celebrate *Pohela Boishakh* and also use it as a means to invite others to their houses or other venues, low-income households in Manikpara, and accordingly a majority of the population in Nasimgaon, cannot afford to celebrate this day intensively. These extra-everyday events are dominated by a re-production of the spatialities of poverty, excluding parts of the population from participating in the celebrations.

6.2.3 Political activities and Bangladesh national celebrations

Political activities on the one hand consist of demonstrations and poster/banner decorations especially on holidays related to the political history of Bangladesh. On the other hand, they are part of the everyday production and negotiation of public spaces which will be discussed in Chapter 1.

On the days related to the national history of Bangladesh, the public spaces in Nasimgaon become a means of expressing the dominance of the ruling party. During all the celebrations discussed below, AL decorations and AL supporters dominated the public space. On *Ekushe,* commemorating the martyrs of the language movement in 1952 (see Chapter 5.1.2), a procession was held at night. When I went to the Nasimgaon *Shohid Minar,* the local replica of the monument for the martyrs on the Dhaka University campus, the next morning I found mainly AL supporters although the day is celebrated by all. Similarly, on 7th March the AL club at Khalabazar played a recording of the speech Sheikh Mujibur Rahman, founder of AL, held on 7th March 1971 and thereby dominated the sounds of the bazar. Victory Day, celebrating the victory of Independence War in 1971 (see Chapter 5.1.2), was celebrated in Nasimgaon with a *michil*, a political demonstration, organised by the AL Ward level with the participation of many Nasimgaon residents (see Photo A-42), among them Shihab who worked as a blacksmith:

> "He [previous Ward Commissioner of AL] brought out a *michil* to show that he is still a powerful person though he is not the current [Ward] Commissioner. That's why he brought out a *michil* by gathering us. Then he helped us by giving us this blanket [for winter times]." (Shihab, 21.02.2010)

This indicates the domination of the AL: although the present Ward Commissioner still is a member of the BNP (see Chapters 5.1.1 and 9.1.3), the public space becomes an arena of representation for the AL activists. While the neighbourhood spaces are thus dominated by the ruling party's supporters, the opposition party resorts to means of collective protest by bringing out road barricades, *michils* and *hortals* in the more public areas, often as an answer to certain events like accidents or erratic electricity and water supply. Many of the Nasimgaon residents regularly participate in either the AL's or the BNP's political demonstrations and, as the quote above indicates, receive some benefits for doing so, indicating the patron-client relationships that politicians tend to build up and nurture.

This was not as explicitly the case in Manikpara which may be due to the fact that politicians are known for drawing (or 'buying') their supporters from the 'slums' rather than the rest of the city. Furthermore, the inhabitants in Manikpara are mostly involved in contractual labour and their fixed working hours do not allow them to regularly join *michils*. I thus found representations of political activities in Manikpara to be more subtly present in the form of posters and banners exhibited in everyday life and on public holidays. On such days, both the leaders of AL and BNP put up their own pictures and send their greetings and/or electoral promises to the inhabitants (see Photo A-43).

The observations in the study settlements have revealed how public spaces repeatedly become arenas demonstrating the power of political parties. While the opposition resorts to expressions of protest, the ruling party celebrates the national holidays with a dominant presence in public space. This includes decorations as well as the mere presence of important AL leaders at central locations of celebration. Some of the national celebrations are furthermore accompanied by extra-everyday educational activities (see Chapter 6.2.4).

6.2.4 Educational activities

Educational activities, while normally part of everyday school time, assume an extra-everyday character occasionally, especially during national or cultural celebrations (see Chapter 6.2.2, 6.2.3), and then they become part of the spatial practice lived and experienced in public space.

On various occasions, especially during public holidays but also on 'weekends', some schools in Nasimgaon hold sports programmes using public spaces. I witnessed such programmes on the Nasimgaon Eid Gah Math and on an open space at the fringe, the only two spaces suitable for competitions in running, 'biscuit running', where between the running a biscuit hanging from a rope has to be eaten without using the hands, and similar games. Most schools in Nasimgaon do not have their own school grounds or only very small ones, and thus shift sports and games to public spaces. The public spaces during such school competitions become very mixed gender spaces with a high number of adult females (mothers) in the gathering crowd – rather contrary to the male crowd watching football tournaments of the local youths (see Chapter 6.1.2) or the *Eid* prayers. During one school competition which I attended there was a competition for adults. While the young men did 'cock fighting', i.e. fighting against each other while jumping on one leg only, the women, including me, played musical chairs. Women of all ages joined in this game, which in Manikpara, i.e. a more restrictive society considering women's movements, would have been unimaginable. The Nasimgaon Eid Gah Math also provided space for an adult educational (and at the same time political) activity, specifically a convention on the International Water Day. On this day, information on the new water supply system that was to be introduced was shared and discussed in a temporary tent put up on Nasimgaon Math[17]. Although the photograph taken by a participant of solicited photography at a first glance suggests a predominantly male crowd, there is a female crowd sitting in one part of the tent.

The only educational activity that took place outdoors in Manikpara was a school class for street children organised by an NGO at Bagan. The existing schools of the area seem to carry out such programmes inside the school compounds and outdoor programmes are not common, indicating a different social environment. Furthermore, it is more common than in Nasimgaon to send children to *madrasas* after a basic primary education.

The educational activities taking place in public space are of considerable importance as they change the normal user structure considerably. Especially the school events draw a large and very mixed crowd to public space, but also the NGO events produce a rather mixed crowd in comparison to many everyday events.

17 I did not witness other awareness campaigns taking place in public space, nor did the interviewees mention any other campaigns.

6.3 IMPACTS OF THE QUALITY OF PUBLIC SPACES

The quality of public spaces has an impact on many of the livelihood activities and everyday life spatial practices discussed above. This discussion will focus on how changes in the quality of the road network and public spaces generally impact on mobility patterns and behaviour. The quality of the public spaces will be discussed with regard to pavements and the width of roads.

This sub-chapter does not cover transformations of public space into stable structures that reduce public space. This is especially evident in the example of Khalabazar, which transformed within a year from a semi-permanent open-air vending place to being built over by stable structures. This change of public space also results in changing spatial practices in everyday life, as indicated by Afsana's reference to *bazar-ghat* in Chapter 6.1.1. However, these transformation processes will be discussed in detail in Chapters 8 and 9 of this research.

Pavement

In April 2009, Embankment Road in Manikpara was surfaced with concrete (see Photo A-44 and Photo A-45). Previously, it had been covered with sand, and during the dry season a lot of dust came from the road, while as soon as the first rains started the road was covered with mud making movement difficult, especially for pushcarts and rickshaws. Accordingly the owner of a plastic storage business located on Embankment Road expressed his appreciation of the concrete coverage:

> "The road was broken before, there was too much sand and dust on the road when the vehicles ran. Now the road is smooth, there is no sand. The road was full of sand and as a result I did not want to work." (Tariq, 27.06.2011)

This appreciation of the infrastructure improvement was repeated by all interviewees who were involved in plastic recycling and needed to transport goods along the road. For the people working in sorting shops along Embankment Road, the working conditions also improved, despite new problems:

> "The road condition has improved. Now we can move easily. But there is another problem that we are facing now. The drains are clogged with waste which creates water logging in the drains of the road and if rain comes the situation gets worse. When the cars and buses move here, this waste water is splashed to our shops. Before the construction of the road, there was a problem of the movement of the vehicles, but there was no problem in the shops. But now our shops and we are facing this problem of waste water splashing." (field note, Tariq, 16.09.2009)

For pedestrians, however, the improvement means decreased safety. Vehicles speeding after the improvement of the road was also perceived as a nuisance by the group of men using the slopes at one *ghat* for their *adda* round. They realised that crossing the street had become dangerous for themselves and for those who wanted to access the *ghat*. In an effort to slow down the vehicles they put up a barrier to separate the lanes (see Photo A-46), however, this only worked for a day before the barrier was destroyed. It was not put up again.

Interestingly, during the re-construction process the new road became an important temporary public space. While construction was going on, the traffic was directed one-way. The other lane, partly finished, was then used extensively to store plastic recycling materials and to load and unload goods (see the left lane on Photo A-45). This immediate and intense use of public space that evolved only temporarily indicates the high demand for public space in Manikpara.

Textbox 5: Improvements of pavement and disruption of livelihoods

The improvement of Embankment Road in Manikpara was not understood as being an advantage for space-based livelihoods by all. Kabin Mia used to operate a rickshaw garage and rickshaw repair shop at the corner of Bagan, where a small triangular space facing Embankment Road existed (see Photo A-47). Due to the improvement of the road changing the flow of traffic, he had to shift his working place:

"I worked at the side of the road before, but now it will no longer be possible. Once the road is constructed there will be too many vehicles on the road. There were not many vehicles before. Now the road is good and vehicles will move with high speed and accidents will occur." (Kabin Mia, 29.04.2009)

As he had negotiated a good arrangement at his location at Bagan, he faced problems in changing his location because suddenly the rent payment for the rickshaw garage increased. This dislocation from his previous location led to the serious disruption of his livelihood activities, and for a long time his wife, who operated a *pitha* stall in the same place where he used to have his repair shop, and his son, who worked in a plastic sorting shop, were the ones solely supporting the family. His wife did not face the same problems due to road construction, because her shop was set back from the road along the stairs leading to Bagan.

In Nasimgaon unpaved roads also become a nuisance during the rainy season, as they get muddy and dirty and, according to the interviewees and my own observation, hamper residents' movements. This also results in losses for some businesses because access to the shops becomes difficult. Accordingly, many initiatives to pave *kaccha* roads take place inside the area, which can also be understood as resulting from perceptions about perceived security of residence in Nasimgaon (see Chapter 5.4.1 for changes in this feeling of security). While some larger activities, including also the improvement of drainage along the road, are conducted with the support of NGOs, the pavements of smaller areas were constructed by the joint efforts of neighbouring shopkeepers.

The Nasimgaon Math received a new layer of sand on the initiative of the MP in July 2009. At first this considerably improved the quality of the Nasimgaon Math and it thus became more attractive for children and youths to play football, cricket and other games there. However, the improved condition did not last long. Mokbul Hossain suggested this was because the money the MP had provided and the amount of sand put onto the Math did not match, suggesting that a large amount of that money had ended up in other pockets. Furthermore, some of the sand was washed away during the rainy season. Thus only a few months later the

Nasimgaon Math became less attractive for playing games because of its uneven surface and standing water, except for small children who enjoyed playing in the water. Thus no permanent improvement of the surface condition could be achieved (see also Chapter 8.3.1).

Road width

The road width was especially mentioned in Manikpara as being an issue for everyday life. Although most of the internal roads still appear very narrow, they are the product of road widening initiatives initiated in the past after severe fire outbreaks and after experiencing the problem that dead bodies could not be carried through the roads. The joint effort was narrated by the Ward Commissioner:

> "At that time people became united to widen the roads like during the Liberation War of 1971 when people became united for the independence of our country. Even someone who had only one *katha* of land contributed some space for the road because he was also a sufferer of fire accidents. Everyone knew that the fire accident becomes severe because of narrow roads. But they didn't take any initiative. We took the initiative first. When we went to break someone's building then he told us 'I will break down my building by myself and leave the land within seven days.' We didn't need to invest anything. Everyone broke down his own building and DCC has constructed the roads. Some people also didn't want to leave land for the road network. But they have also left land because of collective social pressure." (Ward Commissioner, 15.04.2010)

His emphasis on the Liberation War indicates the willingness of the community to together improve the road conditions. Furthermore he referred repeatedly to the atmosphere of social pressure and his initiative to also make very powerful leaders donate land for the roads. According to his narrative, the community had largely united to improve the road network. Till today, however, the electricity poles remain in the road space rather than on the sides, and the Ward Commissioner said that ever since the initiative he had been trying to get the electricity authority to relocate these. While the road widening initiative has improved accessibility, many of the internal roads remain narrow and this was perceived as unsafe by women, for reasons of safety at night but also because of potential body contact with men when passing them in the road.

In Nasimgaon the streets in the bazar areas are used intensively for trading. Accordingly, it is not possible for rickshaws to drive through these streets at a normal speed and without hindrances. In principle, some roads are accessible by car and I witnessed a small ambulance making its way through the settlement, but most roads hardly allow such movement. Apart from that, many footpaths are too narrow for rickshaws, or impossible for rickshaws to use due to the open drainage system. The width of the roads was only mentioned as a concern in the bazar area by Baharul, an AL leader of one sub-area. He proposed that the shops should be constructed with a shutter system in order to preserve more road space. The current system, where the front side of the shop made of CI-sheets is laid onto bamboo pillars, extends onto the road space and thus causes problems for traffic, especially if ambulances need to access the area. At his own compound Baharul experienced a conflict when he built a veranda on the second storey, extending onto the road (overhanging). As a consequence, the road became more difficult to traverse by heavy laden rickshaws because of the height limitations.

The above discussion has shown the surfacing of public space to be an improvement necessitating changes in space-based livelihoods and spatial practices. Furthermore, the improvement of Embankment Road in Manikpara has revealed the general demand for public space in the settlement, confirming what in Chapter 5.3.2 was identified as 'scarcity of public space'. Road width can be regarded as being most important for safety and security, which has triggered road widening initiatives in Manikpara, while similar initiatives have not yet occurred in Nasimgaon.

6.4 A RICH DIVERSITY OF SPATIAL PRACTICES IN EVERYDAY AND EXTRA-EVERYDAY LIFE

The activities explored in everyday and extra-everyday life indicate the many uses of public space as well as the values attached to public space. Public space forms the basis of many livelihoods, be it of the vendors on Khalabazar, of the plastic drying businesses on the embankment slopes or of the businessmen using public space to store the plastic materials to be sorted. But public space also forms a base for many non-economic activities, such as the spending of free and leisure time. Cultural meanings and values find their expressions in public spaces during extra-everyday festivals and public holidays. Furthermore, the discussion has revealed conflicting uses, for example the space-intensive production activities on the Nasimgaon Eid Gah Math and its simultaneous use as a football field and playground and as a place for protection in case of fire, and multiple interests in access to public space – leading to the notion of contested space. Given the space-demands of the numerous activities, the contestations arising in such a dense entanglement of uses can easily be imagined. Similarly spatial practices, such as the use of roofs in Manikpara or the temporary use of Embankment Road when under construction, indicate the high demand for public space. Public space remains a highly scarce and often contested resource in the study settlements.

In performing their everyday life routine and celebrating extra-everyday events, the users of space actively contribute to the production and re-production of space. The evolving spatial patterns depend on the public spaces available, continuously reconstructed by spatial practices, while the spatial practices also provide the basis to understand these patterns. The lived space, in Lefebvre's terms the 'spaces of representations', can then be understood as the practice of making use of the public space available in both the everyday and the extra-everyday. But the discussion has also pointed out that the conceptualisations of space by different actors determine activities, for example the interference of the mosque committee in the organisation of a *mela*.

The activities outlined above indicate that different public spaces are accessible to different user groups dependent on time, events and social and religious norms. Especially the examples of free time activities, religious and spiritual activities, and the issue of going to the *bazar* for grocery shopping indicate the specific importance of gender relations in defining mobility patterns and uses of space. In the following Chapter 7, I thus continue to analyse the factors defining a person's behaviour in public space.

7 WHICH PUBLIC SPACE? THE PRODUCTION AND REPRODUCTION OF SPATIAL PRACTICES

The aim of this chapter is to discuss the existing spatial practices in public spaces within the settlements and in relation to the urban fabric of the city. The discussion in Chapter 6 has repeatedly revealed how spatial practices differ according to gender norms. Accordingly, in the first sub-chapter 7.1 I pay particular attention to what constitutes gendered spaces given the institutions defining access to public space for men and women. Furthermore, the above discussion already hinted at different conceptualisations and multiple hierarchies of what constitutes a 'public space'. I therefore follow up on the discussion in Chapter 2.1.1 regarding the usefulness of Western delineations of hierarchies of public space in Bangladesh. In sub-chapter 7.2 I thus discuss the nuances of 'publicness' dependent on the urban fabric, rural/urban identities, imagined spaces and the inhabitants' perceptions of urban life, and user groups, especially female and male mobility patterns. The third sub-chapter (7.3) analyses how specific events and extra-everyday activities produce spatial practices and spatial hierarchies deviant from everyday structures. In sub-chapter 7.4 I summarise and conclude the analyses of this chapter.

7.1 GENDERED SPATIAL PRACTICES AND MOBILITY PATTERNS

Based on the above discussion of everyday and extra-everyday life and emerging spatial patterns, I would here like to follow up and discuss the specific results and patterns of spatial practices with regard to gender relations. The discussion in Chapter 6 has revealed specific female spatial practices and mobility patterns, as well as hinted at the gender relations and norms producing them. Here I aim to summarise and extend the discussion towards the underlying reasons producing these patterns. The focus of the discussion is on female spaces, although I also add a sub-chapter on male spaces[1].

1 I have not extended this research to include the *hijra* community. In Bangladesh, as in other South Asian countries, the *hijra* or transgender community (male to female) is perceived and perceive themselves as a third gender. In India and Pakistan (but not in Bangladesh), the *hijras* have officially been recognised as a third gender. Commonly the *hijras* collect money, especially from markets and after a child has been born, in exchange for giving protection based on traditional beliefs (and fears) regarding their spiritual power. During the course of solicited photography, Rabeya, a girl of about 13 years of age and resident of Nasimgaon, took pictures of a *hijra* inside Nasimgaon with whom she had a good relation. Rabeya furthermore narrated that a group of *hijras* often played the *karom* boards which she was in charge of overseeing. This indicates how the *hijras*, dressed in female clothes, can venture

7.1.1 Female employment, spatial practices and gender norms

Regarding the acceptability of women's employment, there is a considerable difference between female spatial practices in Nasimgaon and Manikpara, as the discussion about economic activities in everyday life has already indicated (see Chapter 6.1.1). In the following I will analyse how the spatial practices regarding economic activities relate to existing gender norms.

In Manikpara, women's employment outside of their own home is accepted if it is an economic necessity for households belonging to the low-income groups of the local society. As a consequence, moving around in public space is acceptable if it is part of the economic activity in question. For example, Kabin Mia's wife does the shopping for her *pitha* business herself and runs the business independently. The women working in the sorting shops need to traverse public space in order to reach their workplaces. In Nasimgaon, given the fact that most households are socio-economically poor and the local middle class is less pronounced, most women are engaged in income generation activities. A gap, however, exists between the economic necessity of contributing to the household income[2] and the preferences of these working women.

Despite the economic necessity, many women in Nasimgaon and Manikpara described having to work outside of the home as a hardship. As already discussed in Chapter 6.1.4, women in Nasimgaon in expressing their 'dream life' prefer village life, where they could focus on domestic tasks and maintain a *porda*-oriented lifestyle and 'get peace', while their husbands would earn the monetary income. Others do not refer to the village, but express the same wish to stay at home and focus on domestic chores, here expressed by the vegetable vendor Shoma:

> "I feel bad. I am a woman. I could do another work. But since my husband cannot work I have to sit in the shop. He cannot earn. If my husband was ok then I could stay at home and could do household works and it would be good for me. Now lots of men move in the bazar and I sit in the shop. Don't I feel bad? [...] I would like to stay at home with my kids, cook and eat and do household works. I like it." (Shoma, 27.04.2010)

Hamena, who did not work at the time of the interview, related working outside of the home to religious norms, but then analysed that it was not possible to maintain these religious norms under conditions of poverty:

> "It is very hard for women. In our religion, working outside the home is prohibited for women. Leave it. If I want to support myself then I cannot follow the religion. It is not acceptable that only men will work and women will not." (Hamena, 23.05.2009)

beyond restricted female spatial practices. Except Rabeya, no other interviewee mentioned the *hijra* community in Nasimgaon, which shows the social exclusion of the third gender that continues to persist in Bangladesh despite beliefs in their powers (see for example Khan et al. 2009). In Manikpara I did not meet any *hijras*, nor did anyone mention a local *hijra* community during the interviews.

2 In quite a considerable number of cases the husbands did not provide sufficient income for the household, thus increasing the pressure on women to work. One reason was that they spent the money on themselves rather than on the family, but some husbands also had health problems and could not contribute as much as they otherwise could have done.

Many of the economic activities at the focus of this research involved not only working outside of the home, but working in public space. The social rules and norms here can be considered to be stricter, and mobile vending for example is hardly ever carried out by women (although in the past this was done by Roxana's mother who sold snacks together with her children). In Manikpara, the plastic sorting quite often involves sitting in public space, also for women. However, the women involved in these activities did not consider this a problem because of society but only because of the poor environmental conditions. A remarkable economic activity for a woman was Momena's involvement in her husband's plastic washing and drying business, as here she moved around on the embankment slopes, constantly exposed to heat and public scrutiny. While her husband Rokib would not allow her to do any kind of paid work (she had once suggested working in a factory) except for the family business, she did not face problems working on the slopes and in a male domain as her family supported and 'protected' her. Her family seems to be of a higher social status – indicated by education levels and ways of dressing – and the support extended to kin of lower socio-economic status seems to be a classic case of family relations in Bangladesh (see more on this support in Chapter 9.1.3).

Instead of working in public space, the women in Nasimgaon expressed a preference for working indoors, which also includes working inside a house as a housemaid or inside a factory, as opposed to working in open, outdoor space. During the focus group in Nasimgaon, the discussion between Ishrat and Firoza underlined that working indoors is more accepted by society because it provides the possibility of maintaining *porda*:

> Ishrat: "Society [*shomaj*] accepts working inside the house. [...] It is because we can maintain our clothes according to *porda*, we can move in a right way. [...]
>
> Firoza: "We can say our prayer five times a day."
>
> Ishrat: "We say our prayers, our behaviour with people is good. Then society [*shomaj*] calls us good." (Nasimgaon FG, 24.04.2010)

For the women of the middle class in Manikpara this, however, still is not acceptable, and only home-based working meets the requirements expected of them by society (see below).

These self-perceptions and preferences of working women are the result of common gender norms in society, and when discussing the necessity to work the middle-class women in Manikpara underlined that this was not in accordance with the norms of *porda* and maintenance of shame, honour and prestige. This becomes obvious from the following conversation, where Mashrufa said that only 'aliens' could work outside for the whole day, and then the discussion continued that by mixing with men the women working outside did not take care of their honour properly, and that the managers, *mohajons,* and co-workers 'took advantage' of this non-honourable behaviour:

> Mashrufa: "After doing all these [household works], we want to work for the rest of our time. But the *mohajons* do not want that. They will tell us to work for the whole day, from 8am to

8pm. Not everyone can do that. Who can do that? Aliens do it. Maybe they have come from Mymensingh [town and rural area three hours north of Dhaka]. Leaving their family, they have come here for work. [...] Their husbands and children are starving, and they are working here with no cloth on their head, no cloth on their chest."

Hasna: "They are not good, not good."

Mashrufa: "You can see them poking [talking, touching, laughing with the guys] the guys from the factories. They mix up with them, get close to them, and have fun with them. The *mohajons* and guys of the factories want this. They want girls to do this with them. Then they will provide work for them, the girls will work, and come to take money." (Manikpara FG2, 03.04.2010)

The consequence of their perception and the notion of society is the impossibility of leaving the house to pursue economic activities. In the following conversation among women of the middle class in Manikpara, this was related to concepts of shame, prestige and honour:

Azufa: "We are good people."

Mashrufa: "We cannot do such things [working in factories outside], we will never do this. So we remain inside the house."

Hasna: "She [Azufa, Hasna's daughter] will not go to the street [to work] even if she is starving."

Mashrufa: "But we are not girls from those families where girls go outside for work and have food. We want to live with our families inside the home, underneath the fences [physical fences] and have simple food. This is our honour and prestige. This is the reason. Otherwise there are lots of factories in Manikpara, we also can work there. Why don't we?"

Hasna: "Shame [*lojja*]."

Mashrufa: "We cannot do it because of our shame [*lojja*]. We cannot do it because we are afraid of losing our prestige. Our relatives might come and see. They would say 'We have seen his wife working in that factory'. "

Subina: "Yes, yes."

Mashrufa: "The news would spread. Another person would say 'I have seen Mesbah Uddin's wife working in a factory. Does Mesbah Uddin not work?' We are afraid of this kind of discussion."

Hasna: "The husband will get insulted."

Mashrufa: "We could be eating by working in the factory in the presence of our husband. If the relatives get to know this, this would be a matter of extreme shame."

Subina: "We cannot go outside as we are afraid of losing our prestige. Otherwise we also could work outside [if we wouldn't be afraid of losing prestige]." (Manikpara FG2, 03.04.2010)

The discussion shows how working outdoors was not acceptable for middle-class women. However, working indoors would be acceptable and most of them engaged in home-based work from time to time. But their discussion above makes very explicit that working outside of home or even in public space would never be an option for them, also as they were under the constant surveillance of society.

Lojja (and similarly *shorom*, often used in the same context) here refers to feeling shyness, uneasiness and shame and to a certain behaviour of women expected by society, i.e. a woman who has shyness in her expression, is not arrogant in behaviour, and is known as well-mannered is behaving according to the concept of *lojja*.

While at most times during the focus group the women of the middle class expressed that they could not work outside without regretting this, there also was a reference to a 'dilemma'. Being part of middle-class society means that they have to act according to the expectations of society and their husbands. So in times of hardship, when an additional family income would ease the household's everyday life, they are not able to generate this additional income. The way they discussed this dilemma carries a notion of regret about their behaviour being under the constant surveillance of society, which expects them to act in accordance with the concepts of *lojja* (shame) and *man-shomman* (respect perceived by society):

> "Allah gave money to the rich people. They remain inside the house. Those who are very poor, they are really needy. They would be working in the factories. We, the middle-class people, cannot work outside like the poor people; we cannot just sit inside like the rich people. We are always tensed. [...] No one understands our pain. We even do not want to let anyone understand. We cannot ask for money to anyone, like the poor people. This is our real need and this is our real poverty. We, the people of the middle-class society, cannot work outside because of our shame and honour [*lojja, man-shomman*]." (Mashrufa, 03.04.2010)

Thus the obligation to maintain the *porda* lifestyle tied to their status at the same time becomes a burden to them, both socio-economically poor and socio-economically rich women are freer in their movements (see also Mashrufa's comment in Chapter 6.1.4).

The acceptability of becoming involved in economic activities also depends on position in the life cycle. For young women, working outside can become unacceptable as soon as they reach marriageable age. Not all women of marriageable age stop working, but for example Taslima in Nasimgaon stopped sending her daughters to work as housemaids in order to keep their prestige and honour, and Roxana in Manikpara was told to stop joining her mother in mobile vending. On the other hand, the movement of elder women for economic activities (e.g. Ramisa and other women selling rice as mobile vendors, or Roxana's mother working as a marriage broker) is more accepted according to the rules of *porda* (see Chapter 5.1.3). Nonetheless it is seen as a heavy burden, as this means they are not adequately supported by their sons. Riaz took a picture as part of solicited photography which shows an old woman working in plastic sorting. With this picture he wanted to show how the old woman had to suffer. Both Ramisa and Roxana's mother expressed their sorrow and pain that their sons did not provide for them adequately and thus they had to continue working despite their old age (and in the absence of husbands providing for them).

The above discussion indicates how spatial practices related to economic activities are highly dependent on gender norms. The low number of women performing economic activities in public space in Manikpara is a result of the middle class being able to 'afford to maintain *porda*', which has also been observed by Gardner (1994) in her research on rural Bangladesh. The dominance of these gender norms becomes apparent from the preferences expressed by those having to work out of economic necessity, indicating a persistence of gender ideologies. Accordingly, the female production of public space is based on economic necessity rather than on an active formulation of and claim to female participation in public space. In the following sub-chapter, I will focus on how the movement of women, except for going to/from the workplace, is also subject to social norms and a concept of societal surveillance.

7.1.2 Female mobility patterns and determining gender norms

The above discussion on women's employment has already pointed out different mobility patterns, as those who work outside of the home need to move around, at least in order to get to their workplaces. This sub-chapter will follow-up and extend the above discussion to the gender relations and norms determining and explaining female mobility patterns. In Chapter 7.2.4, the female mobility patterns will be discussed further under the concept of nuances of 'publicness'.

The acceptability of women moving around apart from going to their workplace is highly dependent on their socio-economic status, similar to the issue of working outside. This has especially been analysed above with regard to going to the bazar (see Chapter 6.1.3). While in Nasimgaon most women go to the bazar themselves, producing the bazar as a partly female space, this is completely the opposite in Manikpara's middle class, while some of the women of Manikpara's low-income groups go to the bazar, at least irregularly. Interestingly, Momena, working on the embankment slopes, was never allowed by her husband to go to the bazar. This indicates that their status in the locality is perceived as slightly higher than their work and economic necessity suggests, resulting in Momena's movement outdoors only being acceptable as long as it is to work with her husband, but not for any other purpose. Thus socio-economic status is only one dimension of a respected status, and identification of belonging to a certain status group and availability of economic assets do not necessarily correlate (see also Gardner 2000: 136).

The leisure time women in Nasimgaon (and in Manikpara, if they are able to do so at all) spend outdoors (see Chapter 6.1.2) is mostly related to a specific purpose, e.g. spending time with their children. Spending time without such a purpose is rather uncommon. Hamidul, a boatman of Nasimgaon, reported how boys and men from Nasimgaon went on boat rides on the lake just for enjoyment, while women from Nasimgaon did not do so. However, this is a common activity for the women from the high-income neighbourhoods surrounding Nasimgaon. This indicates that Nasimgaon women, although feeling free to move around inside Nasimgaon, would not leave Nasimgaon unnecessarily. They also cannot afford the time and money for such boat rides. While for men from Nasimgaon this may still be acceptable as a pastime (and they are the ones who are 'allowed' to spend money on themselves), women are expected to take care of the household or to be at work.

Similarly, during the women's focus group in Nasimgaon a discussion evolved about whether it would be possible for them to move around. They referred to women who did not have to work and could move around, and they referred to me as someone who could move around with friends:

> Meher: "And those who don't have any work, they can move around. But we cannot. We cannot go out with our children."
>
> Ishrat: "After days of working and working, suddenly we need to have some fun, like we will go out, we will eat some good food [expressed as wishful thinking]." [...]
>
> Firoza: "For example, *Apa* [myself] went to Wonderland [amusement park] the day before yesterday, she goes to Shishu Park [amusement park], she goes to the zoo, she goes to the cinema hall. Am I not right *Apa*?" [...]
>
> Meher: *Apa* has some female friends. They take her to good hotels and feed her. We don't have female friends, so whom will I go with?"
>
> Firoza: "Not female friends, male friends." [this was discussed later on; Firoza's point was that female friends do not treat among themselves and only male friends spend money to invite their friends] (Nasimgaon FG, 24.04.2010)

Here they mainly expressed their dreams and perceptions. In contrast, leisure time movement without a specific purpose such as shopping or taking care of kids is not part of their spatial practices. An important dimension here was the spending of money for oneself. While it is common for men to spend money for themselves, for example on tea and cigarettes, but equally so for moving around with friends (see Chapter 6.1.2), this is not done by women. Women's spatial practices are thus influenced by the patterns of spending the money they earn to sustain the family, i.e. on necessary activities, but not on themselves, i.e. on optional activities. This was also referred to by Rasheda, working in plastic sorting, who explained the salary differences (see Chapter 5.3.3) between men and women doing different kinds of work. Her explanation of why men would not do the easier work of sorting polythene indicates the different ways of spending money:

> "Moreover the work that we do is not good enough for men because we are satisfied if we get 400-500 Tk [per week], but men take tea, cigarettes, so this amount is not sufficient for them. But we do not consume anything, so we can manage with this money." (Rasheda, 29.05.2009)

Additionally women, although often participating in a household's decision making, also depend on the permission of their husbands when they want to move around in public space. This obedient behaviour can be understood as another consequence of the notion of surveillance. Firoza thus mentioned that she could not move around with friends anymore since her marriage:

> "For example, if I want to go to my brother's place I need his [my husband's] permission. He won't let me go. Again, if I go to any other places to move around with my [female] friends, he won't let me go. He will say 'Why, for what? Am I not here? You cannot move around with them'." (Firoza, 24.04.2010)

Similarly the working women in Manikpara mentioned that in most cases they consulted their husbands when leaving the house. Gulshana's rhetorical exclamation "Is it possible to go [outside of home] without the husband's permission?" shows how deeply engraved this gender norm of the husband's power over the behaviour of a household's female members is. She continued that she needed to tell him if she was going to a faraway place "because men can have devil ideas on their minds". This hints at the common perception of women being the initiators of 'trouble' as in the rape cases on Jahangirnagar University Campus (see Chapter 5.1.3). Her mother Salma brought to the point the responsibilities women had for the reproductive tasks within the household:

> "Imagine, when I was coming here, I told him [my husband] that I am going over there, these *Apas* [me and my female field assistants] will come, they will talk. Then he asked what you ladies will talk about. I answered that whatever it is, I want to go. Then he said 'Ok, you can go but tell me what should I do with the rice on the stove?'. I told the other tenants to watch that rice." (Salma, 02./09.04.2010)

Salma's husband's permission was based on her fulfilment of household's tasks – here the preparation of the evening meal.

Contrary to the stories of permission-taking, Fahima also showed some resistance to having to obtain her husband's permission. If he was not there, she just went out with her children, for example to buy small things from the surrounding shops. Furthermore, she stated:

> "It does not matter whether he gives permission or not. This man has gone to so many places in the country or abroad, leaving me. I lived by myself with my children at these moments. Nothing is going to happen if I do it now, too." (Fahima, 02./09.04.2010)

This 'rebellious' statement is all the more interesting considering the behaviour of her husband. As soon as he appeared during the discussion, she completely changed her way of talking, presenting her husband as dominating all her decisions and herself as obeying (see Chapter 4.2.4 for the difficulty of keeping a 'female space' during the women's focus group as her husband found several excuses to enter the room). This dominance of male household members also became visible during the solicited photography, where women had given the cameras to their husbands as a matter of course (see Chapter 4.2.3).

Some women said that their husbands accompanied them depending on where they went and at what times. Nasrin, for example, was accompanied by her husband when going outside of Nasimgaon. Aklima usually picked up her younger sister when she was coming back home from the garments factory. While she did so up to 10pm on her own, if it got later her husband accompanied her.

The story of Aklima picking up her younger and unmarried sister, whom she characterises as good-looking, thus explaining her wanting to prevent anyone being able to harm her sister, indicates yet another dimension of the movement of unmarried women. Unmarried women, as has been noted before, are under stronger surveillance given their perceived vulnerability related to their position in the life cycle. Hamena, an unmarried women working in a garment factory, was not allowed to move around except for going to and coming from work. When she came back to Nasimgaon from the factory after 10pm, she did not go home but to her brother-in-law's shop in the busy bazar area, and was then accompanied home by her brother-in-law or married sister. Here the bazar area with its increased publicness and watching crowd was considered more appropriate to ensure the maintenance of Hamena's honour and reputation (see also Bose 1998: 369) than the dark lanes leading to her home. This was similarly the case with Roxana in Manikpara, whose mother often did not allow her to move around, while the talking I noticed about her indicates how delicate the issue of maintaining an unmarried woman's reputation is (see Chapter 6.1.2). Reputation and the societal pressure to maintain it also encouraged Rehana to marry off her daughter:

> "Then I arranged marriage for my daughter. Everyone said 'Your daughter does crazy stuffs. Arrange marriage for her. Then she will leave the crazy stuffs.' Then I arranged marriage of her. [...] [Crazy stuff means] she used to go here and there like a mad person. She did not listen to what I said, she did not come home properly. [...] She quarrelled with people. People talked about her 'She became crazy seeing boys. Arrange marriage for her.' Don't you understand? Suppose if I say every day without any reason 'I saw *Apa* talking with a man here' then it will be a disgrace. If people say something about you in front of many people again and again how will you feel? Then the mind gets disturbed. I also became disturbed hearing such things from people again and again and arranged marriage for her. [...] Now people cannot say anything. What can they do now? Now if she talks with ten men it is not a problem because she has a husband now." (Rehana, 23.04.2010)

The quote also indicates how certain behaviour that causes gossiping before marriage can be accepted after marriage (although to a certain degree only, as for example the rules of *porda* mean that women should not talk to men outside of the family).

In contrast elder women were relatively freer to move around, such as Roxana's mother, who worked as a marriage broker and Tariq's mother-in-law who did the shopping in Manikpara (see Chapter 6.1.3). In Nasimgaon Hamena's mother Ramisa did much of the family's shopping. Furthermore, Ramisa regularly moved around with her (female) friend. The mother of Ishrat, who earned her and her family's living by begging in the surrounding high-income areas, was also able to trespass gender norms and smoke cigarettes or joke with the men in the bazar by expressing herself using pronounced body language including the touching of men she was joking with.

If women move around in public spaces, the way they do this also carries with it a notion of good or bad manners or behaviour. During the focus group with the working women in Manikpara, Gulshana referred to both the style of walking itself and behaviour towards others on the street as determining good or bad female behaviour:

> "There are some who do not walk in a proper way. When they walk, it seems like they are running. On the other hand, there are some who walk in a very good way. That feels good. [...] [Those who walk as if they are running], they walk like men. There are girls who, no matter how many works they have to do, walk slowly and nicely. Again, there are many who do not even see men or women. They force to pass by. They do not observe that men are getting in touch with their body. They think it is just about walking." (Gulshana, 02./09.04.2010)

Similarly in Nasimgaon, Aklima, a vendor selling in public space, referred to women walking around carelessly without taking care who else was around, while Taslima disliked women walking like models:

> "There are some females who drop their dress [referring to not wearing for example the *orna* properly], don't tie their hair – I don't like them. They walk like they are doing modelling. I don't like modelling. I become angry if I see this. They look like the devil. If you wear a good dress and walk properly, then people will say after seeing you 'This is a good woman'. We are from the village. We always think in this way." (Taslima, 18.04.2010)

The statement furthermore touches on dress codes as part of women's manners; this will be discussed in more detail in Chapter 7.2.4. The reference to the village was also made by Gulshana's mother in Manikpara, who underlined not liking the way women moved in Dhaka, feeling disgusted to see how women tended to walk like men in Dhaka city. This can be understood as a critique of a more articulate presence of women, as opposed to the expected behaviour of moving in public spaces in a more invisible way. Another issue raised above is the way of walking in relation to (accidentally) touching others and thus the possibility of creating a female private space in public. Fahima reported how during power cuts it became difficult for her as a woman to walk amidst men (see further discussion in Chapter 7.3). Generally, the way that women talked about their own mobility and movement suggests that 'moving around' is especially considered a male activity and thus if carried out by women the underlying rules and surveillance are considerably stricter.

In many of the women's narrations, it appeared that they avoided crowded places, especially in order to avoid touching men in passing. This was expressed by women both in Manikpara and in Nasimgaon (for example Salma's mention of her husband not liking her to be present in crowded places, see Chapter 6.1.4). Referring to a picture she took of women (presumably garment workers) gathering in the bazar area, Meher explained that it was not easy to move in a crowd. Hamena, before she joined the work in garments factories again from November 2009, mentioned the necessity of going back home in a crowd at the end of work as one reason why she did not like to work in the garments factories:

> "It is difficult to move on the roads because both men and women move together and our bodies get in touch with men's bodies. When we are walking together then it happens. But this is very difficult." (Hamena, 23.05.2009)

As the discussion on ways of walking indicates, the women tend to be held responsible for avoiding touching. That women in the common perception 'do not belong' in crowded places especially in Manikpara also becomes obvious from Tariq's argument, presented when asked whether women joined the launch picnics: "We did not take them because there was a big crowd gathering inside the launch" (Tariq, 27.06.2009).

The above discussion has at different times revealed notions of surveillance (see also Bose 1998 in Chapter 2.2.2) especially in relation to distance, types of public spaces, character of the surrounding environment, company and time of the day. Both in the discussion of women's economic activities and in the discussion of mobility patterns it was seen that living up to the expectations of society and not provoking negative talking about oneself or one's daughters evolved as a prime female concern. The trespassing of gender norms by Dipa and Roxana (see Chapters 6.1.2, 6.1.4) immediately led to rumours, indicating how the concept of surveillance is always present. In Nasimgaon, Rohima also mentioned that she would not go far out of Nasimgaon on her own, but would rather take someone with her. This can be interpreted in relation to the concept of surveillance, because if others can confirm her activities no rumours could start spreading. The incident of Dipa's illness (but similarly of Rehana's daughter's marriage, see above) underlines how quickly the system of surveillance operates and how it can damage a woman's reputation. When Dipa was seriously sick in 2009 and had a swollen belly, people started talking negatively about her being pregnant without having a husband (see also a similar account in Rashid 2006).

The above discussion has furthermore shown how the crowded areas can become areas for the protection of women's honour, as in the case of the bazar area protecting Hamena or the crowd of garment workers going home that makes it possible for women to go home late at night. On the other hand, crowds are experienced as being male-dominated and having to walk amidst men can result in unwanted touching with men.

The discussion so far has mainly revealed how women are expected to behave in spaces dominated by men. However, the gender norms discussed above also create the explicit female space of the semi-private housing compounds and private rooms. During the day this is a completely female sphere with a few exceptions (see Chapter 7.1.3 on the occasional transformation of 'female space' into 'male space'). In public space, explicitly female spaces do not exist, but at times public space in Nasimgaon is dominated or at least considerably shaped by female presence. This is especially the case in the morning, when the garment workers move towards work, but similarly so during the educational events discussed in Chapter 6.2.4. This is a non-permanent but significant change of the 'normal' everyday and extra-everyday life pattern in public spaces and it signals the importance of women in employment.

The gender norms determining women's employment and related spatial practices similarly influence the general mobility patterns of women. Here the notion of surveillance by other women or men is particularly useful for understanding the spatial patterns. Again, public space can only in a few instances be considered an arena of active female production, while mostly the spatial patterns are influenced by persisting gender norms about women's behaviour defined by society. This corresponds with Bose's statement (Bose 1998; see Chapter 2.1.1) that access to a public setting for women, here the workplaces in public space, does not necessarily mean equal access to power and resources.

7.1.3 Male spatial practices and mobility patterns

While outdoor 'female spaces' are thus rather restricted, the public spaces of both Manikpara and Nasimgaon tend to be male-dominated, and specific spaces can be considered as explicitly 'male spaces'. The discussion below summarises the main points; however, given the absence of similarly detailed gender norms for men, this discussion will be kept brief.

Within both Manikpara and Nasimgaon, explicit male spaces exist, often depending on time of day and specific occasions. In Manikpara, public space is generally rather considered male space, as the female perceptions discussed above also indicate. Explicit male spaces are the Bagan and embankment slopes at night times – although the discussion of shifting borders (see Chapter 7.3) will reveal how they can also transform to include women at specific times and on specific occasions. The launch picnics regularly organised by the community tend to be male spaces, not least because of the large crowd. Mesbah Uddin's statement made during a discussion of male and female salaries in plastic recycling also carries a notion of women 'disturbing' male spaces:

> "Moreover, men do fun [meaning fun in the sense of impertinence, impudence] when work is done involving women." (Mesbah Uddin, 21.05.2009)

As already pointed out above, the women of higher socio-economic status in Manikpara tend to consider more spaces as male space, for example the environment where working women have to move and work. But also some of the working women's statements point at public space as being generally considered male, as Gulshana's appreciation of public spaces as places where 'men can sit'. In Manikpara politics also seems to be a male space, as the knowledge of women about the party contestation remained rather vague, suggesting that there may be less 'spaces' to discuss these issues that are typically discussed among men at tea stalls and in *adda* rounds. The young men also tend to spend their free time moving around without girls, and if one of their friends brings his girlfriend they make fun of him. So this moving around, especially in Manikpara, is the domain of young men – similar female groups are not seen moving around. In contrast, in other parts of the city, especially on the University campus, it is not uncommon to see groups of young women during the day.

On certain occasions, the Nasimgaon Eid Gah Math turns into a male space. This is the case during the football tournaments (see Chapter 6.1.2), but also on the *Eid* days during the main prayers (see Chapter 0). Similarly, the *karom* boards are predominantly a male place where female presence is not accepted according to social norms. The tea stalls during cinema times become male spaces, especially in the evenings. Of the bazar areas, one space was rather a male space in terms of the presence of only male sellers mostly selling bananas (Shaonbazar). The sellers are mainly old men and as such they may be respected for their age and have their own 'community' undisturbed by the 'outside world'.

During the thanksgiving prayers of *milad*, the private homes where these are held – normally female spaces for most of the day – transform into male space as only men can join in these prayers. After a contestation between my landlords and a group of young politicians, the veranda of the compound, during the day mostly accessed by women, suddenly became a male space when my landlord used it to discuss the matter with friends. Suddenly, I was prevented from going there, although during the daytime this place used to be open to me at all times. At night the veranda also sometimes transformed into a male space when my landlord and his friends gathered to smoke cannabis.

The above discussion of male spatial practices and mobility patterns has indicated the relative freedom of men to move around according to their wishes. In comparison to women's spatial practices, men make considerably more use of the public sphere, and accordingly public spaces in both everyday and extra-everyday life tend to be dominated by men. This is especially the case in Manikpara, where women's movements and presence in public spaces are more restricted than in Nasimgaon. Public spaces that in Manikpara are male spaces, like the bazar area, are rather mixed spaces in Nasimgaon.

7.2 NUANCES OF 'PUBLICNESS'

In addition to the Western delineation of public, private and semi-private space (see Chapter 2.1.1), this chapter seeks to explore the nuances of 'publicness'. These refer to a gradual approach to levels of integration, as will be elaborated in the following sub-chapters. The nuances of publicness are explored from various perspectives and differ depending on the perspective taken. Below, multiple hierarchies of publicness in both Manikpara and Nasimgaon are discussed with relation to

- the urban fabric and integration into the cityscape,
- rural/urban identities and perceptions and images of the city,
- mobility patterns of male inhabitants, and
- mobility patterns of female inhabitants.

7.2.1 Urban fabric, recognition and nuances of 'publicness'

The urban fabric of the study settlements and their integration into the urban surroundings are outcomes of the production of city space over time. The emerging spatial patterns can indicate spaces of inclusion and exclusion.

Manikpara is characterised by a high degree of integration into the cityscape. Its urban fabric is, despite a few specialities, a continuation of the structures of the surrounding areas, especially of Old Dhaka and the adjacent settlements along the embankment. The building structures and structures of public spaces and the street network display similar patterns, and borders between the neighbourhoods are rather fluid and indefinite. Manikpara, then, cannot be understood as a 'fractured space' in relation to the surrounding urban fabric. Although it differs considerably from the model town neighbourhoods (see below on the perception of 'residential areas'), it is not isolated from the dominant production of urban space in Dhaka. Thus, in terms of its integration into the city, Manikpara displays a 'familiar publicness', with 'familiar' referring to the similarities of urban fabric that suggest a smoothness of moving in-between as there is no symbolic or mental border to overcome. This 'familiar publicness' in terms of the urban fabric may, however, also be perceived as an opportunity for others to enter the neighbourhood because of the lack of distinctness of the neighbourhood boundaries that renders them easily traversed by 'strangers' (see discussion on women's perceptions of publicness in Chapter 7.2.4).

From the perspective of Nasimgaon, the city rather is the 'other'. The urban fabric of the surrounding neighbourhoods stands in stark contrast to Nasimgaon and thus immediately points at less integration and familiarity. The contrasting urban fabric and the separation created by the lake create a physical barrier between Nasimgaon and the surroundings, which is reinforced continuously by conceptualisations of what constitutes the city from outside Nasimgaon (by planners, politicians and middle and upper classes, see Chapter 5.2). The physical barrier furthermore produces Nasimgaon as a rather homogenous entity with relatively high 'privacy'. In the city context it presents a rather secluded space, one which maintains connections to the city, but the flow of outsiders into the neighbourhood is almost non-existent. On the other hand, the explicit borders and limited extent of the familiar in terms of the urban fabric produce an atmosphere of community solidarity internally, despite the ever-present everyday life contestations. This enclosed familiarity, for example, translates into more flexible movement patterns of women within the neighbourhood (see Chapter 7.2.4).

How the integration of a neighbourhood into the urban fabric re-produces specific spatial practices is related to a branding of spaces, or to a 'blackening' and 'whitening' of urban areas (Yiftachel 2009; see also Chapters 2.2.4, 2.3.2 and 5.2.2). Manikpara is integrated into the surrounding city through its status as an officially recognised neighbourhood where land owners generally have a legal title to their land (despite a few contested land ownership cases) and the area is officially connected to the water, electricity and gas supply systems. In contrast, the insecure tenure status and the non-availability of water, electricity and gas via the official systems create a stark difference between Nasimgaon and the surrounding city. The non-permanent and insecure status of Nasimgaon re-produces a spatial practice that re-enforces the physical, social and mental boundaries between Nasimgaon and the city. Nasimgaon becomes a space of exclusion, and this space is continuously re-produced externally, for example by the threats and rumours of eviction (see Chapter 5.4.1) which, despite phases of relative silence on the matter, are deeply engraved into the inhabitants' perception of their right to citizenship and their spatial practices.

The perception of inhabitants of their neighbourhoods underlines especially the otherness of Nasimgaon in relation to the city, while it adds another aspect to the integration of Manikpara. Nasimgaon is not considered a 'residential area' (*abashik elaka*), neither by its own inhabitants nor by others. The term *abashik elaka* is only used in the context of urban residential areas that are well equipped, with 'good-looking' wide roads and generally a calm and quiet atmosphere. Examples in Dhaka are Banani, Gulshan or Dhanmondi – thus the areas of high-income groups. By referring to *abashik elaka,* Nasimgaon inhabitants distinguish their neighbourhood from the surrounding high-income areas and critically compare the level of urban services and facilities they receive. In Manikpara, Mesbah Uddin understood Manikpara as an 'industrial area' instead of a 'residential area':

> "Manikpara became popular as it became an industrial area. However, our area is an industrial area. Our area cannot be mentioned as residential area [*abashik elaka*]. There is no residential area in this *Thana*. Dhanmondi *thana*, Hazaribag *thana* etc. are residential areas. There is no word like 'residential area' in this area; it cannot be a residential area where there is high density." (Mesbah Uddin, 31.03.2010)

This quote has to be interpreted differently from the use of *abashik elaka* in Nasimgaon. It tells of a predominant image of urban living representing only areas affordable to the high-income groups. However, it does not exclude Manikpara from the city context and the first sentence carries an expression of pride about the importance of Manikpara.

The 'familiarity' of Manikpara with the city and the 'otherness' of Nasimgaon became apparent with the reactions that followed the introduction of daylight saving time on 19th June 2009 as an attempt to deal with the electricity crisis. Locally the new time was referred to as *shorkari shomoy*, government time, while the previous time was referred to as *bangla shomoy*, Bengali time. The adaptation to this change underlines the degree of integration of the two settlements into the city. Manikpara, being more integrated into the surrounding urban structures and with a less distinct border, adapted quite fast to the new *shorkari shomoy*. The plastic factories soon changed their working hours from 8am to 9pm so that the natural daylight rhythm could be maintained. This was not what had been anticipated with the introduction of daylight saving time, however it eased residents' everyday life and supported a fast adaptation to *shorkari shomoy*.

In contrast Nasimgaon for a considerable time remained an 'island' of *Bangla shomoy*, whereas the surrounding city had changed to *shorkari shomoy*. This became obvious in the 'conflict' of timing between those who based their livelihoods on the internal area and accordingly stayed with *Bangla shomoy,* and those who went to work outside, e.g. in garment factories which operated according to *shorkari shomoy*. The change of eating habits – because there was now less time to prepare food in the morning after waking up with sunrise for those going to work – was felt by those vending within Nasimgaon. The banana seller Khalil experienced a 30% reduction in sales. Furthermore, the boatmen experienced the consequences as the movement of people towards the factories shifted to one hour earlier in relation to sunrise. Hamidul, a boatman who was used to getting up at sunrise, thus exclaimed about missing out on the good income that could be earned at the most crowded early morning hours when people were moving towards the factories. Daylight saving time here took a long time to become familiar to all residents, and conversations about time often ended with the question about whether the timing given was *shorkari* or *Bangla shomoy*. This small anecdote symbolises differences in integration into the city.

Depending on the integration of a neighbourhood into urban structures, the bazar seems to become a more or less public space. Public here has not to be understood as one-dimensional; it is my understanding that there can be a hierarchy of 'publicness'. In Manikpara, the bazar is one of the most public spaces, being a meeting point between inhabitants and strangers, and a place of exchange and transfer. This publicness has been produced continuously during the development history of Manikpara. It can be said to influence everyday life and mobility, e.g. fewer women are going to the bazar. This is in contrast to Bose's observation in Kolkata that the market despite its high number of strangers was an accepted place for women, as the presence of strangers countervailed any "improper male-female interaction" (Bose 1998: 369). In Manikpara, the specific Old Dhaka identity and religious and social life further produce the 'strangers' publicness' – it is thus not only a physical construct, but as much an expression of conceptualised space, in this case of societal institutions. Similarly, the higher degree of publicness, together with social norms, may be a reason for the school programmes not taking place outdoors in Manikpara. Being outdoors, also for children participating in a sport programme, means being exposed to a greater public than in the enclosed neighbourhood of Nasimgaon.

In conclusion, the above discussion identified Manikpara as a space of inclusion, and Nasimgaon as largely a space of exclusion. Manikpara is well integrated into its surroundings and also constitutes a transfer space, especially with Embankment Road where traffic from Gabtoli (the starting point of the Dhaka-Aricha highway leading towards Khulna and Rajshahi, the major road out of the city towards the west) traverses towards Old Dhaka and vice versa. Nasimgaon, in contrast, is not a thoroughfare and its relations to the city are not mutual. These conceptualisations are also implicit in the planning and development guidelines of the Detailed Area Plans concerning both areas (see Chapters 5.3.1 and 5.4.1).

7.2.2 Identities, citizenship and perceptions of the city

The familiarity and otherness/strangeness that can be read from the urban fabric is also reflected in the identities of inhabitants, which can be oriented towards the urban/city or towards the rural/village. A feeling of the city being home indicates familiarity, while a perception of the village as the only home indicates a feeling of otherness/strangeness towards the city.

Rural and urban identities in Nasimgaon and Manikpara

The statement of many Nasimgaon residents 'if this area is evicted we will go back to the village' expresses that the city, despite providing a livelihood, remains unfamiliar and strange to them. It furthermore indicates an identification with the village home, while the city is perceived as a workplace. This identity is also visible from the internal organisation of the settlement with its institutions of social life resembling those prevalent in rural communities, especially the *shalish*. These institutions are not necessarily absent in urban areas, indeed some similar institutions also exist in Manikpara, however the way that people interact and the way in which community is organised much more resembles a rural set-up than an urban neighbourhood. Furthermore, the organisation of housing compounds resembles typical rural compounds with their central courtyards (*uthan*), a resemblance that is not common to this extent in all socio-economically similar areas of Dhaka, and which may partly be attributed to the specific history of Nasimgaon (see Figure 11 for the organisation of a rural compound).

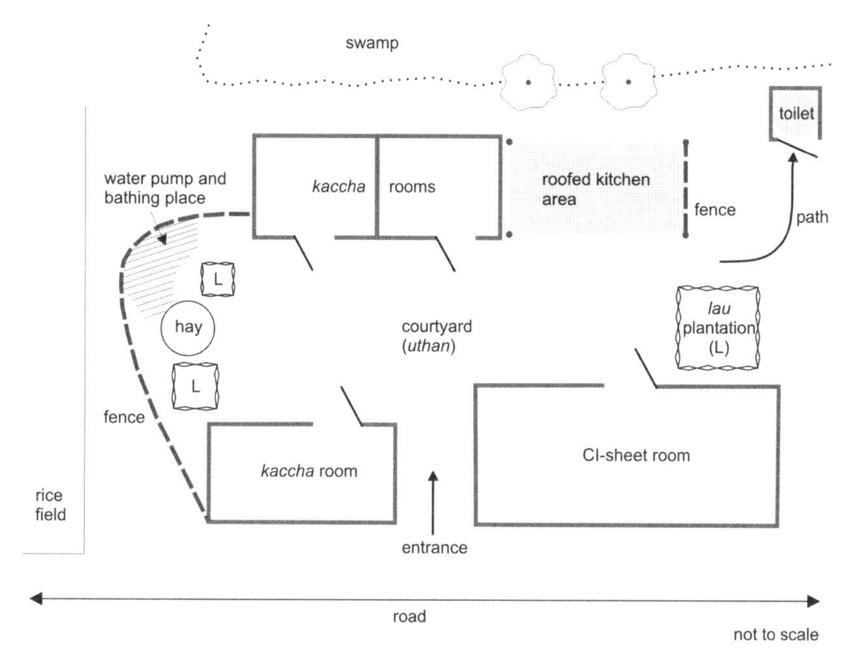

Figure 11: Organisation of a rural housing compound

This sketch depicts the rural housing compound of a family that has mainly lived in Nasimgaon for 15 years. Although the family owned a shop and let rooms in Nasimgaon, they primarily invested their money in the village. Over the years they had invested in their own room, the largest one on the sketch. The floor was made *paka* and the walls erected in corrugated iron sheets with concrete pillars (as opposed to bamboo which is commonly used in Nasimgaon), including a pitched roof. Their relatives who lived in the same compound remained in *kaccha* rooms. Although the family did not own any of the surrounding fields but only the compound, they invested in the village by leasing land for rice cultivation for 85,000 Tk in 2010 – money which they partly received from their in-laws during their daughter's divorce and otherwise covered by taking out a loan of 40,000 Tk in Nasimgaon. This arrangement included a person who cultivated the fields using the family's money and who finally received a share of the harvest. This considerable investment in the village reveals a preference for the village and the permanent feeling of insecurity concerning their Dhaka life.

Many participants of solicited photography in Nasimgaon went to their villages and took pictures (see Photo A-48 and Photo A-49), or if they could not do so, then they verbally expressed their wish to take a camera to the village and take 'beautiful pictures of the countryside'. The interviews and participant observation also revealed how the village is still central to residents' lives, manifested in regular visits to the village home, the number of close relatives remaining in the village, the regular exchanges of both money and goods, and the preference for marrying off daughters in the village rather than the city. Savings are often also invested in the village, mainly to buy land or improve the house owned in the village or the house of the parents (see also background text of Figure 11). Taslima, for example, invested her savings in cows in the village, so that she could save for her daughters' marriages which she preferred to be in the village (see below for a discussion of women's preference for the village, and the related concepts of honour/shame/prestige).

Only two of the interviewees in Nasimgaon expressed an urban identity based on Dhaka being their birthplace. One was Rabeya, a 13-year-old girl who asked: "Can a person whose birthplace is Nasimgaon leave it?" (Rabeya, 05.04.2010). The other one was Firoza, who was born in Dhaka and perceived the village as 'strange' and Dhaka as 'familiar', manifested in her belief in ghosts:

> "[In the village] you cannot go out in the evening because of spirits and ghosts. Isn't it scary? If I cannot go out at night, if I cannot have fresh air at night, do I feel good? […] Suppose now I am in Dhaka city. I can even roam around the whole city at night, can't I? But in the village you cannot move anywhere at night. You have to stay inside the house from the evening." (Firoza, 24.04.2010)

For her, this belief translated into a constraint of movements, most likely being both a result of the spirits/ghost stories and the matter of keeping one's respect according to the social rules of the village society. With her 'urban background' she thus perceived the city as a place of increased freedom of movement, although she also did not move around outside of Nasimgaon at night except to walk from the garments factory to her home, accompanied by her colleagues.

In Manikpara, in contrast, most interviewees felt at home in the neighbourhood and identified themselves with the locality (see also Chapter 5.3.1 on the Old Dhaka identity). Almost everyone regarded the rural home (or place of origin of the family, if born in Manikpara/Dhaka) as a place for family visits especially during holidays, but could not imagine living there. During the solicited photography none of the participants went to their home villages, and the only pictures of a rural setting were taken by Hortem during one of his leisure time excursions with his friends to a place approximately 20 minutes by boat/rickshaw from Manikpara. Among the low-income residents, the identification with Manikpara is related to the duration of stay – those who have only been there for a short time (a few months up to two years) do not yet identify with Manikpara. Although this time period is not necessarily longer than the time period residents of Nasimgaon spent in Nasimgaon, the identification with the area seems higher. Two tenants who had stayed for a long time expressed their relationship to the area accordingly:

> "I like all things [in Manikpara] such as people's behaviour, movement etc. Everyone is like a relative here." (Fahima, 29.05.2009)

> "Until I die I want to live in Manikpara. I don't want to go anywhere." (Kabin Mia, 29.04.2009)

Fahima's quote points at her feeling of familiarity with the area by referring to quasi-kin relationships. Other interviewees of low-income households also revealed how the presence of relatives was central to their identification with the locality:

> "I only stay here on rent. All of my relatives like my brother-in-law, my nephew – all have five or six-storied houses of their own here. That's why I live here, as there are houses of all my relatives." (Rokib, 30.04.2009)

Property ownership, social status and community engagement

Although Rokib himself was a tenant, in the above quote he indicated the importance of owning property, i.e. land and a house. While he and other low-income households could not afford to buy property in the neighbourhood, ownership is a central category for one's identification with the area. Owning a piece of land with a house and/or a business/shop is the aspiration of those categorising themselves as the middle class of Manikpara. During one of the women's focus group discussions, two women, who both grew up in Bikrampur District, expressed their local identity:

> Mashrufa: "Now I have become a resident [*nagorik*] of Manikpara. Now my house is in Manikpara, my own house."

> Azufa [commenting on Mashrufa's statement]: "Now Bikrampur [the district of the village home] is an outsider." (Manikpara FG2, 03.04.2010)

Similarly Mashrufa's husband repeatedly pointed out during interviews that he had an official deed for his two *katha* of land (appr. 140 m^2). In 2009 and 2010 he was engaged in a court case against the government, together with other land owners of his lane, because they wanted to pay land tax to document their ownership status:

> "The government is getting income tax and bills for electricity, water and gas. An annual tax has to be given for the land and actually the government is not taking this tax. We are ready to pay that tax. [...] We are not giving less to the government [means they have no interest in cheating/not paying]." (Mesbah Uddin, 31.03.2010)

This underlines that he considers himself entitled to full citizenship in the urban area of Manikpara.

The economically-based identification carries with it a connotation referring to one's social status within the local society. Ownership can provide access to this social sphere as it carries the notions of long-term residency and belonging to the area, but ownership does not necessarily include someone in this social sphere, nor does non-ownership exclude anyone. Social status in society thus becomes another important category for a local (urban) identity, especially of the middle class. Those residents who have been there for a long time have formed a close community. For example, Mesbah Uddin was considered a '*dada* (brother) of Manikpara' and accordingly his wife Mashrufa was considered the '*bou* [wife, but here referring to brother's wife] for the *mohalla*'. This respected status indicates an urban identity, but at the same time it also carries with it a certain code of conduct, which includes the seclusion of women as a means to preserve status (see chapter 7.1.1).

Furthermore, the interviewees of the middle class commonly engaged in community affairs and social events. Some businessmen of the locality had constructed the Bagan, Mesbah Uddin was engaged in workers' affairs as the president of the workers' union, Ripon had taken over responsibilities connected to the community police, Jahangir regularly organised religious functions at his shop close to one of the *ghats*, and Tariq together with his friends established the small garden at another *ghat*, which he believed should also serve the community one day. Furthermore, the community regularly organised launch picnics as well as a launch to jointly visit the *Bisho Ijtema* (see Chapter 0). This community feeling is expressed by Mesbah Uddin: "Our birth and death – everything is here" (see Chapter 5.3.1). It is also noticeable that the Manikpara interviewees quite often relate to places in Manikpara and the surroundings by using 'our', e.g. "this is our Bagan" or "it [Curzon Hall at Dhaka University] has similarity with our Fort".

In Nasimgaon, community engagement is also not absent. Here the local respected elders have also taken various initiatives serving the community. The very common story, mentioned by almost all inhabitants who lived in Nasimgaon at that time, is the driving out of *shontrashi* (armed gangsters, extortionists) from the area and by doing so stopping a period of *ottachar* (exploitation, including *chada*-payments). But the leaders also claim to have been involved in road improvements and road widening and other infrastructure projects, mostly implemented with NGO support. This seems to be one of the main differences between the two settlements. While previously much of Manikpara's development also depended on NGO and donor programmes, the community today is self-active. In Nasimgaon, however, development work greatly depends on NGOs and donor support or on political initiatives, e.g. of the MP.

Women's preference for village life

With regard to women's lives, the focus group results indicate a general preference for the village by women from socio-economically poor households, as it would be possible there to live 'in peace' and to maintain a lifestyle more oriented to the social and religious norms, e.g. to carry out reproductive/domestic work only and to pray five times a day. Other than economic factors, they did not see many advantages in urban life or in participating in the generation of household income:

> "Men, no matter where they are living, will earn and send the money to the village. Women will take care of the house; they will take care of the domestic animals like chickens, ducks, and goats. They will run the family in a good way. They can maintain *porda*; they can take their prayers five times a day. But we live in Dhaka city, we work, we come back home and cook, we eat and then we sleep. In the village, we can do our household work properly. We can have two ducks, chickens as pets. We can keep the surrounding clean; we can do our work properly. But in Dhaka city, we keep sweeping other people's houses, there is nothing for us." (Gulshana, 02./09.04.2010)

This perception expressed by a working woman in Manikpara is also common in Nasimgaon. During the focus group the women also pointed out the hardship of having to work for others and indicated how their preference was the village life:

> Meher: "Now we have come to Dhaka. People of our village say that those who have come to Dhaka are bad."
>
> Firoza: "When an unmarried girl comes to Dhaka, it becomes impossible to arrange her marriage in the village."
>
> Meher: "It is not possible to arrange her marriage in the village. You see, my daughter got divorced. You should have seen my daughter's father-in-law. We would be considered good if we would feed them continuously. It is a very difficult world out here, you will get problems in your mind after listening to all our stories." (Nasimgaon FG, 24.04.2010)

The village life promises less hardship because women believe they would not have to work for others. But, all of them agree, it was not possible to lead this life in the village because they were not in an economic position to do so (see Chapter 7.1.1). However, their image of village life indicates a rather conservative understanding of male-female roles. Dhaka was furthermore considered as a place where they could not maintain their respect and honour properly, whereas the rural areas were considered places of honour for women, as the following quote impressively indicates:

> "It is said that when you enter into Dhaka city, you should leave your shame [*lojja-shorom*] at Shadarghat [the main ferry terminal]. Again, when you will go back to the village, you will take that back from Shadarghat." (Firoza, 24.04.2010)

In Manikpara, on the contrary, the women who consider themselves as belonging to the middle class cannot imagine a village life and give preference to the city:

> Mashrufa: "Even when we go visiting our village, we come back from there."
>
> Subina: "It does not matter whether we eat or not, we like it here."
>
> Mashrufa: "I mean Manikpara has become known to us. We don't like going anywhere. We don't feel good even when we go to the village. Suppose I stay for a week, I don't want to stay a couple of days more." (Manikpara FG2, 03.04.2010)

This is all the more interesting as these women are also the ones most conscious of maintaining *porda* and socially acceptable behaviour to suit their group's standards. For them, in contrast to what was expressed by Gulshana in the focus group with working women, maintaining a socially accepted life is not a question of urban or rural. This presumably is the case because they do not need to work and thus are able to maintain *porda* wherever they are, and thus can 'afford' to prefer the city life they have grown accustomed to. The city for them does not mean working, while for the socio-economically poor the city means generating an income in the first place. They remain socio-culturally attached to the village and understand the village as naturally allowing a different lifestyle.

Perceptions of citizenship confirming the familiar/strange

The conceptualisation of the city as familiar/strange and of identity being urban/rural is also reflected in the right to the city. Here it is necessary to differentiate between the perception of citizenship by inhabitants and the construction of citizenship by society in general, including the implications these constructs have for the residents. The notion of citizenship displayed by Manikpara's residents is inherent in Mesbah Uddin wanting to pay land tax. With his willingness to pay taxes, he also makes a claim to be recognised as a citizen with civic and political rights and is aware of both his rights and obligations as a citizen. This full citizenship of Manikpara inhabitants is also recognised by society in general. In contrast, Nasimgaon residents are aware that they have less of a right to the city. They are aware of the government ownership of the land and the threat of eviction emanating from this ownership structure. While they claim that they are citizens of the country and also have a right to a place, they nonetheless do not express their citizenship as self-confidently as Manikpara residents. This is also what is produced from outside – political citizenship as voters, but no civic citizenship concerning the right to a place within the city.

Concluding remarks

The discussion of familiarity/strangeness as identified with regard to the urban fabric is continued by looking at the identities of inhabitants of Manikpara and Nasimgaon. The urban identity of Manikpara inhabitants is multifaceted, including their own perception of rural/urban identities, the network of relatives, economic status, social status and respect and religious affairs. In contrast, the inhabitants of Nasimgaon refer to the village as the reference point for their identity. This differing identification may be closely linked to the exclusionary practices that Nasimgaon residents experience, including the insecurity of the permanent right to residence, the notion of a differentiated citizenship and the resulting lack of integration into the urban context. Furthermore, identities are based on the local societal set-up, and the social sphere of Manikpara encourages a much stronger identification, again initiated by the feeling of permanence. The identities of inhabitants of Manikpara and Nasimgaon are furthermore related to the age of the settlements, the generational and life cycle related perceptions and the socio-economic background of inhabitants. All these factors can be considered to influence the spatial pattern of inhabitants and the way they participate or are able to participate in the production of space on a city scale.

7.2.3 Men's mobility patterns reproducing the familiar and strange

The mobility patterns of Manikpara and Nasimgaon male residents again indicate a different perception and conceptualisation of urban space. In Manikpara male residents tend to visit areas outside Manikpara more frequently in their free and leisure time and thus display a higher level of integration into urban space. In Nasimgaon, on the other hand, residents spend less of their free time outside the neighbourhood and mainly leave Nasimgaon for activities related to economic livelihoods. These differences can be explained with urban/rural identities, but also with the integration of the areas (urban fabric) into the city and the socio-economic situation of inhabitants. As a result, however, these spatial practices reproduce the urban space that is inherent in the urban fabric.

For Manikpara's (male) inhabitants, the city is perceived as familiar and home. The familiarity becomes obvious from the mobility pattern of Hortem (see Figure 12). As an unmarried employee in a plastic sorting shop, he spent most of his free time with friends both in Manikpara and in the surrounding areas of the city, among them the Dhaka University Campus which is among the most central public places representing the city of Dhaka. While the mobility patterns of married men and elder men slightly differ because they have to take on family responsibilities and are thus more restricted in their daily mobility (see Chapter 6.1.2), they nonetheless display a similar familiarity with the surrounding city although they may visit places less frequently. For socio-economically poor and married men, familiarity with the city is lower. However, they too identify themselves with the city, also influenced by the duration of their stay in the area and whether Manikpara/the city is their birthplace.

Even for those inhabitants who due to family obligations cannot move around as easily as Hortem, there are opportunities for mobility beyond the neighbourhood. These are, for example, the regularly organised launch picnics (see Chapter 6.1.2) or the day of the *Bisho Ijtema* (see Chapter 0). Furthermore, when I joined *adda* rounds in Manikpara I was often asked whether I had been to a place, both in Dhaka and surroundings as well as all over Bangladesh. This indicated a higher knowledge of the surrounding city and told me that many residents had visited such places.

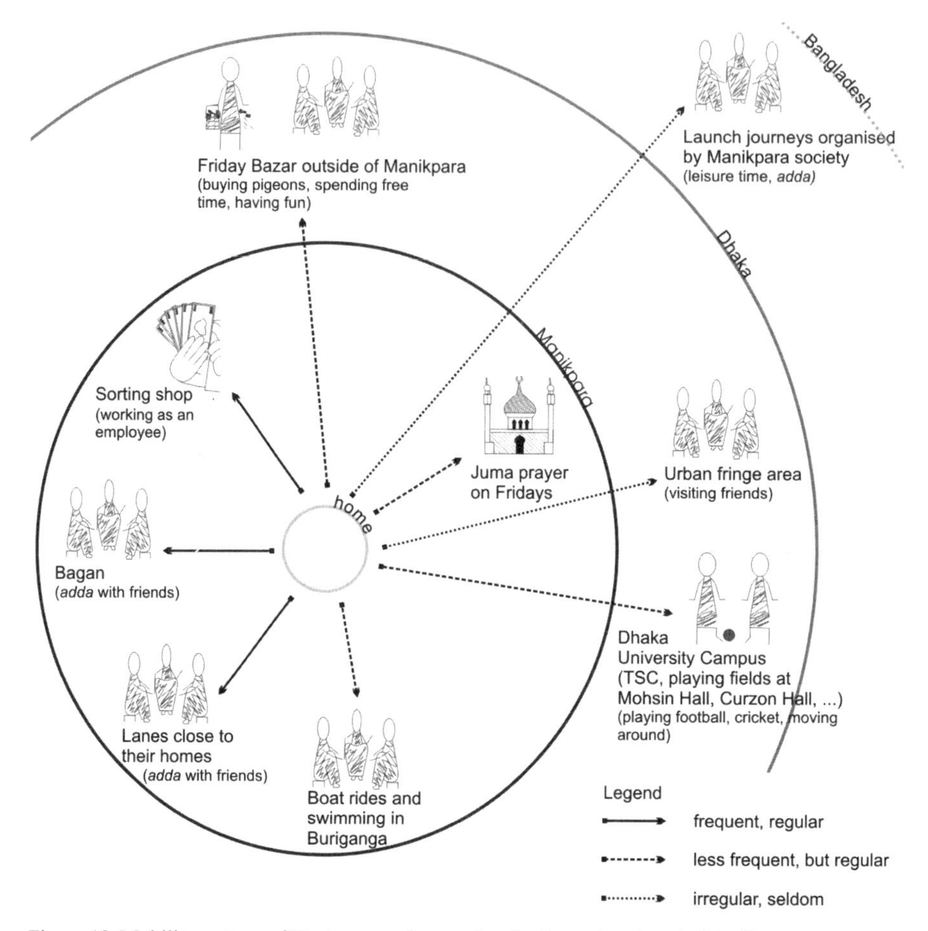

Figure 12: Mobility pattern of Hortem, employee of a plastic sorting shop in Manikpara

> Hortem, aged 23 and unmarried, worked in a sorting shop in Manikpara six days per week. He spent much of his remaining time visiting different places with his friends (indicated in the figure by the icon with three people chatting/spending time together). His mobility pattern was drawn up according to his description of his daily activities during the solicited photography interview on 29.05.2009.

In Nasimgaon the common male mobility pattern revealed by the solicited photography did not include many spaces outside of the neighbourhood. Most men only left Nasimgaon for the purpose of supporting their business, e.g. to go to Kawranbazar to buy fruit and vegetables from the wholesalers, to Gabtoli to buy construction materials, or to Gausia Market to buy cloth. The mobility pattern of Khalil (see Figure 13) and the way he talked about the city and the village, for example, revealed that the city for him only represented a means of earning an income. He had previously been in government service and then operated shops in the village, but he failed to make a profit because too many people bought his products on credit. Thus he decided to run a banana shop in Nasimgaon. While his daily urban routine was to operate his shop and get supplies from the wholesale market if necessary, he visited the village where his wife remained very frequently and had no plans to settle in Dhaka permanently. Other inhabitants also did not leave Nasimgaon commonly, except for journeys to the village or to some religious places, especially *mazars*.

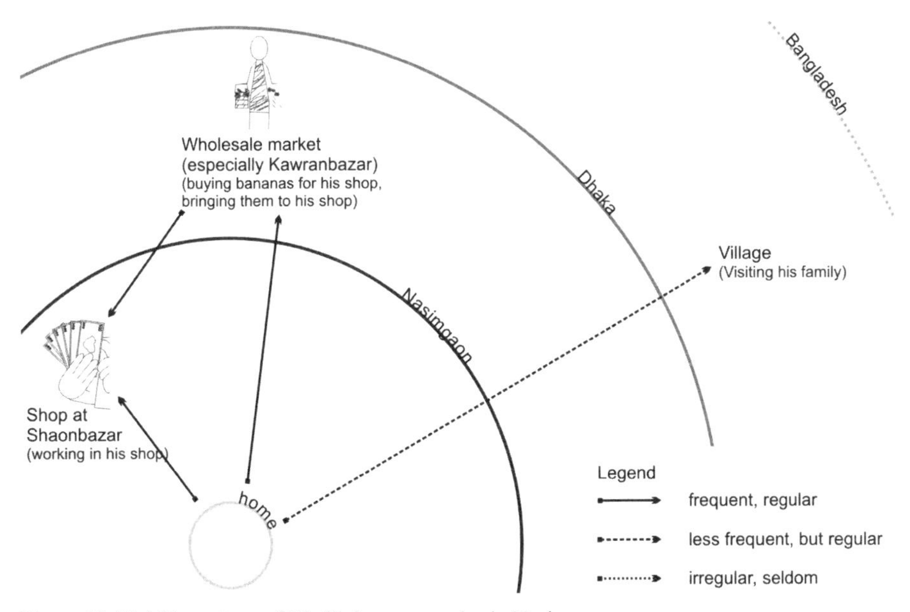

Figure 13: Mobility pattern of Khalil, banana vendor in Nasimgaon

Khalil, aged around 55, sold bananas from a small shop at Shaonbazar in Nasimgaon. His mobility pattern was drawn according to his description of his daily activities during the solicited photography interview on 26.05.2009.

This job-focused mobility pattern, however, cannot only be understood as a result of exclusionary processes and a predominantly rural identity. It is also attributed to the socio-economic status of many Nasimgaon inhabitants which does not allow them to invest a lot of money or time in personal leisure. Only two Nasimgaon participants of the solicited photography displayed a different mobility pattern. The young father Assad used the camera moving around with his friends and visiting places. His flexible working hours as a carrier in the garment factories enabled him to do so, and his position in the life cycle, young age and recent fatherhood, supported his movements. The other person was Abdul, a rickshaw puller, who enjoyed visiting places in the nearby surroundings of Nasimgaon. This he was able to do extensively in 2009, when his family was in the village and he was able to use his time off from driving the rickshaw. But when his family (two wives living in two housing compounds and five children that stayed with his younger wife although she was the mother of only three of them) came to Nasimgaon in 2010, he was more restricted in his movements as well.

With regard to men's mobility patterns, the discussion of publicness supports what has already been analysed concerning the urban fabric and rural/urban identities. Men's mobility patterns in Manikpara suggest a high degree of familiarity with the city and thus a large extent of 'familiar publicness'. In contrast, men in Nasimgaon, and also the working men of Manikpara in their everyday life, do not have the same familiarity with the city, as their movements are more restricted, especially due to socio-economic reasons. Thus the 'familiar publicness', here denoting the spaces that are visited frequently, is of more limited extent in Nasimgaon than in Manikpara. Besides socio-economic reasons, the high degree of otherness in terms of the urban fabric may also contribute to this difference.

7.2.4 Women's mobility patterns, clothing style and hierarchies of 'publicness'

The above findings and interpretation focussed on men's mobility patterns and urban experiences. For women the situation is rather different, as the discussion of female spatial practices and mobility patterns has already indicated. This sub-chapter will thus investigate these women's mobility patterns from the perspective of the underlying hierarchies of publicness. Building on the discussion of gender norms in sub-chapters 7.1.1 and 7.1.2, women's mobility patterns will be analysed by referring to women's clothing styles[3].

Female clothing and nuances of 'publicness'

Female clothing habits support the above observation on the nuances of 'publicness'. Female clothes include the *maxi*, a long loose-fitting dress worn together with an *orna*, *salwar kamij*, *sari* and *borka* (see Chapter 5.1.2 and Figure A-2). All of this clothing can be worn without covering the head, or with covering the head loosely (principally referred to as *ghomta*, carried out with the end of the *sari* or an *orna*) or tightly in *hijab*-style. How a woman dresses is highly related to the familiarity of the area which she is planning to visit. This observation is supported by Tarlo's research on women's clothing in rural Gujarat, India (Tarlo 1996; see also Chapter 2.2.2). In her research, Tarlo established a relationship between the choice of clothing and a person's identity, considering caste, age, but also belonging to a rural-traditional or an urban-modern household. Accordingly, what a woman wears and where and when becomes an expression of her need for representation in public. This at the same time is interlinked with the social gender relations of a neighbourhood's society, and thus the dress code has to be under-

3 As it is rather uncommon for women in Bangladesh not to wear the local dress, I also decided to wear the *salwar kamij*, a combination of loose fitting trousers and a long blouse worn with a scarf to cover the chest (see Figure A-2). This was perceived positively by the inhabitants of the study settlements. In carrying out this research and asking questions concerning respect and honour in relation to gender norms, my 'appropriate' clothing was an important means of being accepted in the study settlements.

stood as an outcome of the familiarity of the public sphere, the social gender relations and the urban fabric itself.

Familiar and strangers' publicness

The internal familiarity of Nasimgaon and the perception of the city as 'other' is reflected in the way women dress. While dress codes inside the housing compound during the daytime when men are absent tend to be relaxed, as soon as women leave their compound they put on an 'acceptable dress':

> "At that time [when leaving the housing compound] it becomes different. Suppose I can put off the *sari* from my head when I feel hot and can stay in this way at home. But if I go outside I have to cover myself properly and put on the *sari* on the head. Otherwise if other men see me isn't it bad?" (Rehana, 23.04.2010)

While for Rehana this means wearing *sari* and using the end of the *sari* to cover the head, this 'acceptable style' also includes *maxi* combined with *orna* and *salwar kamij* (which already includes *orna*). Covering the head is not required, however, many women wearing *sari* or *salwar kamij* use the *sari* or *orna* to loosely cover their heads. Within Nasimgaon all the women interviewed stated that whether they did their work, including selling foods by sitting in public spaces and bazar areas, or moved around in Nasimgaon for other purposes like going to the bazar or visiting friends and relatives, they would always wear this 'normal style' – between the more 'informal' style inside the housing compound and the more dressed-up style for moving further away from home. Also the wife of Mokbul Hossain, who can be understood as part of the local middle class, moves around within Nasimgaon in the same clothing style, and without specifically dressing-up.

The above discussion indicates how the internal area of Nasimgaon is perceived as familiar and not as an area of strangers. It underlines the findings that the specific fabric of Nasimgaon and its lack of integration into the city produce an internally homogenous space with a 'familiar publicness', which, however, stands in stark contrast to the surroundings.

This 'familiar publicness' is less existent in Manikpara. Here, only those women who work mainly in the plastic recycling industry leave their homes to go to work dressed in 'normal style'. As they normally do not go to the bazar themselves, this 'familiar publicness' is restricted to a relatively small area. For women of socio-economically middle-class background such a 'familiar publicness' does not even exist at all, as they seldom leave their house to move around within Manikpara. The higher degree of 'strangers' publicness' may also be attributed to the fact that Manikpara, much more than Nasimgaon, is a space that is 'traversed through', due to its integration into the urban fabric. The limited mobility patterns are shown in Figure 14 for Fahima, who works in the plastic sorting industry. The map shows how her activities did not range beyond the immediate surroundings of

Manikpara to cover the city of Dhaka. Only in the company of her husband did she move outside of the city.

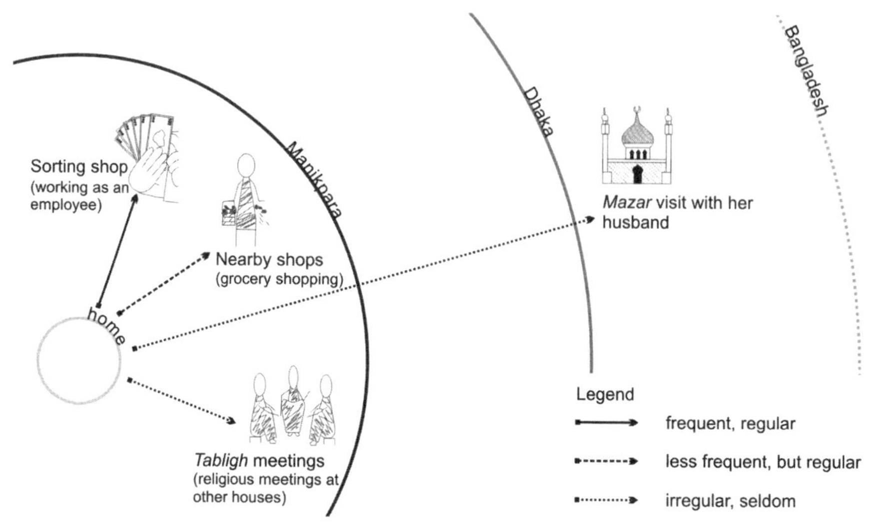

Figure 14: Mobility pattern of Fahima, employee of a plastic sorting shop in Manikpara

> Fahima, aged ~40, worked in a plastic sorting shop along Embankment Road. Her mobility pattern was drawn based on the solicited photography interview held with her on 29.05.2009 and the focus group of working women on 02./.09.04.2010.

Women's movement and dress styles in Manikpara suggest that a 'strangers' publicness' begins almost at the threshold of the house. This is especially the case for women with a slightly more well-off socio-economic background who are stricter in maintaining the rules of *porda* (see also Chapter 7.1). Their mobility map would display even less locations than Fahima's map. Wearing *borka* from the threshold of the house symbolises both a protection from a 'strange' outside (as opposed to family and neighbours in a housing compound or apartment block) as well as the religious decency and social respect of a family. While some women who consider themselves as belonging to the middle class despite being socio-economically poor would leave home wearing a good quality *sari* and *orna* in their immediate neighbourhood, Mashrufa would always wear *borka* when going outside:

> "I always go outside with *borka*. If I go nearby, I take *borka*. If I go far, I take *borka*. I mean I always wear *borka*. I don't go outside wearing *orna*. I don't go outside without *borka*." (Mashrufa, 03.04.2010)

To go beyond the neighbourhood of Manikpara as well as to the village, the women interviewed in Manikpara all referred to the *borka* as the appropriate clothing style. The 'strangers' publicness' for the (female) inhabitants of Manikpara, then, begins in the immediate vicinity of the home, while the extent of the 'familiar' depends on the socio-economic status of the household (which again is related to the general status of respect in society, see Chapter 9.1.3 for a further discussion).

In Nasimgaon this border of 'familiar' and 'strangers' publicness' is more distinct and homogenous for all female Nasimgaon inhabitants. The 'strangers' publicness' begins where the urban fabric changes, i.e. as soon as Nasimgaon is left behind (although the small neighbourhood adjacent to Nasimgaon, which has the same building structure as Nasimgaon, can be considered 'familiar' too). As soon as women left the area of Nasimgaon, they were often (not always) accompanied by their husbands or by female neighbours, and they either changed their clothes into better ones or they took care to maintain a certain decent dress code:

> "Then [if I leave Nasimgaon] I definitely give *ghomta* properly. Everyone knows me in Nasimgaon. Suppose if I don't have *ghomta* at one time, still I can walk here. But now if I go to Sohelpur [neighbouring settlement], nobody knows me there. Then I have to put on *sari* on the head definitely. If it falls then immediately I put it on again. But I wear this *sari* that I am wearing now; I don't change the *sari*. [...] If I go to some other place I wear a good *sari* which I have kept with care." (Rehana, 23.04.2010)

Here Rehana referred to the familiarity of Nasimgaon, while she perceived Sohelpur, where nobody knew her, as unfamiliar and 'other'. This perception reflects Simmel's notion of the anonymity of the city in '*Die Großstädte und das Geistesleben*' (Simmel 1903), and points on the one hand at Nasimgaon as a space outside of this anonymity, and on the other hand at a reproduction of spatial patterns of exclusion concerning Nasimgaon in the city context. While Rehana continued to wear the same dress but took care to present a decent appearance, other women said that they changed their dress and wore a better *salwar kamij* or *sari* once they moved outside of Nasimgaon. Both 'clothing strategies' point at the division of publicness into familiar and strange. The mobility map of Rohima (see Figure 15) underlines the above statements about a familiar publicness in Nasimgaon and a change as soon as the neighbourhood is left. Rohima was highly mobile inside Nasimgaon, while she only occasionally moved outside Nasimgaon. All the locations she then visited were in the immediate surroundings, which were more familiar than the rest of the city (see Figure 16 for the delineation of the nuances of publicness). When she went to these locations she normally was accompanied by others (see discussion of surveillance and company in Chapter 7.1.2).

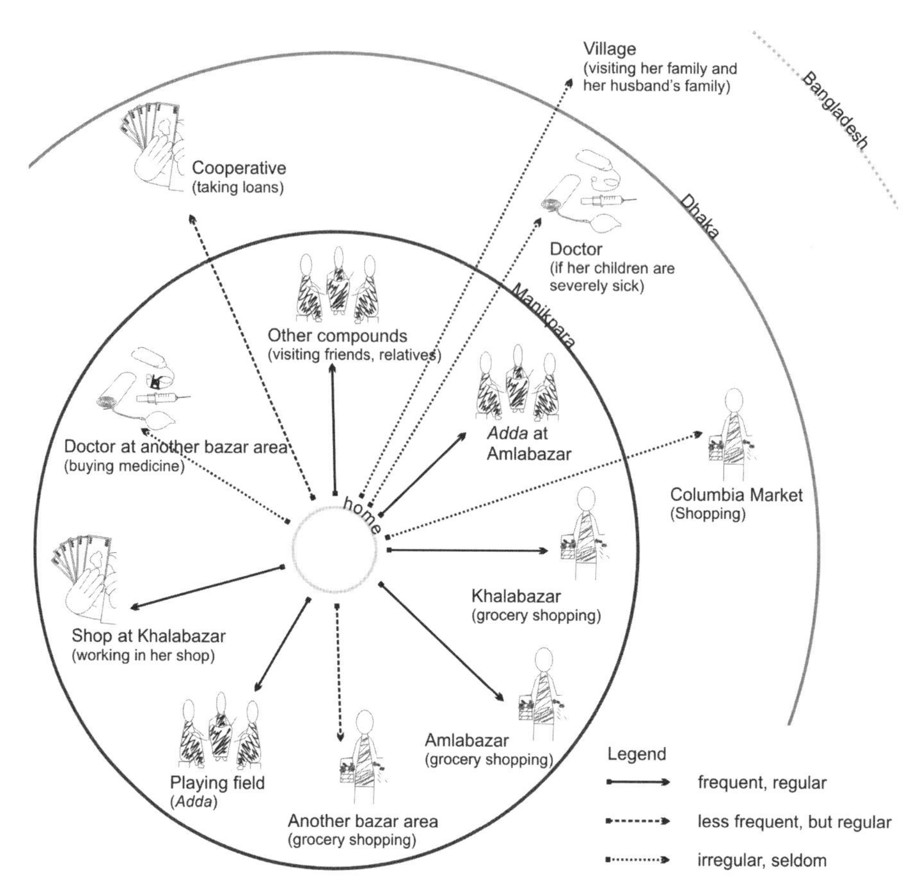

Figure 15: Mobility pattern of Rohima, tea stall operator and fruit vendor in Nasimgaon

Rohima, aged ~25, together with her husband Shahin sold fruits and operated a tea stall in Nasimgaon. Her mobility pattern was drawn based on an interview combined with creating a mobility map on 02.07.2009.

Village dress style

For Nasimgaon residents there seems to be yet another nuance of 'strangers' publicness'. The interviewees separately referred to their dressing style to go to the village. Here the rules of what to wear depend on the village society they come from, as in some rural areas it is common to wear *borka* while in others it is not. To go to the village, most women give more care to their dress than when moving around in the city outside of Nasimgaon:

"At that time [when going to the village] I wear a better *sari*. I have come from my shop now. This *sari* is dirty now. I can't go to outside of Nasimgaon wearing this *sari*. I have a separate

> *sari* and petticoat in my room. I wear those dresses and make myself well looking. After that I go to my village." (Taslima, 18.04.2010)

This dressing for the village symbolises a higher degree of 'strangers' publicness', especially for the journey to the village:

> "Suppose I am going to one place to travel. There are people in the roads; so I wear good clothes." (Aklima, 18.04.2010)

Once in the village, however, strangeness does not work as an explanation. Good clothing then becomes a matter of respect and social status, quite similar to how women from Manikpara would move within their own neighbourhood. This can furthermore be interpreted as the need to tell a 'success story' of city life and to demonstrate its benefits. This then again indicates the urban identity of Manikpara residents and the rural identity of Nasimgaon residents. There is yet another component, however. The need to produce a success story has also been discussed in the context of Bangladeshi women returning home from working abroad (Dannecker 2008). Dannecker found how women on the one hand needed to prove the 'usefulness' of having been abroad, while on the other hand this was expected to be done more 'quietly' than in men's cases. More 'quietly' here means that while men are expected or even obliged to bring home status symbols, women's periods abroad were not equally celebrated and displayed due to the lower acceptability and general suspicion of women's improper behaviour. While I was not able to confirm this for residents of Nasimgaon or Manikpara and their home villages, it may be an issue for further consideration of urban-rural linkages.

Surveillance in 'strangers' publicness'

The nuances of publicness also become apparent from the acceptability of moving around outside Nasimgaon alone. Dipa's frequent journeys to the *mazar* had resulted in gossiping about dishonourable activities (see Chapter 6.1.4). As soon as women expose themselves to the 'strangers' publicness', the surveillance of neighbours can no longer be upheld and thus 'bad behaviour' is assumed. The common pattern of husbands normally going to the wholesale markets outside of Nasimgaon supports this argument. Only those women who do not have a husband or whose husband is not able to go, enter the 'strangers' publicness' of the wholesale market themselves. This is not without surveillance, as Dipa's gossiping about other women going to Kawranbazar by themselves underlines:

> "If I am going to Kawranbazar to buy my vegetables, I am taking care of my dress, I'm going decently dressed up, putting the scarf and wearing the *sari* decently. But the other women are not wearing a proper dress and are showing too much [she shows that they are not wearing the *sari* covering the blouse]. Because of that the other women are having a good access to the market sellers at Kawranbazar and are sometimes getting some materials for free. But I'm buying the materials and am not getting any benefit from the shopkeepers. Because of that I have some loss in my business [earning less than them]." (Dipa, 31.05.2009)

The readiness to react with gossiping and rumours to female behaviour that trespasses certain norms underlines the importance of nuances of publicness for female movement. This aspect is also supported by Bose's observation of the parochial realm (referring to the neighbourhood area, resembling what I here term 'familiar publicness') being the sphere where women could engage in economic activities, while activities that necessitate leaving this realm of the "watchful gaze" were opposed and less accepted (Bose 1998: 367).

Concluding remarks

The above discussion has revealed how female clothing styles relate to the perception of familiarity and strangeness. The 'familiar publicness' women experience within Nasimgaon explains much of the spatial practices and mobility patterns of women analysed above and in Chapter 7.1.2, where women in Nasimgaon are freer to move around within the neighbourhood without this being perceived as a matter of shame by society. The in-between space of Nasimgaon, which does not directly extend to a space where strangers meet, seems to serve as a buffer from the 'anonymity' of the city. In contrast, the surveillance of society prevents women of the middle class of Manikpara from moving around outside the familiar home, as this would immediately trigger rumours among relatives and neighbours and harm the household's reputation.

7.2.5 Multiple hierarchies of publicness

The different perspectives explored in the above sub-chapters are closely interlinked with each other, i.e. the urban fabric influences the mobility patterns, but also seem to contradict each other (see Figure 16). This characterises the multiplicity of perceptions, meanings and uses of space which produce and re-produce specific spatial practices. Inherent in the hierarchies is then a delineation of spaces of movement, freedom and perceived familiarity ('familiar publicness'), as opposed to spaces of restricted movement/freedom and perceived otherness/strangeness ('strangers' publicness').

In conclusion, the idea of multiple hierarchies of publicness, which was introduced in relation to integration into the urban fabric in sub-chapter 7.2.1, is reconfirmed by the hierarchies of publicness expressed in urban/rural identities and in male mobility patterns. The mobility patterns of women in Manikpara, however, propose a different hierarchy. While socio-economic and socio-cultural reasons influence the different nuances of publicness, it nevertheless seems that there is an interrelation between the integration into the urban fabric, the perception of publicness, and the spatial practice. Secluded spaces, like Nasimgaon, present urban niches for an everyday life that is not as strictly ruled by social gender relations as the more 'public arenas'. Similar findings were discussed by Wedel (2004; see Chapter 2.2.2), who found women's mobility in Turkish gecekondus to be restricted to the internal settlement and vicinity perceived as an extension of private spaces, as opposed to men's mobility patterns characterised by distance. On the other hand, seclusion and urban niches are an expression of homogenous space indicating dominated spatial practices rather than a counter space for personal freedom.

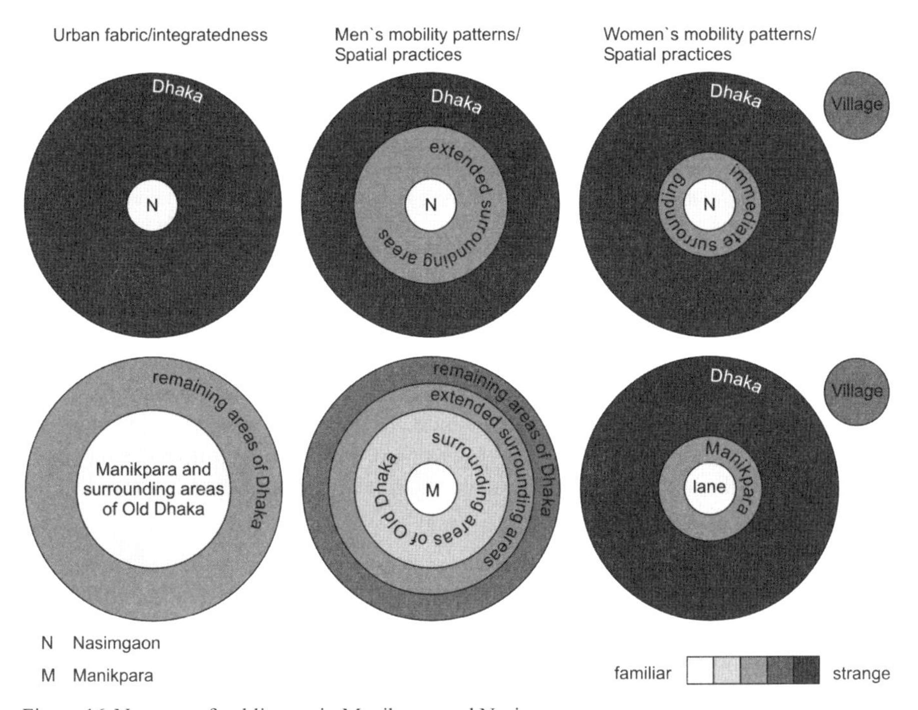

Figure 16: Nuances of publicness in Manikpara and Nasimgaon

7.3 SHIFTING BORDERS OF FAMILIAR AND STRANGE

The borders of what is familiar and what is strange identified above are, however, not static. Instead, specific events and seasonality initiate changing conceptualisations of familiar/strange. Thus the multiple hierarchies of 'publicness' do not only have to be understood as a compilation of layers depending on perspective, but also as a hybrid and fluid construct with shifting borders.

Extension of the 'familiar' I: summer months and electricity cuts

With the beginning of the hot season, usually from March until June[4], the frequency of electricity cuts increases due to the necessity of load shedding[5]. These electricity cuts change the borders between familiar and strange, especially for women. In Nasimgaon, many of the female interviewees indicated that they went out to enjoy a fresh breeze late at night if the power was cut (see also Chapter 6.1.5):

> "When electricity is gone then people sing and do *adda* [on Nasimgaon Math]. […] I also go. […] The house owner and all of us who live here sit and do *adda* on the Math. Later, when the electricity comes back, everyone goes to his/her house and sleeps. [If the electricity goes after 12 at night] then I stay in the house. I don't go outside at such late night time. Electricity goes many times during the night, and then everybody sits on the road taking his/her own chair. At that time we cannot stay in the house due to the hot condition." (Nasrin, 11.05.2009)

Without the power cuts, Nasrin normally stayed at home after 8pm. However, this also differs based on location: in the bazar areas, which are usually busy until 11 or 12 at night, women still stay outside for *adda* after 8pm even without electricity cuts. But here as well, the time when it is acceptable for women to stay outside is prolonged by power cuts. Even spaces which are normally 'reserved' for men late at night suddenly become frequented by women until late. Ishrat, however, indicated how the distance the women would move away from home became shorter with the progressing hours:

> "If it is 11pm [when electricity goes], then we stay closer to this gate [of the housing compound]. Many of the people are awake until the electricity comes back. Again, if it becomes 12 or 1am, then no one stays. Everybody goes to sleep in their own room." (Ishrat, 24.04.2010)

The increased familiarity of this staying outdoors until late at night was also expressed by Rohima as she narrated how she enjoyed moving around in these night time hours:

> "There are lots of people who stay till 12 or 1am in the night. The shops are also kept open. […] The people become tired due to too much heat and can't sleep. At that time if the shops

4 The summer time of the Bengali year starts on 14th April and lasts until 14th June. However, temperatures are usually high during the daytime from March onwards. While nights at that time are still cool in the countryside, the dense urban fabric of Dhaka results in high temperatures at night as well, at least from mid-March. The monsoon usually begins on 15th June, but for example in 2009 it did not begin until two weeks later and temperatures remained relatively high despite occasional rains.

5 In 2009 and 2010, Dhaka experienced serious load shedding from the beginning of the hot season onwards, as electricity production did not meet the increased demand. For example, the Dhaka Power Distribution Company was short of almost 600 MW on 26 April 2009, both during the daytime (demand 1,800 MW) and evening hours (demand 2,000 MW) (The Daily Star 2009a). In most urban quarters until late evening the electricity was off for one hour after one hour of supply. Later on the cycle was changed to two hours on, followed by two hours off, which was more favourable to industries which needed to warm-up machines first, e.g. the plastic recycling in Manikpara.

> are open, they sit there and take tea as well as have some *adda*. Thus the time passes and electricity comes back again. [...] Women also stay. It is really nice to walk around in the night because the place becomes silent and there is a breeze. Then I feel really good." (Rohima, 02.07.2009)

The familiar can also be understood as the way both men and women enjoy interacting with each other late at night, doing *adda* at the tea stalls. The presence of people extends 'familiar publicness' into the night hours and thus changes the everyday spatial practices and spatiality of livelihoods. But also in Manikpara, where women especially of middle-class households tend not to leave the house, especially not in the evening, the border between strange and familiar changes during power cuts. For example, Mashrufa, a middle-class woman who normally would not leave the house at all, repeatedly talked about how she went outdoors late at night:

> "Mashrufa tells us [research team] about the power cut last night, after 1am. It was very hot when the power was gone, and no one was able to sleep, so everyone in the area went out of their homes. Together with her husband she went down to the road, but many people went to the Bagan. She did not feel like going to the Bagan, because there were too many people, and rather preferred to stay in the lane close to the house." (field note about conversation with Mashrufa, 19.06.2009)

The above examples show how seasonality leading to power cuts also alters the everyday spatial practice, and for brief moments what is acceptable behaviour, especially for women, changes. This indicates again how multifaceted the notions of publicness and the expressions of norms in public spaces are. Accordingly, the next sub-chapter analyses how the same event of electricity cuts can at the same time lead to an extension of the 'strange'.

Extension of the 'strange': busy hours and crowding in public spaces

Above, I have outlined how electricity cuts can lead to an extension of the familiar. However, these can also result in an increased 'strangers' publicness'. This was reported by the women working in plastic sorting in Manikpara:

> "We face problems while walking in the lane if there is load shedding. The lanes are filled by men, only men. It is difficult to walk in midst of men. The load shedding always starts when our work is done. Then we have to walk back home in the midst of men. We suffer for that. You cannot fight with men for walking. [...] People get out of their house whenever there is load shedding in the lane that we use. All the people in the industries of that lane come to the lane. That's why there are more sufferings." (Gulshana, 02./09.04.2010)

For the working women the route between home and the workplace normally belongs to the immediate surrounding and is thus familiar (see Chapter 7.2.4). But the increased presence of men during load shedding results in a perception of this usually familiar place as a 'strangers' publicness'. Walking amidst men in a crowd, and in darkness, is a stressor for the women and makes them feel highly uncomfortable as they are eager to avoid physical contact. Thus not only electricity cuts, but also the crowded hours of the day when people move to the bazar or back home from their workplaces are considered as occasions of an increased 'strangers' publicness', both in Nasimgaon and Manikpara. Women tend to feel uncomfortable in crowded roads due to the increased potential closeness to men (see Chapter 7.1.2).

Furthermore, within Nasimgaon, which generally can be considered familiar, 'strangers' publicness' is also created in secluded spaces off the main roads. For example, women referred to some narrow lanes where 'bad boys' stayed and which were then avoided in everyday life mobility patterns, e.g.:

> "I like this road. I don't move by the other road. I don't like to move by that other road. That road is narrow and local boys take cannabis and wine there. [...] I don't come by that narrow road. Bad boys stay there. Don't you understand? So I don't come by that way. (Shoma, 27.04.2010)

This 'extended strangeness', however, is also dependent on a woman's position in the life cycle (see also Chapter 7.1.2). While for young and unmarried women the rules are rather strict as they have to take care not to harm their honour and dignity, older women are freer to move around in the darkness or crowded areas. This was expressed by Roxana, an unmarried woman in Manikpara, with reference to the different rules of movement valid for her mother:

> "There is no anxiety for the elder women [to go outside after dusk], for example: there is no anxiety for the women who have my mother's age. But there is anxiety for women of our age. Suppose we went at 8am and came back at 10pm. At that time there is anxiety about how we can come back by road since there are lots of bad boys." (Roxana, 08.05.2009)

While strangeness thus can extend depending on the hour of the day, the necessity to react to this increased strangeness depends very much on a woman's position in the life cycle. Women of old age, who are also more free in choosing their dress styles, can also move around more freely. On the other hand, for young women of marriageable age or who are recently married, public spaces are characterised by extended 'strangers' publicness' at specific times.

Extension of the 'familiar' II: Bengali New Year

A considerable disruption of women's everyday life routines is the Bengali New Year *Pohela Boishakh* (see Chapter 6.2.2). Women of the middle class, who normally would not leave the house without wearing *borka* or at least not go far, show up at the Bagan to enjoy the programme (see Photo A-50 which depicts women belonging to the middle class of Manikpara wearing red and white *Pohela Boishakh saris* on *Pohela Boishakh* in 2009). This is in stark contrast to the everyday life routine during the summer. On a normal working day or a Friday, only men frequent the Bagan to get some fresh air and/or engage in *adda,* and children use the space to play, while women are absent (see Photo A-51). Accordingly, the

Pohela Boishakh marks a day of extension of the 'familiar publicness' and increases women's mobility range accordingly[6].

In Nasimgaon, *Pohela Boishakh* did not change the range of familiarity considerably. This may be attributed to two factors. Firstly, as discussed above, within Nasimgaon there is already a high degree of 'familiar publicness'. Secondly, many of Nasimgaon's residents continue to work on *Pohela Boishakh*, e.g. those operating shops and stalls within Nasimgaon and the rickshaw pullers. While many residents wear new clothes of *Pohela Boishakh* designs, the day is mainly celebrated by children and those whose workplaces are closed, e.g. the garment workers. Similarly, also the working population of Manikpara does not change their activity patterns considerably. Thus for them, *Pohela Boishakh* does not lead to the production of a different space in extra-everyday life.

Islands of the 'familiar': secluded spaces

Secluded spaces indicate how certain norms are translated into public spaces. Here this refers to two very different types of seclusion. On the one hand, seclusion points to a consensual religious norm during *Ramadan*. On the other hand, seclusion can be an exclusionary practice employed by a group to conceal certain activities which are not characterised by social consensus or generally accepted.

The fasting month of *Ramadan* with the food stalls covered by cloth curtains produces a specific type of secluded space changing the appearance of public space considerably (see Chapter 0). Here a new boundary of public-private is drawn: the curtains give privacy to those who do not fast while the public sphere is maintained according to religious norms. This public-private boundary is drawn by a religious norm but is further enforced by statutory authorities: in Nasimgaon, Sania, a restaurant owner narrated that the special police force RAB would scold the respective shop owners in cases of non-conformation with these norms.

A different case is the choosing of a secluded space for a higher level of privacy as a voluntary strategy. In Manikpara, an *adda* round of elder men regularly met on the embankment slope to smoke cannabis around noon and in the early evening. This is not a respected and acceptable behaviour, and thus the men had chosen the location on the embankment slopes as it was relatively shielded from public view because it was walled-off. Seclusion here is used to protect behaviour that is not according to the norms of society. However, this group is not considered as 'anti-social' like the activities going on along the embankment slopes at night. Rather, this group consists of respected men who shield themselves from

6 During my stay in Bangladesh, I was not able to participate in the two most celebrated Bengali national events, the Victory Day celebrations (16th December) and the *Ekushe* commemoration (at night on 21st February). Thus I cannot confirm whether this finding of an extended familiarity also characterises these two days. Furthermore, I did not witness any of the *Eid* celebrations in the study settlements. Although I got some information through interviews, I cannot derive findings on the nature of the border between familiar/strange on these days either.

the public only for this purpose, but otherwise are actively engaged in community matters (e.g. politics, community police and religious functions). Nonetheless by doing so they potentially exclude others from using this public space. This exclusion is not explicit for there are no rules prohibiting access. But the group of men sitting in front of their shop and doing *adda* represents an invisible border of belonging/not belonging to the group.

Similarly, such a shielding from publicness can be employed by women. This was not so evident from the interviews, but one discussion with two female university lecturers suggested this direction (Hafiz and Shafi, 26.04.2010). They narrated how women can use an umbrella to shield themselves from publicness and thus create their own private place within the public. This option of creating a private sphere within the public and thus shifting the borders between the two on demand was also observed by Tarlo in her study of women's dresses in a Gujarati (Hindu) village (Tarlo 1996: 160). This shifting of borders was performed by 'doing shame' (*laj* in Hindi), a concept also known in Bangladesh, which in Gujarat refers to pulling the end of the *sari*, which is worn in a different style than in Bangladesh, across the face.

Extension of the 'familiar' III: shifting levels of privacy

While the above examples have indicated how public spaces are characterised by familiarity or strangeness, there are also cases where private spaces undergo such transformations. To understand this, it is necessary to briefly discuss the multifunctionality of private spaces common in both Manikpara and Nasimgaon, where in most cases one family shares one room. This room then becomes multifunctional as it is used by the family members to sleep, eat, take a rest and sit. In this private space of the room the social gender relations are inscribed in the everyday rhythms. While the male head of the household and guests sit on the bed to eat, the women squat on the floor. During meal times, the private space then becomes separated into a male and guest space on the bed and a female space on the floor. The space of the compound in Nasimgaon or the corridors of the houses in Manikpara is semi-private, as it is generally accessible to all residents of a compound or house, but normally not to outsiders unless invited.

Especially the cinemas in the evenings (see Chapter 6.1.2) transform the private space of the family owning a TV set into a semi-private space of the compound, when other women join to watch, without necessarily being invited. In the compound where I lived there was only one room with a TV. Though this was actually the private room of my landlords, in the evenings many of the women of the compound gathered in front of the TV. This was not by invitation, but nonetheless the women entered the private room. What remained the only private space during these female (and children's) cinema sessions was my landlords' bed, which was hardly ever occupied by anyone but the family. The threshold to the room, however, no longer served to differentiate private and semi-private.

That the idea of privacy is rather different from any Western conceptualisation was also indicated by my landlady's behaviour towards me when I started renting a room from her. She felt that it would not be possible for me to sleep alone in my room, and thus without asking me shared the bed with me during the nights. When her daughter came back from her husband's village, the family of my landlords did not ask me whether she could now sleep at my place, it was simply taken for granted.

The bathrooms in Nasimgaon are often only separated from the courtyard space by a CI-sheet or a curtain of cloth. Thus there is no real privacy to take a bath. However, there is a social consensus that men would not move towards the bathroom area and the toilets (if those are next to each other, as is commonly the case) while women are taking their bath and vice versa. Due to this social consensus, women, who would normally always take a bath in full clothing, felt safe from unwanted glances and removed their clothing above the waist when bathing. Thus, although the space is understood as semi-private, the social norms create a more private space. In fact, the compound itself becomes a female space during the day, as men seldom spend much time there except for the meals or to take a rest. This was also reflected in the behaviour of omitting the *orna* during the day, while men are not in the compound, and putting it back on in the evening, when the men come back home (or as soon as men visit the compound for other reasons). Rizia also narrated about a time in the past when women had to wait for the men to leave the compound in order to be able to use the toilet – a situation which according to her had been solved when NGOs started to assist in the construction of toilets. When the compound where I lived was still open towards the lake, my landlord used to sit on the 'veranda' facing the lake and smoke cigarettes with his friends, but ever since the compound courtyard became a closed unit he no longer spends these times inside the compound but in his shop facing the roadside or on the nearby public space.

7.4 A MULTIPLICITY OF FACTORS PRODUCING AND RE-PRODUCING SPATIAL PRACTICES

The above discussion has pointed out how spatial practices are the outcome of social norms, especially concerning social gender relations, of the perception of the urban environment and of urban/rural identities. The discussion has indicated how the everyday life activities and mobility patterns of women are considerably shaped by the concepts of shame, honour and prestige depending on the socio-economic status of a household. Furthermore, there seems to be a correlation between the urban fabric and the nuances of publicness this brings about, and the range of female spatial practices. The internal homogeneity of Nasimgaon and the distinct border towards the surrounding presents a familiar space, especially for women's movement. At the same time, this has to be understood as a process of othering and exclusion from city-scale urban life. So the in-between space of the neighbourhood of Nasimgaon seems to present both a potential and a hindrance, depending on the perspectives of easing movements and claiming citizenship. On the other hand, Manikpara is characterised by easily fitting into the urban context, without a distinct border producing a more internally homogenous sphere. Here the 'strangers' publicness', at least for women, lingers at the doorstep of the house. Nonetheless, residents of Manikpara can be considered as being more integrated into the city context, as expressed in men's mobility patterns and recognition of the settlement. Finally, the discussion has pointed at the fluidity of the concept of multiple publicness and continuously shifting boundaries, which can change especially during extra-everyday events, but similarly based on time of day and seasonality.

CONCLUSION: THE SPATIALITY OF LIVELIHOODS

In this short chapter, I would like to summarise and bring to a point the main findings of Part II of the empirical research. The findings will be outlined by bullet points and then discussed in each subsequent paragraph.

- Livelihood activities always carry with them a spatial expression and experience, and the pursuance of livelihood activities in public space actively produces and re-produces spatial practices.

The analysis in Chapter 6 has uncovered a rich diversity of spatial practices carried out by the ordinary and elite groups. Public space in everyday life is not only the arena for economic livelihood activities, but also for recreation by engaging in leisure time activities, for carrying out reproductive activities or for expressing and living out religious and spiritual beliefs. These everyday activities are complemented by religious, cultural, political and educational extra-everyday events. The value of and demand for public space for livelihood activities is inherent in the intense use of the (scarce) existing spaces in the study settlements.

By understanding a person's spatial strategy and 'spatiality of livelihoods', much can also be understood about a person's embeddedness in social relations and institutions. Hence the 'spatiality of livelihoods' provides a useful extension of the as yet not explicitly spatial livelihoods framework. Furthermore, the rhythms of everyday and extra-everyday life produce and re-produce spatial practices by inscribing themselves in (physical) space in a repetitive manner. This underlines the understanding of space as a product of practices, of space as allowing a multiplicity of parallel experiences and of space as a temporal condition (Massey 2006; see Chapter 2.2.1). Breaking out of such practices becomes difficult and, as the next part of this research will demonstrate, challenging the dominant modes of the production of space is an ambitious endeavour.

- Spatial practices and mobility patterns are strongly influenced by the social norms of a society, resulting especially in manifold restrictions in using public space for women.

As analysed in Chapter 7.1, the persistence of traditional gender norms, manifested especially in the concepts of *porda* and honour, prestige and shame, especially influences the spatial practice of middle-class women in Manikpara. The women of socio-economically poor households participate in employment and economic activities, however the persistence of traditional gender ideologies is here manifested in the 'dream life' of a *porda*-oriented lifestyle. Rather than an empowerment, employment is seen as an economic necessity, while preferences indicate the presence of the very same gender norms that shape the life of Manikpara's middle class. In any case, women's movements tend to be observed by society, underlining the importance and usefulness of the notion of surveillance (Bose 1998) to understand female spatial practices. Acts of resistance to traditional gender ideologies tend not to be easily accepted but result in the spreading of rumours and a bad reputation. An exception, as will be discussed in the next part of this research, are women entering the political sphere. In consequence women's share in an active female production of public space is comparatively low.

- The Western model of public, semi-private and private space needs to be expanded to include a notion of the multiple hierarchies of publicness as found in this research. Depending on perspective, public space offers different spheres of 'familiar publicness' and 'strangers' publicness'.

The analysis in Chapter 7.2 revealed how nuances of publicness change between familiarity and strangeness/otherness. The urban fabric and integration into the city of Dhaka of the two study settlements suggest that the internal area of Manikpara is exposed to a 'strangers' publicness', while the internal area of Nasimgaon constitutes a 'familiar publicness'. From the perspective of integration into the city, however, Manikpara can be understood as integrated, while Nasimgaon remains excluded due to its distinct border to the surroundings but also with regard to full citizenship. These urban structures can be understood as being actively produced by urban planners and politicians as much as by the practices of urban inhabitants. The perceptions of being part of the city or separated from the city continue to be manifested in the identity of inhabitants, moving between urban and rural. Furthermore, the nuances of publicness differ when looking at male and female mobility patterns. In terms of women's perceptions of 'strangers' publicness' and 'familiar publicness', the choice of clothing has revealed the distinctions between different public spheres. Overall, the findings indicate the existence of multiple and partly competing public spheres, produced and re-produced by a diversity of actors and processes both external and internal to the settlements.

- The borders between 'familiar publicness' and 'strangers' publicness' are not fixed but fluid and can shift especially in the case of extra-everyday events, but similarly depending on changes of everyday life spatial practices based on time of day, and seasons.

The analysis in Chapter 7.3 has illustrated the shifting borders of familiar and strange. These shifting borders are most relevant for women, as they can either result in an increase of the socially accepted spatial activity range for women, or in a decrease of these spaces. The increases tend to be related to extra-everyday events or the electricity cuts related to seasonality. As such they cannot be understood as expressions of resistance challenging the everyday life spatial practices, but as mere exceptions 'granted' by society. The decrease of 'familiar publicness' at the most crowded times of the day has to be understood as a re-production of common spatial practices.

- The analysis has indicated public space being a contested resource both concerning its use in everyday and extra-everyday life and in the field of social norms.

The contested nature of public space has been explored, for example, in the discussion of the changes of the style of economic activities over time at Khalabazar in everyday life, and the organisation of the winter fair in Nasimgaon as an extra-everyday activity, revealing the conflicting interests in the use of public space. But contestations about the use of public space are also inherent in the specific gender-based rules applying to female spatial practices and mobility patterns. Here the contestations are between dominant practices and norms and those who trespass such norms eventually. The notion of contested space is discussed further in the following Part III that is concerned with the process-oriented negotiations of access to space, where contestations and conflicting interests play a major role.

PART III – THE NEGOTIATIONS OF ACCESS TO PUBLIC SPACE

In Part III of this research, I will analyse the negotiation processes of access to public space and how these contribute to the (domination/resistance-based modes of) production and conceptualisation of space in society. In the first chapter (8), I outline several narratives of the production and negotiation of public spaces in Manikpara and Nasimgaon. The subsequent chapter (1) then identifies the actors, their aims and motivations in making spatial claims and their potential sources of power. Based on these, I will analyse the strategies actors apply in the negotiations of access to public space with a view to reaching certain aims. I will then conclude with specific regard to the outcomes of negotiation processes and to dominant and resistance conceptualisations of public space negotiations.

Main characters appearing in Part III

This overview serves to introduce the 'main characters' appearing in the stories and analyses of this part of the empirical study. It serves as an orientation for the reader, while the complete list of interviews can be found on page 377.

Manikpara

Foyez	leasehold operator of Manikghat from July 2008 to August 2010, afterwards leaseholder, age ~35, married, 1 child
Jahangir	previous leaseholder of Manikghat, now operates a plastics recycling shop on the embankment slopes, age ~45, married, 3 children
Mesbah Uddin	plastics businessman of the area, well connected and respected in the locality, age 46, husband of Mashrufa, 2 children
Momena	together with her husband Rokib operates a plastics washing and drying business on the embankment slopes, takes work on commission from *mohajons*, age ~30, married, 2 children
Rokib	Momena's husband, see above, age ~55, 2 children
Tariq	young plastics businessman, together with his friends established the small garden at Dokkin Ghat, age ~25, married, 1 child

Nasimgaon

Afsana	operates a vegetable shop at Khalabazar (renting one unit), in November 2010 she shifted the location of her shop, female political leader of AL (involved at city level), age~35, married, 2 children
Azad	owner of a rickshaw garage in the south of the Eid Gah Math since BNP times, thus part of the conflict that evolved in March 2010, age ~35
Baharul	political leader of AL Sub-area Committee (overseeing Khalabazar), age ~60, married
Dipa	in 2009 operated a vegetable stall or tailoring unit (depending on weather conditions) at Khalabazar, had to leave the place in June 2009, struggled for a long time to settle at a different location, but eventually in 2010, with NGO support, started to successfully operate a tailoring shop and in November 2010 even joined AL politics in the neighbourhood, age ~37, divorced, 1 child
Hatem	previous inhabitant of Nasimgaon, political leader of AL (several levels, involved in most organisations within Nasimgaon, but also beyond), age ~50, married, 2 children
Mokbul Hossain	operates a construction materials shop (shop owner) and uses Nasimgaon Math for production activities, age ~45, married, 2 children
Rizia	previous inhabitant of Nasimgaon, female political leader of AL (especially *thana* level), ~40, married, 1 child
Rohima	operates a fruit shop and tea stall (renting two units) at Khalabazar together with her husband Shahin, age ~25, married, 3 children
Shahin	Rohima's husband, see above, age ~30, married, 3 children
Taslima	operates a *pitha* stall on Nasimgaon Math, age ~40, married, 4 children

8 NARRATIVES OF THE NEGOTIATIONS OF PUBLIC SPACE

As a basis for the following analysis, the stories of three public spaces in Manikpara and Nasimgaon which evolved as major arenas for the negotiations of access to space (see also Chapters 5.3.2, 5.4.2) are told in this chapter. These narratives are based on my own observations and several interviews and informal discussions which I use to cover the stories from different perspectives.

8.1 THE EMBANKMENT SLOPES IN MANIKPARA

Access to the public space of the embankment slopes (see Chapters 5.3.2 and 6) in Manikpara is subject to negotiations between various actors. The general setting and events are outlined below[1].

8.1.1 The statutory *ghat* leasehold and sequence of arrangements

The management of the respective *ghats* along the embankment is contracted to an individual through an annual tender procedure by the Bangladesh Inland Water Transport Authority (BIWTA). The leasehold contract in Manikpara includes a series of smaller *ghats*. The year of the leasehold lasts from 1st July to 30th June of the following year. According to the contract, the leaseholder has the right to collect charges from boats landing at the *ghats* as well as passenger charges for using the *ghats*. These charges are fixed in the regulations of the authority (see Table 5) and the leaseholder is not permitted to collect any other or higher charges nor can he set up new toll collection points that have not previously been part of the agreement. The contract further prohibits the leaseholder from making use of the slope land: "He [the leaseholder] is also not permitted to do any agricultural cultivation on the slope land of the water body. He has no rights on the slope land of the water body" (mutual agreement between the authority and the leaseholder). The duties and responsibility of the leaseholder by contract include the maintenance of the *ghats*, the reporting of any accidents to the authority and the provision of safe drinking water and sufficient lighting at the *ghats*.

If the leaseholder fails to adhere to the rules and regulations or does not fulfil the payment requirements in due time, BIWTA reserves the right to cancel the leasehold. For example, in April 2010 it was announced that the leasehold system of Sadarghat, the largest passenger river port of Bangladesh, would be terminated

1 Parts of this chapter have been published in disP (see Hackenbroch 2011).

in July 2010 because of mismanagement and harassment of passengers due to the regulations established by the leaseholder (The Daily Star 2010a).

non-motorised boat transfer	
passenger fee	0.5 Tk
under 2 years passenger fee	0.25 Tk
loading and unloading of goods/cargo	
general goods for each quintile or less	2 Tk
ducks, chicken baskets	2 Tk / basket
goats or sheep	2 Tk / each
domestic animals	4 Tk / each
other animals	5 Tk / each
bicycle	2 Tk / each
motorcycle/tempo/rickshaw	3 Tk / each
products which are measured in cubic metres	4 Tk / m^3

Table 5: Fees to be charged officially
(Source: BIWTA List of fees 2010; Hackenbroch 2011: 62)

In 2008/2009, Foyez operated the leasehold in place of his brother, who had won the tender by placing the highest bid of 1.1 million Tk (incl. VAT). The tender process for 2009/2010 was won by a person from another area for 1.3 million Tk (incl. VAT). This person started to operate Manikghat on 1st July 2009, however, as an outsider and unexperienced with the business, he faced strong opposition from the local community. After less than two weeks, the ongoing contestations made him realise that he could not operate the *ghats* by himself. Consequently he sub-contracted the operation of Manikghat to Foyez on the basis of a daily fee of 4,000 Tk (about 1.46 million Tk per year, mostly paid on a monthly basis) in order to recover his costs. All the additional revenue from the *ghat* was then at Foyez' disposal.

The bidding process for 2010/2011 did not take place and Manikghat was re-tendered to the same leaseholder for unknown reasons but in line with the statutory provision which reserves the right of re-tendering. Subsequently, Foyez and another person competed to win the leasehold operation from the leaseholder. By drawing on AL leaders' support, Foyez finally managed to win the sub-contraction from the leaseholder for the *ghat* operation. At the end of July the leaseholder, however, had still not received confirmation of the re-tender by BIWTA. The re-tender was furthermore contested by others, who offered to pay higher amounts to BIWTA to get the leasehold contract. It then turned out that the re-tendering was only valid for a month, and from August 2010 onwards a new leaseholder (two persons from Manikpara) was in place. At the beginning of August, Foyez was still not sure whether he would be able to keep operating the leasehold, as one of the two new leaseholders was on his side, while the other was thinking of giving the daily operations to someone else. Eventually Foyez got the sub-contract for 5,000 Tk per day. As Foyez preferred to be the leaseholder himself, he decided to 'play tricks', his own expression for the following story:

> "To run a *ghat* is a dangerous matter. This profession is full of problems. The police can demand money. Various leaders, *mastans* can demand money. I always called *Bhai* [brother, one of the leaseholders] if I faced any kind of problem, even in case of a small problem. I did it deliberately so that he becomes annoyed about the *ghat* matter. I knew he was busy with his politics. Moreover, there are some costs to maintain a *ghat*, like construction of staircases, lighting, repairing and others which should be carried by him. I called him almost every day to inform him about these problems. As a result he became annoyed. Then he called me to his place and told me 'I don't want to look after the *ghat* anymore. You do manage everything of the *ghat*. I remain busy all the time, you know. I get another 170,000 Tk from BIWTA [*Bhai* refers to the money that he had already paid to BIWTA for his leasehold duration, indicating that if he gave up the leasehold now he would claim this money back]. You can give me this amount and take the *ghat*.' After that I paid him 166,000 Tk instead of 170,000 Tk. I gave a bribe of 7,000 Tk to the chairman of BIWTA to change the name of the leaseholder officially in the BIWTA office. Now I am the leaseholder of the *ghat*. I have all the documents. You can see these documents in my house." (Foyez, field note, 25.11.2010)

Instead of consulting the BIWTA in the first place, Foyez consulted the AL Ward committee to negotiate his claim of becoming the registered leaseholder:

> "To get the *ghat* I also had to sit in meetings twice with AL leaders from the Ward Committee. I gave 10,000 Tk to the president and the secretary of AL Ward Committee so that they won't demand money from me anymore. I also gave another 10,000 Tk to two other leaders. […] I helped many AL leaders. I took part in the election campaign of the MP. That's why I am a leaseholder now." (Foyez, 25.11.2010)

8.1.2 The daily operation of the *ghats* and embankment slopes

Acting against the provisions of the statutory contract, the leaseholder extended his area of influence to include the embankment slopes, where especially the plastics washing and drying takes place (see Chapters 5.3.1, 6.1.1), and he also changed the charges to increase his income.

For Foyez, the daily charges collected from the businesses involved in the washing and drying of pieces of plastic were his major source of income. Rokib and Momena had to pay 4 or 5 Tk per bag of plastic processed on the slopes to the leaseholder. The same amount was due from any other business drying plastic on the slopes. Rokib passed this fee on to the *mohajons* who gave him work on commission. He thus did not feel that he was paying an extra amount to the leaseholder. Rokib occupied one particular part of the slopes almost every day. In his perception he had a 'right to this place' as he was the one who had cleaned this part of the slopes about eight years ago, when it was a garbage dump and the smell was unbearable. Nonetheless, he has experienced contestations of his spatial claim by other plastic drying parties. Until today, however, he has been able to maintain his position, something he attributed to the support of his powerful relatives.

The leaseholder did not charge for plastic drying on the embankment slopes in front of the Bagan. Foyez explained that he gave these collection responsibilities to the Banganbari Committee which collected 5 Tk per bag of plastic which was then used for the maintenance of the garden. According to Saifur, a committee member of the Bagan, this was the established practice, an unwritten agreement, between all leaseholders and the Bagan. Foyez referred to these concessions as voluntary and in June 2009 explained, conscious of his powerful status, how he would also be able to keep this money for himself if he wanted to. Concerning the *ghat* in front of the Bagan, however, he experienced some conflicts with one AL leader who also wanted to take over the leasehold. He solved these by offering the jurisdiction of that *ghat* to this leader:

> "There were some problems in the past. We sat to solve them. After that I asked him 'Which side do you want? You can take this side [Bagan]. Why do you want to take my position? You will get nothing if you don't agree with me'." (Foyez, 06.04.2010)

The example of the *ghat* in front of the Bagan indicates how Foyez secured his influence. In order to secure his power in the locality, Foyez established and maintained good contacts with central actors. This he did by providing 'entertainment', i.e. tea and cigarettes, topping-up mobile phones, giving presents during religious holidays or making concessions to well-established, powerful persons of the locality:

> "We don't take charges from some boats of well-known persons. I can understand from whom I should demand money and from whom I should not from long experience in this sector. As an example, we take 100 Tk instead of 180 Tk per boat if those materials belong to Uncle A. He is a big leader of Awami League. As a result the relationship among us is good." (Foyez, 23.01.2010)

In this way, Foyez gave a minimum of 20,000 Tk per week to the police, the mosque, local marriage ceremonies and local political leaders. Furthermore, Foyez also changed the charges collected at the *ghat* points (deviating from the charges set out by BIWTA, see Table 5), both with regard to who was charged and how much was being charged. Once during his leasehold operation in 2008/2009, he wanted to establish a system of collecting a fee from each passenger using the *ghats,* as provided for in the formal charges. However, the local inhabitants protested the introduction of the fee and succeeded:

> "Last time we thought to collect 50 paisa [1 Tk equals 100 paisa] per person for crossing the river. We also built toll-houses for that purpose. But we could not implement that because local people didn't want it and made problems. If I take any initiatives [for change] then I should not do such kind of initiative. This may cause lots of trouble." (Foyez, 23.01.2010)

He did not start such an initiative again until the end of 2010. When the statutory authority raised the total leasehold amount, increasing the daily lease payment by 100 Tk, the leaseholder considered it necessary to raise charges for boats and plastic drying activities. Because of his close relations with and payment of bribery to the BIWTA's officials, he achieved formalisation of the increased charges according to his wish.

On 8th August 2010 BIWTA carried out an eviction drive on the embankment slopes in Manikpara. Jahangir's shop (the place of the *adda* round, see Chapters 6.1.2 and 7.3) as well as the neighbouring shops were evicted without prior notice. Jahangir expressed his happiness that everyone had been evicted, as he did not appreciate his 'neighbours' who had constructed their rooms using political power, while he claimed to have a deed and was confident of re-establishing his shop on the slopes. However, he opened a new shop along Embankment Road and until November 2010 only installed a temporary shop on the embankment slopes for his social gatherings. Rokib and Momena were not evicted from the embankment slopes, nor did the drive cause any problems for Foyez.

8.1.3 Tariq's garden and the temporary mosque

A scene from everyday life: In the afternoon before sunset, men come to a temporary mosque erected on the embankment slopes at Dokkin Ghat for their prayers. They use the makeshift washing place for the ritual washing. Opposite the temporary mosque, the reconstruction and improvement of the main mosque is in progress, which is why a temporary one has been constructed. Adjacent to the temporary mosque, a group of young businessmen had planted trees on the embankment slopes and now they gather in their small garden to enjoy the breeze coming from the river around the time of sunset.

Tariq's garden at Dokkin Ghat (see Chapters 5.3.2 and 6) was cut off from its previously easy access to the Ghat by the construction of this temporary mosque in April 2009 (see Figure 17, the first sketch displays the situation before the construction of the temporary mosque, the second sketch shows the situation after the temporary mosque was established). From then on, access was only possible from the direction of the more distant *ghat* north of the garden, crossing piles of rotting garbage on the way. Despite this decreased accessibility due to the mosque, Tariq and his friends continued to take care of the garden and sequentially enlarged it by extending the earthen layer and fixing it towards the river with wooden logs. They had established the garden by collecting money among themselves. Knowing that it was government property, they justified their project by referring to the contribution it made to public welfare by increasing the quality of life in this place that used to be a garbage dump. Nonetheless, they experienced contestation by some local elders who started to spread the word about unsocial activities going on in the garden. To restore their reputation, Tariq and his friends gathered other friends and went to the MP. Thereafter, the local elders and members of the mosque committee discontinued their protest of the garden activities because quarrelling with apparently 'young boys' would damage their social reputation in society. At the same time, Tariq and his friends stopped extending the garden to avoid new problems and decided to wait until the temporary mosque was removed. They stated about their future plans:

> “If the mosque is removed then we will get at least 70–80 *hat* [feet] of land. We will open a nursery within this place. That will be very beautiful. […] This will be a nursery type garden. People will come to enjoy. Seedlings will be produced here. You can sit here in the rainy season when the river will get its life.” (friends of Tariq, 19.04.2010)

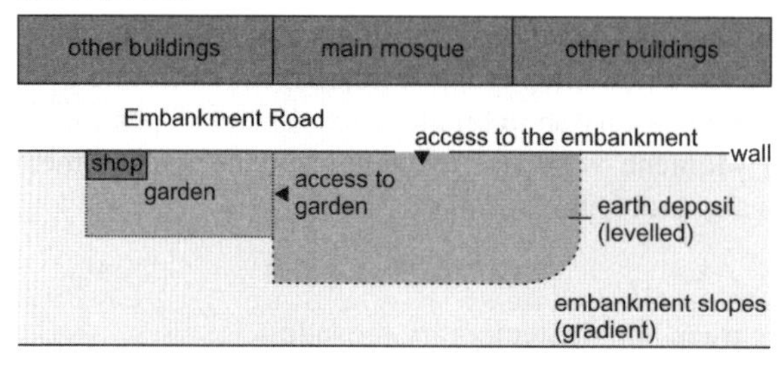

from April 2009 August 2010

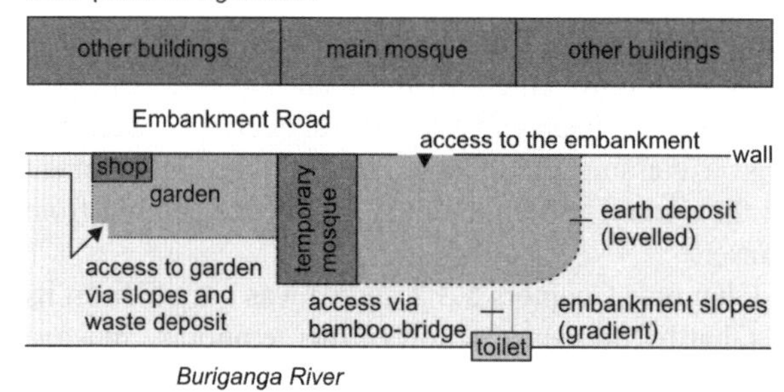

after August 2010

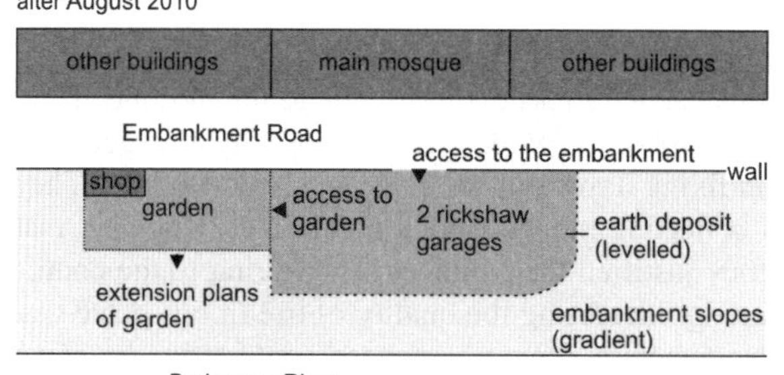

Figure 17: The developments at Dokkin Ghat (Source: Hackenbroch 2011: 65)

By June 2010 the mosque reconstruction had been completed. However, the temporary mosque was not removed before the end of August. During the BIWTA eviction drive (see above) at the beginning of August, the mosque committee members managed to prevent the demolition of the temporary mosque by promising to remove the structure the next day. In contrast, a CI-sheet structure in Tariq’s garden which he and his friends had only recently constructed to sort and store plastic was removed during the BIWTA eviction drive. When the mosque committee decided soon after the eviction drive to remove the temporary mosque from Dokkin Ghat, Tariq and his friends hoped that now they could finally enlarge the garden. But the Mosque Committee decided to rent out the place to two rickshaw garages in October 2010 (see Figure 17, sketch 3, for the situation at Dokkin Ghat after removal of the temporary mosque). Each garage owner paid 2,000 Tk per month to the mosque and used the place for 20 rickshaws, rented out for 100 Tk per day.

8.2 KHALABAZAR IN NASIMGAON

In the following the story of the Khalabazar will be outlined chronologically, reporting the main events (for a general overview of the spatial situation, also see Chapter 5.4.2). The spatial changes that occurred over the course of this research are graphically presented in Figure 18. At the end of this sub-chapter, I will briefly outline the story of contestations at the adjacent Shaonbazar.

Before the election of the Awami League government in December 2008, the Khalabazar had been accessible for all vendors based on mutual understanding. The only regulation applicable to everyone was the daily payment of 5 or 8 Tk[2] for the sweepers and night guards. Among the vendors at that time were Rohima and Shahin, who sold seasonal fruits from the common *chouki*, a wooden bed used as a vending table. Shahin had been the one who had taken the initiative to fill the previously low-lying waste dump so that it became an open space where vendors were able to sit (see Photo A-3).

Very soon after the new government resumed office, however, access to the market became contested. In January/February 2009 the local leaders of AL evicted the vendors from the market place by telling them it should be maintained as an open field. Instead, after some days they established *paka* selling platforms (see Photo A-4) without consulting the original users, i.e. the established vendors – first at the location of the fish market on the north-eastern corner, then on the western part of the main field. The *bhits* created in this way were distributed among the leaders, who registered their rights of possession in the local AL office adjacent to the market. The vendors now had to pay an initial deposit of 2,000 Tk and a monthly rent of 1,000 Tk for the place that they previously accessed free of monetary cost except for the contribution to the sweepers and night guards. For many vendors the new rent payment required a significant proportion of their monthly household income, between 15 and 25%[3].

2 Different interviewees reported different amounts – thus the range of 5 or 8 Tk. This may either be because parts of the area were served by different night guards and sweepers who charged different fees, or because the exact amounts of payments made two years back could not be remembered. Whether the amount was 5 or 8 Tk does, however, not change the general outcome for this research.

3 An income of 4,000 to 6,000 Tk was common among the vendors in 2009. Of this, 1,000 Tk were usually spent on house rent, plus an additional 300–400 Tk for water and electricity. The

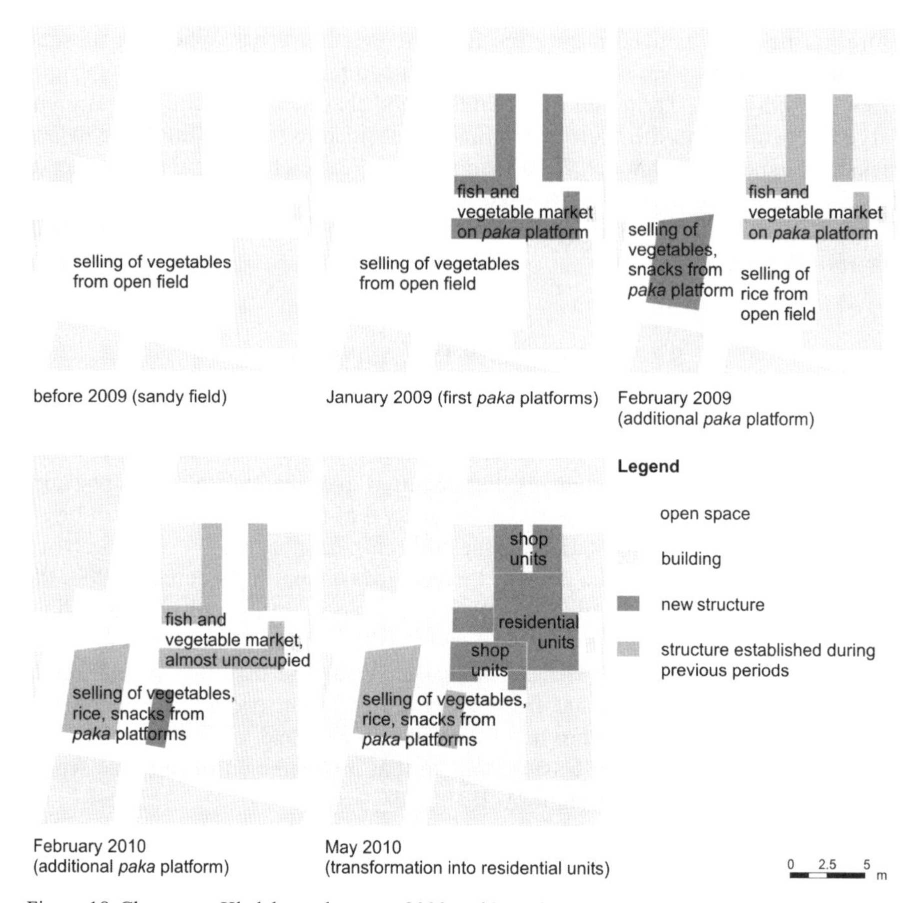

Figure 18: Changes at Khalabazar between 2009 and May 2010

At that time, two women contested the leader's claim by protesting against the establishment of the *paka* platforms. Independently of one another, they requested the leaders to spare their unit as they would prefer to continue selling from a wooden bed. Neither women succeeded in their resistance to the establishment of the platforms, but the outcomes of their contestation of the local leaders differed tremendously. Dipa, a divorced woman who had been living in the area on her own for years, argued with the leaders, but eventually had to accept their 'dictate':

> "I used to do my business over there for seven to eight years. It was a low place before and I repaired it by spending money and did my business. So I have a right. Now, when I was asked to remove the shop to make the place *paka*, I told them to do it except for my place and I would make my place *paka* myself. Then they [local leaders] said 'No, we will make it *paka*. Later you will give us the money for the expenses.'" (Dipa, 25.04.2009)

remaining money was spent on children's education, repayment of loans and daily household consumption. Thus an extra 1,000 Tk per month to rent a *bhit* constituted a considerable challenge for many households.

Consequently, she started to pay the rent and the deposit, however, her business turnover was not enough to sustain her increased monthly expenditures. Although she tried to negotiate with the leaders for the postponed payment of the outstanding amounts, she finally had to leave Khalabazar to establish her shop elsewhere. Afsana, who due to her political affiliation with AL had become a registered owner of one *bhit*, on the other hand, was absent when the local leaders finally constructed the platform. She got angry when she realised that they had included her unit despite her objection. She thus protested and her determination not to pay the deposit becomes obvious from the following statement about her conflict with the local leader Hatem:

> "After coming back, I found that all [*bhits*] were made *paka*. Now Hatem's sleep has gone after seeing me and he was trying to find a way to get me to give 2,000 Tk in the name of the *bhit*. Now my word is that I will not give even 2 Tk to him instead of 2,000 Tk!" (Afsana, 23.03.2010)

After her refusal to pay, Hatem decided to inform the secretary of the AL Ward office, the lowest administrative level of local government. Afsana, however, reacted by consulting the Ward chairman and after another round of back and forth the chairman 'rebuked' Hatem for intended deception while Afsana did not have to pay the money.

Besides the struggle of the vendors to maintain their positions at Khalabazar, the management responsibility of the local leaders had also been negotiated between different local organisations, all backed by AL party support, at the beginning of 2009. The organisational landscape included the Amlabazar Committee whose area of jurisdiction was the adjacent Amlabazar. The Amlabazar Committee, having been established a long time ago, used to be dominated by the members of the ruling political party – thus in the beginning of 2009 AL supporters took over the lead of this Committee. At the same time, the Sub-area Committee was formed as a sub-Committee of the administrative Ward level with its members being AL-supporters. The office of this Sub-area Committee – also the office of the Local Police Committee headed by AL leaders – was the Awami League Club that faces Khalabazar. The different committees had a meeting and it was decided that the Sub-area Committee (which if the matter had been decided by hierarchies would have supervised the Amlabazar and Khalabazar areas) should have jurisdiction over Khalabazar in order to generate income for their activities. A leader involved in the Amlabazar Committee understood this negotiation as a concession made by the Amlabazar Committee which he perceived to be more socially accepted and to have a better reputation:

> "The Sub-area Committee informed the Ward Committee and the *Thana* Committee and they came to sit all together in a meeting. They [Ward Committee] told us [Amlabazar Committee] 'What should we do? As they [Sub-area Committee] are running a club now, they need some money for that.' We told 'What can we do?' Then they [Ward Committee] told us 'You can leave this space [Khalabazar] for the AL club.' After that we left that space." (Sayed, 23.04.2010)

On the contrary, the perception of the Sub-area Committee was that the jurisdiction was given by the Sub-area Committee to the Amlabazar Committee.

After the initial conflict when the leaders constructed the *paka* platforms, the physical transformations of Khalabazar and the changes of access rules, i.e. rent and deposit payments, continued. In June 2009, the market committee planned to construct more permanent market stalls. For this, the deposit was raised to 5,000 Tk. Rohima and Shahin protested the increase, but their complaints did not have any effect, rather the committee members told them to either pay the money or leave the shop:

> "Total advance money now is 5,000 Tk. Then we told them [the local leaders] about our present situation, that we have faced a huge loss because fruits of about 10,000 Tk perished. Then they told 'We don't want to hear anything. You have to give the money or otherwise you'll leave the shop.' Then we said 'We are poor people. How will we survive if we leave the shop? Now you have taken possession of the land since Awami League came into power. Before this we did not give any rent. Now we are giving the rent as you have asked. If you keep talking like this, then it is not fair.' […] Then they said 'Whatever you say, you have to give the money'." (Rohima, 18.01.2010)

After paying rent for more than a year, two vendors tried to buy the *bhits* they occupied from the owners in order to decrease their vulnerability to sudden changes. The amounts the vendors planned to invest – 20–27,000 Tk – represented a considerable amount of their yearly income and they would have only been able to manage this by taking out loans[4].

Confronted with the permanent insecurity of other shop owners and the fate of those who had lost access to Khalabazar in the process of elite groups' claim-making, Afsana eventually decided to try to re-establish the 'old system'. Through her political power, she planned to 'evict' those people who had claimed the ownership of the space and wanted to return the 'right to space' to those who had been there for years:

> "Now Shoma's husband who has one hand only, used to operate the shop along with me. My husband has some problem in his leg; but I had power and so I was able to take the shop forcefully. He [Shoma's husband] was not able to operate the shop using power like me. He gave 5,000 Tk to Hatem in advance and operates the shop by paying 1,200 Tk as rent. […] He is operating the shop by paying rent because he did not have the ability. Then isn't a person like him supposed to get a shop in the market? If he operates the shop by paying rent then what will happen if all these shops do not exist? Then everyone should be able to do their business like before. My chairman and secretary [of AL Dhaka *Mohanogor* Committee] will help us for this. And they told me 'Collect two, four or five people like you and call a meeting and then call us. We will build a hawker's market there.'" (Afsana, 23.03.2010)

Afsana had planned this initiative for after the Ward Commissioner's elections had been held. However, the elections had not been held by the end of 2010 and thus nothing happened, despite her repeated narration of her plans.

4 The above description of events focused on the vendors Afsana, Dipa and Rohima/Shahin. Other vendors not included in these opening stories, but also interviewed in the course of this research, include Shoma (vegetable stall together with her disabled husband), Rehana (rice stall together with her sick husband) and Ramisa (selling rice from the ground).

Instead, in May 2010, one part of Khalabazar where previously a fish market had been located was transformed into residential units and only four shops remained. The fish vendors had long left the place because they were not able to afford the rent payments for the *bhits*, and since summer 2009 the space had been mostly unoccupied. One of the local AL leaders involved in the AL Sub-area Committee and owner of some *bhits* at both the fish market and another part of the market bought most of the *bhits* and constructed the residential compound. From 1st June, he let four rooms for 1,200 Tk per month, including electricity and water. Four *bhits* remained in other political leaders' possession and were let as shops. One and a half years after the AL had come into power, the public space of Khalabazar had thus changed considerably – the previously open area had decreased tremendously and most of the space was under the possession of political leaders who over time transformed it into built-up land.

The adjacent Shaonbazar (west of the map above) was not so much subject to contestations between the ordinary people and political actors. Instead, it was a scene of conflict among political actors. The banana shops had been established before the AL came into power, and the shops belonged to the banana sellers individually – all of them male, and most of them of old age. During the time of this research they never faced problems in their activities – but they already occupied permanent units and thus their spatial practices could not be contested as easily as those of the mobile vendors in Khalabazar. Except for the banana stalls, the Kolabazar was an empty space, not occupied by any mobile vendors. The management of the area, i.e. the organisation of night guards, sweepers and electricity, was the responsibility of the Amlabazar Committee of the AL and not the Sub-area Committee at Khalabazar. However, in 2009 contestations evolved between these two political actors. According to Sayed, the following incident happened:

> "The Sub-area Committee wanted to take possession of the Shaonbazar. I saved that place. A case was filed in the *thana* against me and I am the second most wanted person. On that day, about 2–3 persons were injured severely. They [Sub-area Committee people] put a signboard in the middle of Shaonbazar and wanted to construct a club for Jatiyo Party [coalition partner of AL] people. At that time I said 'There will be no Jatiyo Party in this place.' No club should be located in the middle of the bazar. [...] I will give them a position to construct the club beside the road or within the slum. But it is not possible to give them a position in the middle of the bazar. If we allow them to construct their club room then the businesses will be hampered. I told them 'No' but they wanted to construct the club room forcefully. Then we, [my friend] and me, went to the bazar and uprooted the signboard. They hit me. After that I hurt them severely. Then a case was filed. [MP] *Saheb* [honourable address] himself support the case filing. There was a misunderstanding that they told to [MP] *Saheb* 'BNP people came and demolished our club.' But there was no club room and there will never be a club in the bazar. After that we went to the higher level leaders. We said to the former Ward Commissioner [AL candidate for elections] 'The case has been filed against us and we have broken the signboard. There was no club at that place. They tried to build a club room but we stopped them.'" (Sayed, 23.04.2010)

Although Sayed and others tried to get their cases withdrawn at the *thana* police, they could not succeed. However, this contestation between party leaders of different committees did not affect the everyday life spatial practices of the banana vendors. But despite the protest about the JP club construction in 2009 and Sayed's outrage that someone wanted to take possession of the land, the Shaonbazar transformed considerably: two generators and three more banana stalls were constructed, reducing the open space by 50%.

8.3 NASIMGAON EID GAH MATH

Based on the brief history of the Nasimgaon Eid Gah Math (see Chapter 5.4.2), the more recent developments are outlined below in a mostly chronological sequence of events.

8.3.1 Conflicts, improvements and plans for the Eid Gah Math in 2009

According to various interviewees, the open space of the Eid Gah Math used to be larger in the past. During the previous AL government period (1996–2001), a market had been established in the west, reducing the size of the Math, although in 2009 that area was no longer a market but was occupied by rickshaw garages and housing compounds. During BNP times, a low-lying portion in the south used as a waste dump had been filled and three shops and six rickshaw garages were established (see below). After the new AL government came to power in 2009, both the Sub-area Committee and the Mosque Committee supervising the Eid Gah Math were taken over by AL leaders.

I witnessed the first contestation concerning the usage of the Eid Gah Math for livelihood activities on Friday, 13th March 2009. During observation of the Math at 1:40pm, 20 active vending and production units used the public space. Another 60 uses included storage, parking of rickshaws and extensions of business premises. Only three hours later, the activities on the Math had come to a halt. Almost all the vendors and producers had left the place and most of their vending stalls had been removed.

The planned removal of the shops had been announced, but the local party cadres executed the drive two days earlier than the ultimatum. The justification given for the clearance was to clean the Math for children to play on. Mokbul Hossain protested against the clearance because his business of producing concrete materials required the use of open space in front of his shop. During the protest, some of the concrete materials under production were broken and he was injured. Mokbul Hossain became a target of the eviction activities because he did not comply with the 'rules of the place' established by the mosque committee, and because he also criticised their way of executing power:

> "They [the committee] ask for money from the people who run shops on the Math. I don't give any money and that's why they give me a lot of trouble. They already broke my shop and materials three times, and once they also broke my head while I was protesting. [...] I am working and earning my money and they should also work for their money" (M. Hossain, 05.05.2009).

In May 2009 a conflict about a water tank made the vendors and production units fear eviction again. One person had started to construct a water tank to sell water on the eastern side of the Math. This immediately mobilised some young AL supporters who protested the tank construction and said that they would take possession of the Math and start their own businesses of letting rooms. To demonstrate their seriousness, they subdivided parts of the Math by installing bamboo poles and ropes (see Photo A-52). Mokbul Hossain, for example, immediately reacted and moved the materials and products he usually stored outside to a safe location, fearing that following the conflict all structures and materials would be removed from the Math. But the conflict did not escalate this time, as the water tank construction was stopped and the subdivisions of the Math were removed accordingly. Until 2010 the water tank issue remained suspended (see below for its continuation).

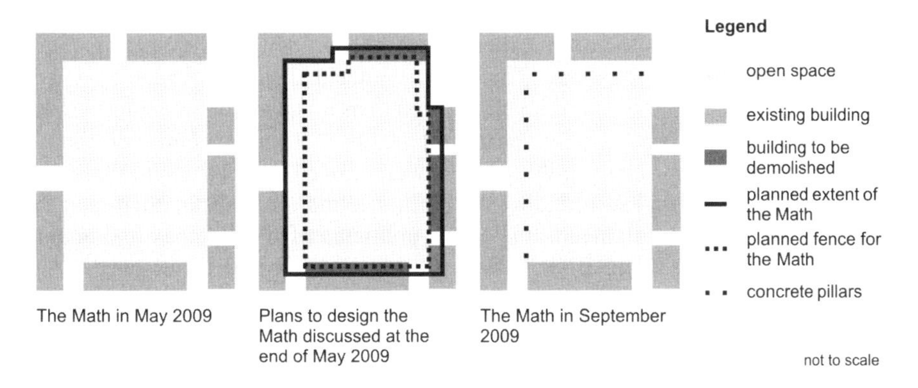

Figure 19: Plans to fence Nasimgaon Math and implementation

From the middle of May 2009, a rumour about eviction and reconstruction of the Math started to spread. The plan of the committees included aligning the Math, designing an eight foot road all around and fencing off the inner part with concrete pillars and barbed wire (see Figure 19, central map). For this purpose a number of shops, including that of Mokbul Hossain as well as the rickshaw garage constructed during BNP times, were targeted for removal. For some time it was very uncertain whether Mokbul Hossain would be able to continue his business, but finally some shop owners made a joint complaint to a high level leader who decided against the eviction. Once more Mokbul Hossain was able to continue his business, although the fencing plan provided yet another threat for it would mean the loss of his outdoor production space. The continuous chain of contestations had also increased the alertness of Mokbul Hossain to the possibility of any evolving conflict:

> "In the meantime police came here two times […]. I thought they came for eviction and so at that time I was very busy moving my materials from the Math" (M. Hossain, 25.07.2009).

In July 2009 the Math's surface was levelled with a cover of sand provided by the MP as part of an election promise to make improvements (see Chapter 6.3). In September 2009 concrete pillars were put up along the road west and north of the Nasimgaon Math (see Photo A-53 and the right-hand map in Figure 19). Afterwards, however, no further fencing initiative was ever started, and the pillars started to deteriorate and fall down.

8.3.2 The conflict about rickshaw garages and school construction

At the end of the BNP period, six rickshaw garages established on the previously low lying waste land in the south of the Math were sold to the operators. The garage owner Azad, for example, paid 25,000 Tk for his garage space. The buyers of the garages still operated the garages at the time of this research.

The first signs of a conflict evolved in October 2009. On 7th October, one participant of solicited photography called me to tell me that the rickshaw garages would be evicted on the next day. This had been decided and confirmed by a Jatiyo Party (JP) leader of the Ward level, referred to as a *mastan* by one interviewee, who also told the garage owners not to protest. On 8th October, however, the removal did not happen. Instead, the garages were moved back by about three feet to make space for a road surrounding the Math. I found out, only in 2010, that the garage owners had complained to the MP (a politician of JP) that they had bought the garages from the same BNP leaders who had appropriated the space. Thus they claimed that evicting them would not be fair and would destroy their livelihoods. Accordingly, the MP started to support the rickshaw garage owners.

At the beginning of March 2010 the conflict about the rickshaw garages escalated. After the conflict in October the MP had agreed to support the garage owners and provide them with new and more stable structures. In February the reconstruction of the rickshaw garages with a *paka* floor and CI-sheet walls had started (see 2nd sequence of Figure 20). At the same time, the leaders of the AL Sub-area Committee had demarcated the portion of the Eid Gah Math north of the rickshaw garage with bamboo fencing and expressed their wish to construct a school, a hospital and a community centre (see Photo A-54). On 7th March, the AL leaders of the Sub-area Committee blocked access to the garages by putting up a bamboo fence (see 3rd sequence of Figure 20, and Photo A-55) with a signboard:

> "By the name of the almighty Allah – Date: 07.03.2010 […] – This place is selected to build a primary health care and education service centre for the children of slum dwellers."

In the following night some of the rickshaws which could not access their garages were parked within the neighbouring settlement. During the night some of the rickshaws were vandalised and, accordingly, the next day one of the rickshaw garage owners filed a case against the AL leaders.

On 8th March I witnessed part of the conflict solving procedure by an AL leader of the Ward level who had come to Nasimgaon. He first expressed his and the AL leaders' anger that the rickshaw garage owners, who could be considered relatively well-off within Nasimgaon, were supported by the MP instead of the poorer households. He gave his account of the potential solution and expressed his anger about what had happened as follows:

> "You can enter with your rickshaw of course, but not like this, this road will not be used for the garages. Find another road! [The MP] provided the garages and he should provide the road in front of the garages anyhow, but not by using the space of the Math. That's why we put the bamboos! We're not against the general people, the rickshaw garage owners, we're just forcing to create another road in front of their garages. To do so, the community police office would need to be partly demolished. So we are negotiating in this way that these three shops leave one foot each and we can also give another foot from the Math, I mean there should be a rearranging of the shops to create the road. But the garage owners misunderstood us. They filed cases against us, against twelve people; I'm also one of them. They filed the cases yesterday. But I was not against them, so why did they file a case against us with my name in first position. That's why the AL people were angry and asked to remove all the rickshaws from the Math. We said 'You have garages, put your rickshaws into the garages by using your own road [the one to be constructed], you are not supposed to keep the rickshaws on the Math'." (field note of the Ward level AL leader's speech, 08.03.2010)

Afterwards some of the garage owners came to see him and they discussed the matter. They clarified that not all of them had filed cases and that they had wanted to talk to the AL leaders previously, but that at this time the police were vandalising their rickshaws. Then the AL leader suggested that the garages should receive their own access road which should no longer cross the Math. Instead it was decided to reduce the width of two shops and the office of the community police (all established during BNP times, and the way the AL leaders talked about the person who had set up the community police office indicates that this was also an act of revenge, further detailed in the analysis in the following chapters) by approximately one foot each in order to make way for an access road three feet wide. This was agreed upon by all conflicting parties, and a few days later the new access road had been constructed (see 4th sequence in Figure 20, Photo A-56).

The rickshaw garage owners were satisfied that they were able to keep their garages, and to receive the improved structure from the MP. However, the space they now had to store rickshaws was reduced and while previously they were able to store about 30 rickshaws, they now could store only 18 rickshaws. Nonetheless, they appreciated the outcome of the negotiation process.

As had been anticipated by critical observers, the school and hospital for which the AL leaders had started the conflict with the MP and the rickshaw garage owners were never built. Instead the bamboo fence deteriorated and the space was used for various temporary production activities and by playing children. From August 2010 onwards, the fenced space was occupied by new rickshaw garages under the supervision of AL leaders. By November 2010 five new rickshaw garages had been constructed with two roofs held by bamboo pillars within their garage space. Adjacent to the garages, a new tea stall of a similar roof-structure was opened (see Photo A-57, see the 5th sequence in Figure 20). A five to six foot corridor was left for access between the new rickshaw garages and the sheds constructed by the MP. Each of the new garage owners now paid 2,500 Tk per month to the AL Sub-area office.

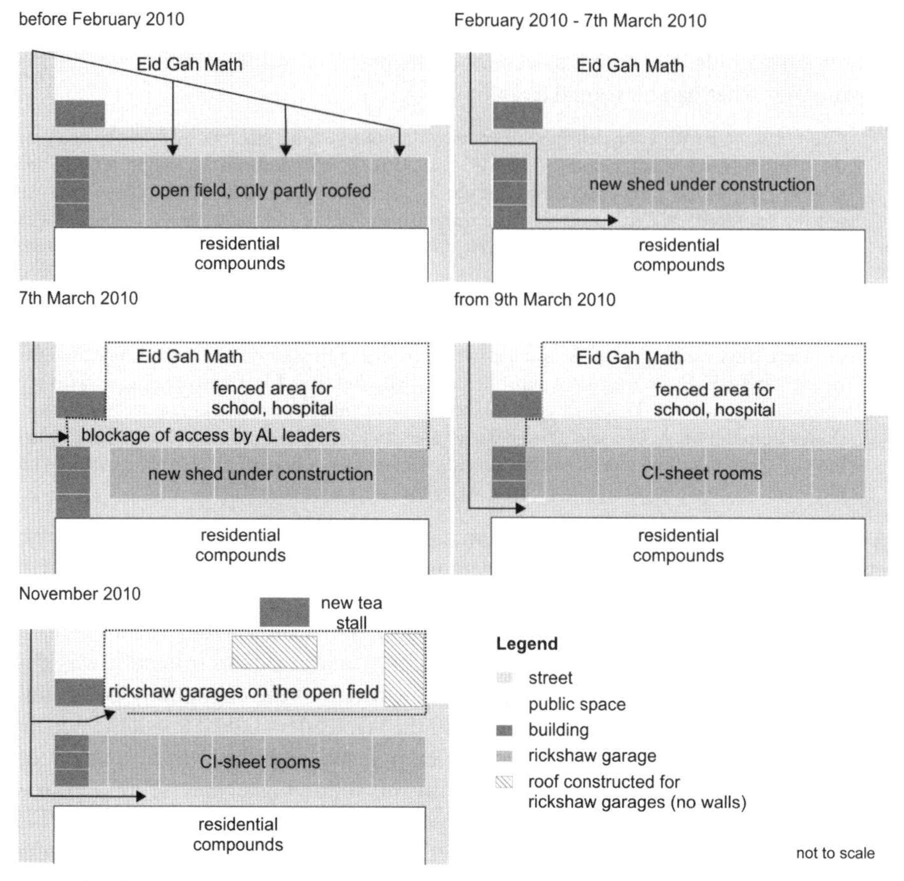

Figure 20: The conflict about the rickshaw garages

8.3.3 Claiming space on the Eid Gah Math in 2010

Besides the conflict about the rickshaw garages, three other attempts to take possession of parts of the Eid Gah Math captured my attention.

After the conflict surrounding the water tank in the east of the Eid Gah Math (see Chapter 8.3.1), construction activities did not resume for some months. In February 2010, a CI-sheet room was constructed next to the unused water tank by members of the AL sub-organisation *Shechhashebok* League (Volunteers' League). At the same time the tank construction was continued by people from the *Sromik* League (Labourers' League, another AL sub-organisation). Harez, one of the local AL leaders, said that this was 'personal work' and not permitted by the committee which wanted the Math to remain without encroachments. Accordingly he stated about the continuation of the tank and AL *Shechhashebok* office construction:

> "Those who are building these things are undisciplined people. They are not our [AL, Sub-area Committee] active people. They use our name to run [benefit] themselves. They are not our people and not anyone of our committee. They are undisciplined. Now if we want to tell something to them, then they will create different problems in the slum. Since already a moral decision has been made about the cleaning up of the Math, it will be executed with the help of the police. If we want to protest directly, then problems will occur because they are a bad type of people. There are lots of problems. I am talking with you; but if we call them they will not talk. They will talk such ambiguous words that you will feel bad. That's why these people are bad. Their task is to spoil a good thing. They think about how to spoil a nice work." (Harez, 16.03.2010)

Despite Harez' statement, the developments were never contested and the tank was finally transformed into a small *paka* room in August 2010, and in the beginning was used by AL activists. By November 2010 it had been transformed into a room used by the followers of a *pir* from Faripur.

South of here, Taslima operated a *pitha* stall in the mornings and afternoons/evenings. For this she used to put three clay stoves, a bamboo-supported polythene roof and a bench for customers onto the Math (see upper map of Figure 21). According to her, she was always supported by local leaders in setting up and operating her business because they understood her poverty and the family's hardship with a husband not contributing regularly to the household income. At one point, presumably at the beginning of 2010, she decided to raise the level of her *pitha* stall by putting debris on the ground, with a view to improving her selling unit. However, some AL leaders immediately told her to remove this so that she would not create an example for others, especially with reference to one person who had started to put debris down as well:

> "He [another person] threw 50 bags of debris over there. Then all people came and told me 'You are a poor person. You are living through this shop. If you throw debris on your place then other people will throw debris on the Math following you. In this way, the Math will be getting smaller. Now do some work. Remove some of the debris from your shop.' Then that person [who had put 50 bags of debris] came to them. But they did not allow him to open the shop on the Math. They told him 'You can't open the shop on this Math. She has been running this shop for five years. She will run this shop but you can't open a new shop.' They didn't allow him to open the shop even though he threw debris on the Math." (Taslima, 18.04.2010)

In February 2010 a sign was put up between her sitting place and the nearby tea stall advertising for *Sromik* League, and soon this space was covered by a CI sheet roof on bamboo pillars. In the beginning this was beneficial for Taslima, as the young men from the League became her regular customers. In May 2010, however, the sign was replaced by a new room, covering the space in between the tea stall and her shop, as an extension of the tea stall (see lower map of Figure 21). Taslima, however, assumed that this room would be used for political purposes. She also expected her business to be hampered because of the new room, as it left less space for her and her customers. Subsequently, she changed the orientation of her shop to face the south (she used to sell to the east, but the new room was constructed there).

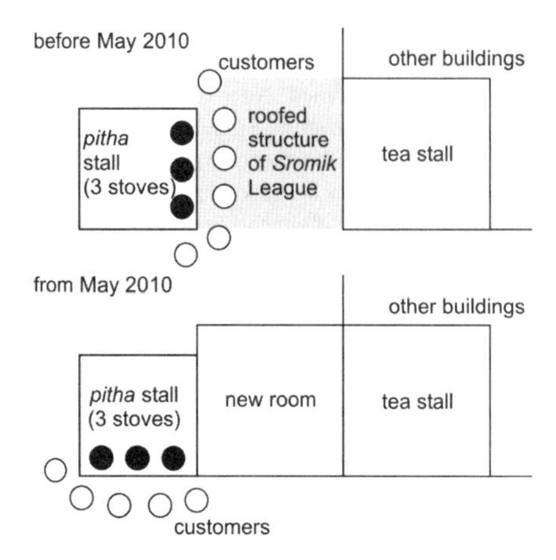

Figure 21: Taslima's *pitha* stall

While the above spatial claims happened over a long period of time in incremental steps, another encroachment took place 'over night'. On 27th April 2010 the brick foundations for the construction of six shops in the north-west corner of the Math were laid. Among those who made this spatial claim, and were present on 27th April, were the AL leader who had contributed to solving the conflict of the rickshaw garages on 8th March and some other AL leaders. However, the construction, uncontested on that day in April, did not continue due to a conflict among AL leaders about the distribution of the rooms. Besides some AL leaders, the freedom fighters' office also wanted a room to support their activities. Due to the conflicting situation, high level leaders suggested stopping the construction. By the end of 2010 the construction had not been continued and the space with the foundation walls was used as a storage space and for parking rickshaws.

9 EXPLORING THE ELEMENTS OF THE NEGOTIATIONS OF ACCESS TO PUBLIC SPACE

In this chapter, I seek to explore the negotiation processes of access to public space with particular reference to the narratives outlined above. In the first sub-chapter (9.1), I will identify the actors, the reasons behind claiming access to and jurisdiction of public space (spatial claims) and the power sources actors draw upon. Sub-chapter 9.2 will then discuss in detail the strategies with which actors preemptively secure spatial claims, react to contestation of their spatial claims, offensively contest the claims of others and resist spatial claim-making and contestations in what can be referred to as more 'subtle modes'. Based on these two sub-chapters, the third section (9.3) will summarise the different legitimations actors draw upon in negotiation processes. The subsequent sub-chapter 9.4 then discusses the outcomes of the negotiation processes and claim-making strategies with regard to access arrangements to public space and the outcome for the condition and availability of public space within the urban fabric. Finally, sub-chapter 9.5 discusses the conceptualisations of public space inherent in the negotiation processes and the specific modes of the production of space, while referring the results back to the theoretical departures concerning contested space and the negotiations of space outlined in Chapter 2.2. The sub-chapters 9.1–9.4 are based on the emerging categories of the empirical analysis and will be related back to the most relevant literature in 9.5.

9.1 ACTORS, CLAIMS AND POWER SOURCES IN NEGOTIATION PROCESSES

This sub-chapter, based on the stories above and the side-stories that could not be narrated in detail, will identify the relevant actors and their motivations in making spatial claims. Subsequently the sources of (dominating/resisting) power on which actors can draw in negotiation processes will be discussed.

9.1.1 Differentiating the actors in negotiation processes

This short sub-chapter aims to differentiate the actors in the negotiation processes of access to space appearing in the above stories.

The statutory actors playing a role in the stories outlined above include the Ward Commissioners on the lowest tier of local government and the Members of Parliament (MPs) as elected representatives at the national level. In the case of the embankment slopes in Manikpara, there is the BIWTA as an administrative statutory organisation and accordingly its officers. Furthermore, the *thana* police and the special police force RAB (Rapid Action Battalion) are statutory actors of the executive. The leaseholder of Manikghat is not a statutory actor, but via his contract with the BIWTA he has a statutory mandate to manage the *ghats*. Despite this mandate he remains a 'private citizen'.

All the other actors can be considered non-statutory and 'private citizens'. However, given the above stories it seems important to differentiate those involved in decision-making via political and other management committees from the ordinary everyday users of space.

The ordinary everyday users are primarily those accessing public spaces in pursuit of their economic livelihood activities (see Chapter 6.1.1), the customers of such businesses (however, their perspective is not the focus of this research) and finally those seeking recreation and pursuing social activities in public spaces (see Chapter 6). The first group comprises the vendors operating stalls in public spaces, the producers of goods using public spaces for their activities and also those using public spaces to store goods. For example, this includes Rokib and Momena operating the plastic drying business on the embankment slopes, Rohima and Shahin operating their fruit stall and later tea stall at Khalabazar, Afsana operating the vegetable stall at Khalabazar and Mokbul Hossain using the public space of Nasimgaon Eid Gah Math to store his materials and for production. The third group includes those creating public spaces for recreation such as Tariq and his friends at Dokkin Ghat in Manikpara. But it also denotes the visitors of festivals and programmes, for example the winter fairs held on Nasimgaon Eid Gah Math or the religious function of the *Eid* prayers (see Chapter 6.2).

On the other hand, the members of political and other management committees, including the traditional *shalish*, exercise considerable influence on the production of public spaces. These committees do not fulfil statutory directives, but voluntarily administer the neighbourhoods. They include the political committees of AL, BNP and JP on different administrative levels, i.e. *thana*, Ward and *mohalla*/Unit/Sub-area, and the political sub-organisations (see Figure 8 in Chapter 5.1.1). Furthermore there are committees of the local mosques, *shalish*, schools and NGO-related committees. Most of those active in the local committees are members of the ruling political party. For example, the Sub-area Committees in Nasimgaon are operated by the political leaders of the AL during the AL government periods. If there were a change of government, the local BNP leaders would take over the operation of these committees. Within the settlements, those forming the management committees can mostly be considered elite groups (see analysis 9.1.3 on power sources), in differentiation to the ordinary. Despite their relative powerlessness, the members of committees of the opposition party also belong to the elite groups.

The differentiation between the ordinary everyday users of space and the elite consisting of members of political and other management committees is, however, fluid, as actors may assume roles in both 'spheres'. On the one hand, those who are involved in management committees may also make their 'private' spatial claims in pursuit of their livelihoods. A political leader claiming public space and gradually transforming it into housing units, like in the example of Khalabazar, clearly does so to support his livelihood; while his involvement in a management committee probably has provided him with the capacity to claim public space (see analysis of power sources in Chapter 9.1.3). On the other hand, Afsana, selling vegetables at Khalabazar, was at the same time involved in political and NGO management committees, and hence was moving in both spheres of actors. The leaseholder of Manikghat also claimed jurisdiction over public space that went beyond the statutory contract in order to generate income. His involvement in managing the embankment space in Manikpara, however, suggests considering him as part of the elite group. Despite the ambiguity of ordinary and elite, I will use the terms to differentiate between actors' general positions. Thus there are the small-scale users of public space whose focus is to provide a minimum living standard for the family on the one hand, and the leaders who exercise influence on the social sphere of a locality and use their position to make spatial claims beyond a survival-oriented level, on the other hand. It is important to acknowledge that a person may assume different roles and positions in the course of a negotiation process, so she/he may be part of the ordinary at one point, and part of the elite at another point.

On a final note, I would like to indicate the non-applicability of the term civil society here. As analysed in Chapter 2.2.3, Chatterjee (2004) in the context of India refers to civil society as being available only to elite groups, while political society is where the urban poor can make their claims to resources. Similarly, this can be transferred to Bangladesh, as the civil society in an empowering sense is out of reach for many socio-economically poor inhabitants.

9.1.2 Aims and motivations in making spatial claims

In the following the spatial claims made by actors will be differentiated according to the aims and motivation for claim-making. These spatial claims can be understood as the basis for negotiation processes and contestations between actors and different claims.

Spatial claims for shelter and basic income generation by the ordinary

A spatial claim commonly made by the ordinary is access to public space for shelter and basic income generation.

The stories of public spaces in Manikpara and Nasimgaon describe a number of spatial claims for income generation activities which are subject to negotiations with various actors (see also Chapter 6.1.1). Such spatial claims may be temporary or permanent, i.e. a person could claim access to public space as an exclusive use right or could make such claims temporarily, allowing other uses at times of non-occupation. For example, Rokib and Momena accessed a particular part of the embankment slopes almost every day from early morning until the afternoon hours. In case they did not have work, the place could also be accessed by others, but their spatial claim can be considered almost permanent. The other plastic drying activities of the factories normally only access public space to wash and dry once a week, thus their spatial claim is made only temporarily. In Nasimgaon the vendors on the Eid Gah Math mostly access public space on a rather permanent basis – although they mostly do not possess stable vending units and this considerably limits their ability to sustain spatial claims. This was the case when the semi-mobile, semi-permanent and even permanent vendors were repeatedly evicted and could not maintain their access to public space for income generation. Similarly, when Taslima wanted to raise the place from where she sold *pitha* on the Eid Gah Math, this spatial claim towards a 'stable vending structure' was immediately contested (see Chapter 8.3). In contrast, the vendors at Khalabazar, e.g. Rohima and Shahin, Dipa, Afsana, Rehana and Shoma, always occupied the same place for which they paid a monthly rent to the local AL Committee. The mobile rice vendors on the other hand only came in the afternoons and accessed space on a temporary basis, but also in negotiation with the local AL Committee.

Spatial claims for shelter are only relevant in Nasimgaon, where land for the extension of housing compounds or construction of new housing units is available. A spatial claim for shelter was made by Dipa, the vegetable vendor/seamstress who could not sustain her position at Khalabazar. She was offered a piece of land along one of the roads leading out of Nasimgaon (Amlabazar Road) in May 2009 by some young local political leaders. The road was constructed as a dam leading through the Lake but was until this time unlined by buildings. In June 2009 Dipa was among the first to start constructing a house along the road crossing the lake. Her motivation was access to shelter and income generation, as she had left her rental accommodation and wanted to operate a shop in front of her new home.

Spatial claims to generate additional income

The need for 'basic' access to shelter and income generation as a motivation to make spatial claims was discussed above but spatial claims can also be motivated by the aim of generating an additional income. Additional income here refers to an income that raises a household above a minimum living standard of 'survival', e.g. when a household besides operating a small-scale business also lets rooms to tenants. Spatial claims to generate additional income can be made in existing public spaces, or by extending into new spaces.

In all three stories outlined above, actors took possession of existing public spaces and generated income by letting these places. These claims were all made by elite groups, indicating the need for a specific power position to be able to secure such claims (see Chapter 9.1.3). Furthermore, such claims in most cases serve both a motivation of income generation by politically backed individuals and a motivation of sustaining and affirming one's power to dominate (see below).

In the above stories, the leasehold operator Foyez claimed jurisdiction of the embankment slopes and thus generated an additional income by charging the parties drying plastic. At Khalabazar, the local AL leaders took possession of the previously open field to establish shop units they could let to vendors. On the Nasimgaon Eid Gah Math, AL leaders also made various attempts to take possession of existing spaces – for example in the case of the water tank conflict in May 2009 and in April 2010 when some leaders started to construct new rooms. While not all these stories ended 'successfully' for those making spatial claims, the interest in taking possession of public spaces was motivated by the aim to generate an additional income.

While spatial claims for basic shelter are a motivation of the ordinary, spatial claims to expand housing compounds must be considered a source of income only available to those who already have a relatively stable source of income, as it necessitates investments in construction materials. Such landlord claims are especially common for elite groups who are also able to exercise the power to sustain these claims, however, it is also a claim made by the more affluent ordinary – here the border has to be considered fluid.

Such claims are primarily made in Nasimgaon, where space for extensions is still available, while the already very dense urban fabric of Manikpara scarcely offers such spaces. As already indicated in Chapter 5.4.1, extensions of housing spaces in Nasimgaon became common practice after the new political government came into power (see Figure 10 for the extensions of one housing compound between March 2008 and March 2010). At that time not only this compound was extended considerably onto the wetland, but there was practically a 'race for space' with several housing compounds competing for the space for extension. The landlords were primarily motivated by the wish to generate income by letting additional rooms. During various stages of the development process, however, these spatial claims were contested by political actors.

But spatial claims were also made on an even larger scale along Amlabazar Road. This development was triggered by local party cadres and *mastans*, who in part offered land to the ordinary (see above, the case of Dipa), which can be considered a testing-strategy, but also made their own spatial claims. Most of the land eventually went into the hands of the elite groups. This was also expressed by Sayed, a local leader of the Amlabazar area, who distributed land to the poor expecting them to contribute to the local mosque fund on a monthly basis, but finally found that most of the land went into the hands of leaders and no one contributed.

Spatial claims to generate income for committee operations

Claiming space is furthermore motivated by the intention to generate revenue for the activities of management committees. In their area of jurisdiction, the committees collect money to recover the costs for services extended to the vendors and other shopkeepers, for example provision of security by organising night guards or keeping the environment clean by organising sweepers.

On Nasimgaon Eid Gah Math, the AL Sub-area Committee collects the contributions for night guards and sweepers from the vendors using public space for income generation, and from the other shopkeepers in their area of jurisdiction. From this fund they also pay for their own consumables in their office. The mosque committee, officially supervising the Eid Gah Math, furthermore collects daily subscriptions from the vendors for thc mosque fund. At Khalabazar (collection by the AL Sub-area Committee) and Amlabazar (collection by the AL-based Amlabazar Committee, including the collection for the mosque) the collection practices are similar. Sometimes extra collections are made to upgrade infrastructure, for example the Amlabazar Committee collected 1,000 Tk per shop owner (of stable shops) to pave one of the main access roads. All these contributions are based on the committees' spatial claims to supervise a defined spatial entity which is also subject to negotiations, as the discussion between the Sub-area Committee and Amlabazar Committee about the jurisdiction of Khalabazar showed (see Chapter 8.2). In Manikpara, a similar example is the mosque committee which started to rent out the place at Dokkin Ghat to two rickshaw garages to generate additional income for the mosque fund. The claims made along Amlabazar Road, according to the local leader Sayed, were also meant to contribute to the mosque and *madrasa* fund, however, finally no one paid as there was no official arrangement:

> "We expected that the room owners will pay rent for the *madrasa* and mosque committee. The rent was 50 Tk per *hat*[1]. 300 Tk would be the rent if someone builds a shop on six *hat* land. They were committed to pay the rent to the mosque. But after getting the position no one pays the rent. This is the matter of his/her belief in religion that someone is not giving money to the mosque. His/her belief in religion has been lost. We told them to pay the rent to the mosque or *madrasa*. If someone pays 300 Tk as the rent then the mosque will get 150 Tk and the *madrasa* will get 150 Tk. We told them to do so and we gave the position to them on this condition. We are not able to give them any pressure to pay the rent as we did not make any deed." (Sayed, 23.04.2010)

Furthermore, the committees are often involved in saving schemes and redistribute part of the income generated from spatial claims. For example, some vendors paying rent at Khalabazar participated in a saving scheme with part of their rent payments. However, the conditions for participation were non-transparent as only a few vendors participated. The leaders emphasised the welfare orientation of their social work (see below). Furthermore, the committees contributed from their funds to support inhabitants for special purposes, for example in case of illness or for marriage ceremonies. Here, however, support went mainly to their own employees (e.g. night guards) or to AL supporters.

1 One *hat* normally equals one foot as a measurement for length. In Nasimgaon, however, it is used as a local measurement and refers to the depth of the shops/rooms, while the width is pre-fixed.

While spatial claims made by the committees serve to recover the cost of services provided, the introductory stories have also demonstrated how claims made via committees may also serve the benefit of individual leaders, and how individual leaders can make use of the power of committees. The strategic use of committees by individual leaders to generate power and make spatial claims will be discussed below.

(Spatial) claims to demonstrate (political) dominance

Even when the main motivation of spatial claims may be the generation of additional income for committees or individuals, spatial claims often also follow a motivation to demonstrate (political) dominance and may be used to sustain and underline this (political) dominance. In a number of instances, spatial claims were used to confirm the power positions of the ruling party and its sub-organisations against the opposition, the minor coalition member JP or other competitors within AL organisations. But claims to demonstrate political dominance do not have to be physical spatial claims. Repeatedly, such claims to demonstrate dominance were made within the social sphere.

The construction of two rooms by the *Shechhashebok* League and *Sromik* League on the Nasimgaon Eid Gah Math (see Chapter 8.3.3) was one such demonstration of political dominance. It can be considered as symbolic of the AL's dominant power that these organisations were able to take possession of such a prime site within the neighbourhood. The conflict between the AL and JP about the rickshaw garages and fencing for a school, hospital and community centre (which were never constructed) can also be understood as being politically motivated. The AL leaders protested against the rickshaw garage because it had been constructed during BNP times and now they received support from the MP of JP. The contestation that followed was at least partly motivated by the AL committee's need to show their power to dominate the social sphere and thus they made parallel claims by fencing parts of the Eid Gah Math (see Chapter 0).

Spatial claims for recreation and organising functions

As has been analysed in Chapters 6.1.2 and 6.2, spatial claims are also frequently made for recreation and leisure time purposes in everyday life (for example organising football tournaments on Nasimgaon Eid Gah Math or Tariq's garden), as well as to organise religious functions, cultural or political programmes as part of extra-everyday activities (for example the *orosh* at Jahangir's place on the embankment slopes or the winter fair on Nasimgaon Eid Gah Math).

9.1.3 The multiplicity of power sources for negotiations

The examples of the negotiations of access to public space have indicated a range of factors determining a person's ability to display and use power. In the following, different ways of acquiring power are discussed, while it has to be noted that these are always manifold and in many cases multiple sources of power are combined to make a (dominant) claim to public space. Sources of power, as indicated below, can thereby be either direct and obvious or subtle, in that an actor generates power via achieving a position of social respect in the local society over time. An overview of actors' sources of power is displayed in Figure 22, while Table 6 at the end of this sub-chapter summarises the importance of sources of power.

Additionally, statutory processes also present a source of power, especially for the leaseholder in Manikpara (see Chapter 8.1.1). This has been outlined in Chapter 9.1.1 (identification of statutory organisations and actors), and thus will not be discussed again in this chapter.

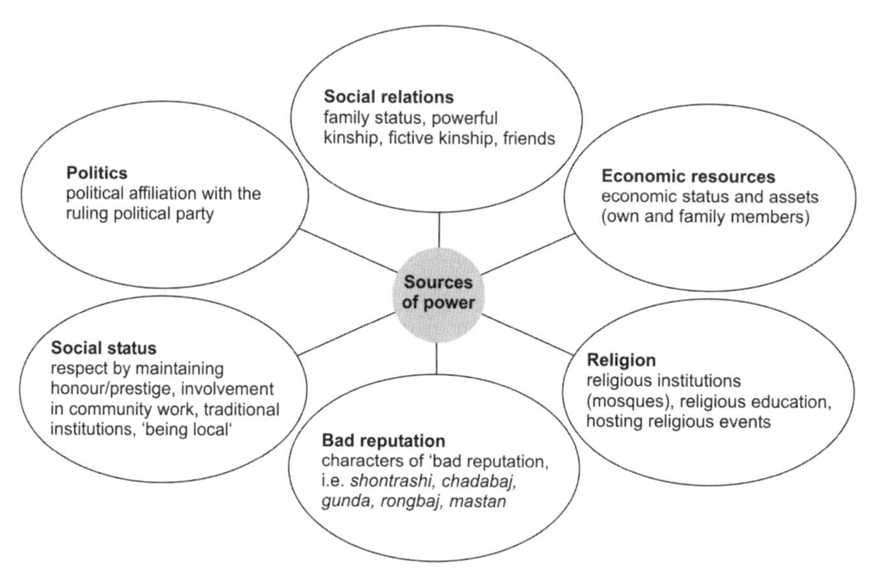

Figure 22: Overview of actors' sources of power

Political system and party affiliation

The political system and party affiliation are the main source of power in all the negotiation processes inherent in the stories narrated above. This strongly supports the analysis of the political system (see Chapter 5.1.1) as being patron-client based and confrontational with regard to the relations between the ruling party and the opposition.

Doing something by 'using the political power' (political power translates as *khomota* in Bengali, while involvement in any kind of politics is *rajniti*) of the ruling party was a very common expression during interviews, especially in relation to the construction of houses in places which were not supposed to be built upon. In Manikpara, Jahangir complained how since the AL government was in power its leaders had constructed rooms on the embankment slopes using their party's power (while he considered his own room in the same location as being legal and claimed to own the land by deed). Nabin, a local leader of AL in Manikpara, who in 2010 also tried to win the leasehold for Manikghat, considered that claim-making by the power of politics was an adequate reward. His statement, made after the BIWTA had evicted his shop from the embankment slopes in August 2010, underlines his expectation that his political activities should be rewarded and that he should be able to use politics as a source of power for claim-making:

> "All these houses were evicted by BIWTA people. They did it without giving us any notice. We came back again and constructed these rooms. What can we do? We do politics. We are very small leaders comparatively. If we don't get even this slope to run a business then what did we get by doing this politics? Do you understand? BIWTA people have done their duties and we are doing our duties." (Nabin, 25.11.2010)

But it is not only individual spatial claims that are made by drawing on politics as a power source. Party affiliation is also used by local leaders as a source of power to dominate the social sphere of a neighbourhood by taking over the management committees[2]. In Nasimgaon, AL politicians operate all local management committees (see discussion of Figure A-3 in Chapter 9.2.1) and Sayed, a local AL leader, perceived it as 'natural' that AL organisations dominated. He also indicated how support structures went to higher level party organisations:

> "The organisations which are closely related to the emerging government do most of the activities in Nasimgaon. People also come towards this type of organisation. This is the slum area, a government's place. So people go to those organisations which are related to government's politics. The members of these organisations also go to those people who are active in higher level politics, like the Ward level. If someone goes to them [political organisations of AL] then his problem becomes simple and hurried up. Suppose, I go to an MP of AL about a *thana* case matter. If I go to AL people about this matter, they would help me, but BNP people would not do anything regarding this matter. Various organisations of the emerging political government are doing most of the activities now." (Sayed, 21.05.2010)

The emphasis of the area being a government's place underlines yet another aspect. Opposition against the government by inhabitants who occupy government land is something that cannot 'easily be afforded' given the insecurity of tenure arrangements and the incalculable risk of action against the government. This re-

2 The fieldwork of this research was conducted during an Awami League government period. The confrontational nature of party politics as discussed in Chapter 5.1.1 as well as the statements made by interviewees about spatial claims during BNP government times, however, indicate that the general practices of the political elite can be assumed to be very similar regardless of which party is in power at any one time.

ality of dependency can be considered to further strengthen politics (in relation to the ruling party) as a source of power.

The ordinary residents also referred to the domination of the social sphere by AL leaders and followers. For example, Hamidul, a boatman living in Nasimgaon, when explaining about the rules of fishing in the lake mentioned the power structure of the socio-political sphere, indicating how deeply engraved this was in everyday life spatial practices:

> "Now the Awami League government is in power. The people of Awami League could not catch fish and could not eat during the BNP regime. Now, Awami League is in power. Now, the people supporting BNP cannot catch and cannot eat." (Hamidul, 07.06.2009)

The stories of Nasimgaon Eid Gah Math and Khalabazar also tell of the interchanging dominance of social spheres between electoral periods. Decisions and actions taken during BNP times immediately became contested when the AL came into power. The dominance of the ruling party's supporters over the social sphere of a neighbourhood is also manifested in their dominance as members of traditional institutions of arbitration, local committees, NGO[3] and other development activities (see below). The low opinion and reputation of politicians that many interviewees expressed is based on their domination power and abuse of this power for personal benefits – in this context the expression 'to eat' [*khaoa*] was repeatedly used in Bengali, indicating how politicians would seize land or collect *chada*, i.e. carry out illegitimate activities by political power.

Hamidul's statement above indicates how the leaders of the opposition are practically powerless and cannot draw on politics as a source of power before another change of government. In Chapter 5.1.1, I discussed the confrontational politics between the ruling party and the opposition and indicated with reference to local government how the power of elected politicians of the opposition was limited. This also became evident from the analysis of political activities in public spaces (see Chapter 6.2.3) which indicated the dominance and visibility of AL politicians. Both in Nasimgaon and Manikpara the Ward Commissioners as elected representatives of local government are BNP members, but the important and powerful persons are the AL candidates for the WC elections. Those candidates are the ones that local leaders consult, rather than the elected representatives. Accordingly, the Ward Commissioner of Manikpara expressed his powerlessness by indicating the AL leaders' dominance in the social sphere (and their abuse of this dominance):

> "I can perform my works without facing any problem. But I don't need to do *bichar* [judgements]. AL people of this area are doing this. The MP is also the representative of people. He also does *bichar* works. They are eating money while doing *bichar*. Though I am a local representative, they don't call me [for *bichar*, disappointed expression]." (Ward Commissioner, 15.04.2010)

3 The dominance of political leaders is also reflected in the education of their children: either they visit the best local NGO school (free of cost and designated for the 'ultra-poor'), or they send their children to schools outside of Nasimgaon which are comparatively expensive.

In Nasimgaon the AL leader Sayed, involved in the water supply business and the Amlabazar Committee, was very conscious of the limited power of BNP followers:

> "They [BNP people] permit us to do the work. They don't try to forbid us though they may not like the work. They say 'Do as you wish'. They have nothing to say as AL has formed the government. They can't raise their voice. If I support BNP then I will not be able to say 'Don't do this work'. The decisions of AL people are the most grantable as AL is the government power holder." (Sayed, 21.04.2010)

The word 'grantable' was used by Sayed to indicate the political power, i.e. that the decisions of AL would remain whether right or wrong, and independent of what might be accepted, rejected or respected by the local society. The statement carries the awareness that AL politicians are the ultimate decision makers in their locality. As an exception to the rule, the Amlabazar area, however, was at the same time the only area where a *somiti*[4] of BNP leaders continued to operate during AL times and was considered a powerful opponent also by the local AL leaders. Nonetheless, the overall picture shows the powerlessness of the opposition and the opportunities for members of the ruling party to use politics as a source of power.

'Doing politics' is furthermore a **source of power for women**. Although female politicians enter a male-dominated domain, they are not considered to violate traditional gender rules. Here, the fact that two women are the heads of the two main political parties, and that female quota exist on the various levels of (elected) local government, seems to contribute to the acceptance of female politicians. Thus, when they join politics women can exercise considerable domination power even in opposition to strong local leaders.

In Nasimgaon[5] the three women politicians I talked to during interviews and informal discussions based their power on both political connections as well as NGO involvement. In the political sphere they held positions in various AL committees, especially in the *Mohila* League, the women's sub-organisation of the Awami League. The most powerful woman politician in Nasimgaon was Rizia, a former inhabitant who was still involved in many local NGOs and political activities. Her successful generation of power via political involvement was reflected in her high profile. When I wanted to visit her house but could not find the location, the rickshaw puller immediately knew where to go when I mentioned her name. She was the president of the *Mohila* League of one Sub-Area Committee in Nasimgaon, a member of the AL Ward Committee and the office secretary of the *thana* level *Mohila* League. Her self-perception also showed that she was aware

4 A *somiti* is a cooperative generally based on locality offering saving schemes and giving loans. The *somiti* at Amlabazar (Amlabazar Somiti) is an organisation of the businessmen of the area and includes saving schemes and micro-credits.

5 In Manikpara I did not come across a woman active in the local political sphere. While this picture may be incomplete, it nonetheless matches the analysis of the societal structure of Manikpara with regard to gender relations. The adherence to the rules of *porda* might also prohibit many women from involvement in politics as a rather public-oriented activity.

of her political power and had been in a powerful position for a long time, as the following account of her protest against AL leaders' plans to evict the then residents of Nasimgaon in 1998 shows:

> "At that time [of the eviction plans of the AL government in 1998], a lot of AL leaders suggested me to leave it [the protest movement] and they told me 'You are a supporter of AL. How could you do it?' I said to them 'My back is against the wall. There is no place to move back. Definitely I will push you if you stand in front of me. Because I need to go, you should better give me a way to go. Then I would not have any need to push you, would I?'" (Rizia, 27.04.2010)

The above discussion shows how political involvement in the ruling party is a source of power to dominate the social sphere of a neighbourhood. Politics furthermore enable women to generate considerable power within a neighbourhood, despite the potential challenge to social gender relations. In contrast, the leaders affiliated with the opposition will only be able to use politics as a source of (dominant) power after a change of government.

Social status

A respected social status within the community can also be a source of power. Social status may be generated by acquiring a reputation of respect and honour, by involvement in community affairs, by involvement in traditional institutions and finally by local origin. Many of these issues are closely interlinked, as the following discussion will demonstrate.

There are various ways of achieving a **respected and honourable status**. For a respected family, the dignified and honourable behaviour of its members is highly important, especially the socio-culturally appropriate behaviour of female members of the household, as discussed in Chapters 7.1.1 and 7.2.4 on female mobility, clothing styles and 'familiar'/'strange' publicness. Especially in Manikpara, this notion of respect via female family members is of utmost importance for a household to sustain power by social status. Furthermore, persons of old age and long life experience are highly respected in society and are thus able to generate power via their respected status. Both in Nasimgaon and Manikpara, interviewees referred to *bishisto murobbi*, notable elders, who are generally characterised by knowledge, age, a philanthropist attitude and impartiality. *Bishisto murobbi* are obeyed and widely respected by society and their counselling is generally appreciated and not easily contested (see also Chapter 5.3.1 for their importance in Manikpara). In Manikpara, the businessman Mesbah Uddin referred to one elder person as 'respected elder brother', indicating that he was a wise senior and a respected advisor whose decisions were obeyed by juniors:

> "Under the leadership of the Commissioner and our respected elder brother Haji [...] this work [land filling between houses built on bamboo platforms above the water] was going on. We call him 'Chairman', but he is not a public representative. We honoured him by giving him the title 'Chairman' after seeing his activities, service and work for the people. Still now he gets 1,300,000 Tk [probably refers to a due amount the 'Chairman' paid from his private

accounts in advance]. He is not a high government official; he is a public ['private citizen']." (Mesbah Uddin, 31.03.2010)

In Nasimgaon, in the negotiations about Rohima and Shahin continuing to run a stall at Khalabazar, an elder person was the one finally heard by the shop owner, which enabled them to keep their position. This indicates that even the political power holders – here the shop owner who was affiliated to the AL – pay respect to elders.

In the above statement, Mesbah Uddin additionally referred to another source of respect and honour, namely **involvement in community initiatives** and contributions to the community. In Manikpara, those already respected in society were also the ones investing in and working for the development of the neighbourhoods. This included the *bishisto murobbi*, the *matbors*, the MPs, the Ward Commissioners and other local (political) leaders, but Mesbah Uddin also referred to his own involvement in the community, see his reference to chasing out foxes and dogs, killers, thieves and criminals in Chapter 6.1.3.

It is difficult to differentiate between the power someone already possessed beforehand due to societal position and the power acquired by being involved in community works. Furthermore, the perceptions of contributions differ according to whether they are told from a power-holder's perspective or from that of an inhabitant. The narrative of Manikpara's social structure is considerably more consistent and widely agreed upon than that of Nasimgaon. This may be due to the fact that in Manikpara many of those contributing to the founding and development of the community have died and are honoured post-mortem[6]. In contrast, in Nasimgaon this first generation of community leaders is still alive and is involved in political everyday life, which earns them respect only from those who benefit, while others are highly critical of their work. But this difference also seems to relate to the specific environment and socio-economic diversity of Manikpara, with its tradition of businessmen sharing their resources with the community, while in Nasimgaon initiatives are mainly based on NGO contributions[7].

Accordingly, the relevance of community involvement as a source of power differs in the two neighbourhoods. In Manikpara the Ward Commissioners and MPs together with *bishisto murobbi* are praised for their initiatives in land filling, road widening and other development works, and the businessmen's contributions to the Bagan. In Nasimgaon community development works are closely tied to the

6 For example, all interviewees who talked about the history of Manikpara referred respectfully to one person (Haji) who built a mosque and a school for the community.

7 While NGOs and donors have also been involved in Manikpara, for example during the SIP (Slum Improvement Programme) of UNDP in the 1980s, no one referred to any external forces. Instead, the narratives completely 'internalised' the development initiatives and ascribed them to local persons of social, economic and political status. Basing my work on local interviews, I cannot verify whether 20 years ago Manikpara was as equally 'flooded' by NGOs as Nasimgaon is today. Given the recent trend of NGOs mushrooming all over Bangladesh and the NGO popularity of Nasimgaon in the low-income urban field, however, I assume that there were comparatively less NGOs and donor organisations active in Manikpara in the 1980s and 1990s.

NGO sector. Accordingly, many local leaders are also members of the local NGO committees and this again gives them a chance to dominate the social sphere and generate power. The local political leaders are not only members of one committee, rather some of them are involved in almost all NGO committees of the neighbourhood; there is thus a small group of elite power holders who generate additional power by dominating the NGO scene. Apart from the NGOs, the management committees get involved in community work by supporting the poor. For example, some of the money collected from the shops as daily market fees, or even in the form of a purpose-based additional collection, is used to help the poor in cases of funerals, marriages, illness, or other unexpected and high expenses (see also Chapter 9.1.2). However, the tendency is to reinforce existing patron-client relationships, and the beneficiaries are the people affiliated to the ruling political party. Accordingly, this involvement mainly serves as a source of power via the respect generated among beneficiaries, while those excluded do not change their perceptions of local leaders.

The **involvement in the traditional judgements** of *shalish* is both a source of power and an expression of power. Being a member of the *shalish* enables a person to dominate the social sphere of a neighbourhood considerably by involvement in solving conflicts (see the example of a *shalish* in Textbox 6, where a conflict between neighbours was solved in a wise judgement; however, the example also shows how the *shalish* members can exercise considerable power over local spatial processes as members of a respected traditional institution). Involvement in *shalish* underlines a differentiated power structure between local leaders and the ordinary. By holding judgement, the *shalish* members possess the power to conceptualise the social field of society in a dominant way. The exclusion of the WC in Manikpara from doing *bichar* (*shalish*, used synonymously) and the dominance of AL politicians in doing *bichar* (see above) indicate how this is an instrument of the powerful and an expression of dominance.

In Nasimgaon, the patronising power generated by being a member of the *shalish* is implicit in the statement of one local AL leader who differentiated between the 'intelligent' and the 'poor' and 'lower class people':

> "There are some intelligent people in this slum. They are intelligent like a magistrate [judge]. These people can make people understand easily, can attract people, can solve any problem easily and can stop any case. We did never allow anybody to file a case. If any person goes to file a case then we call both parties and try to solve the dispute among them. All people of this slum are poor. They can make mistakes but they should not go to file a case. The police administration will eat money. This is a kind of loss for them. That's why we always try to call for a *shalish* looking after the lower class people of this slum." (Baharul, 01.03.2010)

The statement furthermore carries a dimension of securing power and influence by discouraging the use of statutory institutions, here the police administration as part of the executive.

While the AL supporters thus generate power by being involved in *shalish*, the BNP supporters cannot make use of this source of power, and some *shalish* verdicts favour AL supporters and de-power BNP supporters. In another statement, Baharul explained how those shaping the social sphere of Nasimgaon as local leaders today also were part of the *shomaj*, social society organised hierarchically and defining social norms and codes of conduct, in their rural homes (see Chapter 5.1.4). This suggests a continuation of traditional institutions as sources of power. The above narratives and analysis show how traditional institutions become an instrument for those already powerful.

'Being a local' is a significant source of power especially in Manikpara. Those who can be considered respected and powerful in the local society thus refer to their local origin as a reason why they should not be contested, indicating a notion of power generated by origin. Both the current and the previous leaseholder of Manikghat thus stated:

> "Yes, my birth place is Manikpara. That's why it is seen that I can mingle with everybody, every big to bigger person or small to smaller person know me. Again, we are from *Bepari* family [family of businessmen, see below]. So anyone can easily recognise our *ghat*." (Foyez, 26.06.2009)

> "There were lots of *gunda* groups ['destructive' local leaders] and *shontrashi* [armed gangster] in this area. They demanded money, like someone used to come and say 'You have to pay me 2,000 Tk per month'. Another one came next day and said 'You have to pay me 500 Tk per week'. This is my birthplace. To whom should I pay money!" (Jahangir, 11.02.2010)

The latter statement carries a notion of perceived injustice in not treating Jahangir, in his own perception a respected person in the locality, with due respect but extorting money. Being rooted in the locality was also referred to by others. For example, Mesbah Uddin and Mashrufa were highly respected and called the *dada* and *bou* of the *mohalla* (see Chapter 7.2.1), indicating how their local identity at the same time generates power and respect. In Nasimgaon this matter differs somewhat, as most inhabitants are rural-urban migrants and perceive themselves as having a rural-based identity (see Chapter 7.2.2). Nonetheless, the local leaders refer to their long time of serving the community. In an informal discussion with two persons who had just recently appropriated land along one road leading out of Nasimgaon (Amlabazar Road) and were in the process of constructing houses, feeling local was referred to as an explanation of why they did not pay money to anyone for appropriating the space, while some of their neighbours had to pay money and could not refuse to do so.

Social status especially in Manikpara can be generated by honourable and prestigious behaviour, especially of female family members, and by acquiring a status of 'notable elders'. This is often closely tied to involvement in community initiatives resulting in a respected status. In Nasimgaon such a common perception of respect is absent and today's leaders are rarely appreciated for their achievements. Instead, involvement in community initiatives and traditional institutions in Nasimgaon is a source of power, but not necessarily of respect. Further sources of power are the institutions of traditional judgement and a local identity.

Textbox 6: A *shalish* on a conflict between house owners in Nasimgaon

On 16.02.2010, I joined the members of a *shalish* on their site visit to investigate a conflict among neighbours in Nasimgaon. First we visited the compound of the *bibadi*, the accused. He was busy restructuring his housing compound and had torn down most of the previous CI-sheet building. Now he was busy constructing a brick-wall just next to the CI-sheet wall of his neighbour's compound. The neighbour had called the *shalish* because due to the reconstruction, he was expecting an increased amount of rain water on his side which would destroy the CI-sheets and make his rooms more susceptible to flooding. The conflict was solved by the local leaders and accepted by both parties in the following conversation among:

- three members of the *shalish* committee (Afjal – a retired army officer, Ali Hossain – the president of the local Mosque Committee, and Harez – a local leader; all of them have lived in the locality for between 20 and 25 years)
- the complainant, who brought the issue in front of the *shalish* (*badi* in Bengali)
- the accused, who is constructing the brick wall (*bibadi* in Bengali)
- Nazrul, a neighbour who was called as an independent witness by the *shalish* members

Afjal [explaining to us]: "There is a dispute among two families about this land. There is the *badi* and *bibadi*. The one who accuses someone and files a case against him/her is the *badi*. He/she against whom the case is filed is the *bibadi* against whom the *badi* files a case."

Harez: "This home [indicating the neighbouring CI-sheet structure] belongs to the *badi*. This is the new brick wall constructed by the *bibadi*. Once it will be completed, it will create a problem to the neighbour [the *badi*] as rain water from the roof will pour onto the neighbour's roof. That's why the *badi* has filed a *shalish* case against these people, in order to achieve a proper solution. If this person [*bibadi*] had constructed the wall leaving some place in between, then there wouldn't be a problem."

[Another neighbour comes, Nazrul, who lives in the house just beyond the *bibadi's* compound.]

Ali Hossain [to Nazrul]: "Nazrul, you are staying here as a neighbour. Please tell us about this happening, as you are impartial. Is everything alright here? Has someone done anything wrong?"

[Nazrul agrees that a problem will be caused by the construction of the wall once a new roof is placed above]

Harez [to the *bibadi*]: "Your roof will definitely go above their roof. So your roof water will pour onto their roof."

Bibadi: "OK. I will manage to drain the roof water to the outside."

Harez [to the *bibadi*]: "Your roof will go above the other roof, and this is totally wrong. So when they [*badi*] will reconstruct their house then they will merge their roof with your roof. You have to share your wall. They don't need to construct the side wall [meaning that the *badi* will remove his CI-sheet wall to share the same brick wall with his neighbour]. You will arrange the gutter to drain the roof water. You both share the cost for the gutter equally. We have solved this matter now from a social perspective. If the police came here to solve the problem then you would have to remove this wall. Then you would be harmed. Do you want this? Stay together! You will plaster this side of the wall and he will plaster the other side of the wall."

Harez [explaining to us]: "This is government's land. There would be a measurement of land if these lands would be privately owned. As this is public land, people always try to take more land by depriving others. If there were deeds for the land pieces then people could not do like that."

After the matter had been solved through the advice of the members of the *shalish* and both parties had agreed, the *badi* invited the local leaders and us for a cup of tea.

Religion and religious norms

In the context of religion and religious norms, three sources of power are discussed below, namely the mosque as a religious institution, religious education and the hosting of religious functions.

According to religious norms, the mosque is an important and **protected religious institution**. In both Manikpara and Nasimgaon, the mosque committees are respected and powerful actors exercising influence on the production of (public) space. In Manikpara, the example of Tariq's garden (see Chapter 8.1.3) and the road widening initiative of the Ward Commissioner (see below) best exemplify how affiliation with the mosque can be a source of power. Tariq and his friends were not able to protest against the decisions of the mosque committee, who had decided to transform the space of the temporary mosque into rickshaw garages:

> "We thought that our garden would be enlarged once the mosque is removed from the slope. But the Mosque Committee has rented out that space as a rickshaw garage. They didn't tell us about anything. We have also left the matter because it is a religious institution. Otherwise we would never leave it."(field note, Tariq, 25.11.2010)

The mosque as an incontestable religious institution here presents a source of power to claim the rickshaw garage space. During the road widening initiative in Manikpara, the Ward Commissioner made use of this power in a reverse way. The fact that he agreed to break down a mosque which is normally not to be destroyed, strongly underlined the seriousness of his initiative:

> "Some people denied to break down their buildings and said to me 'There is a mosque. Why are you not breaking down that mosque?' I then told them 'Will you leave aside land if I break down a portion of this mosque?' They said 'Yes'. Then I ordered to break down the mosque." (Ward Commissioner, 15.04.2010)

In Nasimgaon, power generated via religious institutions is for example manifested in the jurisdiction over space and the ability to then collect subscriptions, in Bengali known as *chada* in a positive sense, for development activities of the mosque. The above examples all indicate how the mosque with its symbolic function as a religious institution can be used (and abused) in various ways to generate power. In both Manikpara and Nasimgaon, most members of the mosque committees are also in central political positions within their neighbourhoods, especially in the ruling party AL.

A further source of power can be a **religious education** or the taking on of **religious positions**. Religious education is highly valued especially in Manikpara. Mesbah Uddin sent his son to a *madrasa* after basic education in a government primary school and wanted him to become *Hafiz*, an Islamic scholar who can recite the Quran and who occupies a position of high social respect. Foyez' religious family title is *Mollah* (Islamic scholar) which means that his forefathers were *Mollah* and thus of a respected religious status. Such a title evokes respect and power, as also the example of one political leader in Nasimgaon who is *Moulana* (another type of Islamic scholar) indicates. This leader is not only *Moulana*, but he is involved in most religious positions available within Nasimgaon and beyond

– among others he is the president of one local mosque and the adjacent *madrasa,* and the religious affairs secretary at the AL Ward level. Despite the respect paid to those of religious education, this is comparatively less important in Nasimgaon, where many local leaders have not acquired a distinct Islamic education level. This difference can again be explained with the specific Old Dhaka tradition of Manikpara and its close relation to the *Tabligh* movement.

Hosting religious programmes and inviting a large number of people to join can also be a source of power as it serves to maintain a person's respected status in society. In both Manikpara and Nasimgaon, religious programmes are held by groups of people and individual households regularly. Jahangir was very active in the religious sphere of Manikpara which upheld his respected status in society (see Chapter 6.2.1 for his annual organisation of an *orosh* at one *ghat* which included provision of food to the boatmen of the *ghats* while he was the leaseholder, his regular visits to *mazars* and the *Bisho Ijtema,* and his own construction of a *mazar*). Mashrufa, Mesbah Uddin's wife, proudly told us of the women's *Tabligh* meetings held at her house (see Chapter 6.1.4), and it seems that holding these meetings at one's own house increases one's prestige in society, as the household will be considered honourable and religiously respected, especially if the guests are treated with food after the ceremonies. Mokbul Hossain in Nasimgaon also organised an annual *orosh* together with the devotees of a local *mazar*. Although he did not display his power very openly, these activities may also contribute to him being considered a person it is better not to contest, as the story of the developments on Nasimgaon Eid Gah Math and the persistence of his shop despite repeated conflicts indicates (see Chapter 8.3.1).

During my presence in the field I also witnessed three *milads*, religious thanksgiving ceremonies, both in Manikpara and in Nasimgaon. While these prayers held by the Imam of the local mosque at the thanks-giver's home are clearly a product of religious beliefs, they nonetheless are also social events and can become a means of reconfirming relationships to acquire and maintain power. After the prayer held by a group of men, food is served according to the capabilities of the thanks-giver. After a *milad* at Rokib's and Momena's place, all the people important for their business came along to have lunch – which was prepared for 150 persons at a cost of about 14,000 Tk. This included all their (powerful) relatives, the *mohajons* who gave them work on commission as well as the leasehold operator Foyez[8]. Besides the religious purpose of thanking Allah and asking him for his blessing, holding a *milad* also involves an (actively or passively promoted) component of securing power and influence by maintaining relationships.

8 Less than a month later, on 14th April 2010, Rokib and Momena served lunch on *Pohela Boishakh*, although to a smaller number of people. Again, this can be understood as a way of reinstating their power position, this time not in relation to a religious ceremony but to the most important Bengali cultural event. For a family of their socio-economic position they invest rather a lot into maintaining relationships via holding social functions. In November 2010 they also celebrated the *Mussolmani* of their son and again gave a relatively big party.

It is impossible to separate a person's religious belief and his/her use of religion as a source of power. While I experienced a deeply rooted genuine belief in religion, my observations from an outsider perspective, however, suggest that religion is also a means of power, whether subconsciously so or openly displayed. This becomes especially obvious in the power that mosque committees have in conceptualising (public) spaces. Religious belief and practices thus cannot be considered irrelevant to the social sphere of a neighbourhood, as religion can be an expression of power in society in various ways, whether anticipated or not.

Economic resources

Generation of power is also closely interrelated with a person's economic resources, manifested in possession of land/houses and/or business ownership. On the one hand, possession of economic assets can be a consequence of power generated via other sources, on the other hand, holding economic assets can be a direct source of power, especially when it translates into a respected status in society.

In Manikpara, those in a powerful position in society are also known for owning assets, and those **doing well in businesses** are well respected in local society, building on an Old Dhaka tradition of trade and commerce. Accordingly, those considered powerful in Manikpara can also draw on economic resources as a source of respect and power. The lease operator/leaseholder Foyez was economically well-off, owning a house, a plastics recycling business and 70 of the 300 boats plying his *ghats*. Jahangir, one of the previous leaseholders, owned a house in Manikpara, a plastics recycling business and invested in land in a less developed area. Mesbah Uddin owned a house (which his wife proudly mentioned, see Chapter 7.2.2) and a small plastics recycling business. Especially his narrative indicates the respected status in the local society that goes with operating a plastics recycling business. For some time in 2009, Mesbah Uddin was financially unable to operate a plastics recycling business and instead operated a tea stall. During that time he felt that his status in society was compromised as he had to accept harsh words from ordinary labourers, while in plastics recycling he considered himself in a professional business. The economically successful businessmen of the locality regularly spent money on the community, e.g. for schools, *madrasas*, mosques and also the Bagan. The way interviewees appreciated the contributions of businessmen in the past shows how the sharing of their economic resources was also a source of respect and finally power (see also above on 'social status').

Economic resources of other family members may also be a source of respect and power. Mesbah Uddin also based his status on the economic success of his brothers, who owned multi-storey *paka* houses and big plastics recycling businesses, while he still suffered financially from previous business losses. Foyez' power was not least based on the economic status of his elder brother, who was the leaseholder of two *ghats* that were highly valuable in terms of business turnover[9]. Rokib did not possess large economic resources, but he benefitted from his nephews and his sister's husband owning houses (see below on 'powerful kinship').

In Nasimgaon, **power** is rather **a precondition for the generation of economic resources**. The elite groups are the ones with stable income sources generated e.g. from letting rooms and shops and from operating their own businesses. Their permanent sources of income again allow them to invest their time in politics and community work whereas with insecure and uncertain income sources this remains a challenge. The interlinkages between economic resources and elite positions become evident in Afsana's statement, which characterised those who took possession of the vending units at Khalabazar:

> "[…] those who have *bhitas* [demarcated selling places] over here possess at least 60–70 rooms in Nasimgaon and get 60,000–70,000 Tk every month. Hatem has at least an income of 1 to 1.5 *lakh* Tk [100,000–150,000 Tk] from room rent. [Another political leader] also owns a water line and electricity line." (Afsana, 23.03.2010)

My own observations of the economic resource distribution in Nasimgaon confirm this statement. The perception of a differentiated power structure and the interrelation of power and economic resources was expressed by many interviewees. The ordinary almost automatically established linkages between a person's economic assets, e.g. owning houses and letting rooms, and his/her domination power. This also led Shihab to refer to the economic differentiation in the local society:

> "Those who are house owners […] they live like kings; they buy big fishes and eat. And those who live on rent, do work all day, break bricks, work as housemaids, are managing to eat somehow. They are not able to make any progress." (Shihab 07.06.2009)

Contrary to Manikpara, in Nasimgaon economic status was not tied to a similar notion of respect as a source of power. Instead, the perception of economic resources was rather connoted negatively in the sense that those in political power made use of their power to accumulate economic assets. In Nasimgaon, economic status and assets thus rather have to be understood as a consequence of power instead of a source of power. In Manikpara, in contrast, the analysis has shown how economic status/assets translate into respect and consequently a source of power – although the analysis in Chapter 7.1.1 has indicated that belonging to a status group and availability of economic resources do not necessarily fully correlate.

9 This is also expressed in the high bids achieved in tendering the *ghat* leases: in 2009 one of these *ghats* was leased by BIWTA for 99 lakh Tk (9,900,000 Tk) while Manikghat was leased for 12 lakh Tk (1,200,000 Tk). This is partly due to the fact that the other *ghat* is also a launch *ghat* and thus achieves a much higher turnover of passengers and boats.

Social relations

A person's acquisition and possession of power can also be based on social relations maintained with others which here include family status, kinship, fictive kinship and friendship.

Family status and descent are a source of power as they determine a person's respected status in society. The maintenance of respect and prestige (*ijjot* and *shomman,* discussed in Part II of this research) are central to those who have acquired power in Manikpara. The leasehold operator Foyez emphasised that he came from a *Bepari* family, referring to their profession, and a *Mollah* family, referring to their religious title (see above). *Bepari* means that his father was a businessman and indicates a family continuing their family business through successors. Furthermore, a *Bepari* family is generally considered a rich family. Such a rich family is respected, especially in a socio-economically poor locality, and thus *Bepari* indicates the prestigious and well known family status of Foyez' family in Manikpara and beyond. Mesbah Uddin's reference to his prestigious status and the necessity to maintain this, which includes not sending his wife to the bazar (see quote in Chapter 6.1.3 and also the analysis in Chapter 7.1.1), underlines the importance of family status to maintain a powerful and prestigious position in Manikpara.

The discussions in Manikpara further revealed how the **power of relatives** – generated for example via political affiliation – can serve as a source of power for a family member. Foyez mentioned several relatives who were in central and powerful political positions in the neighbourhood and who were highly important for his leasehold operation, among them his maternal uncle and his elder brother:

> "My own elder brother is also a big leader of AL, [...] from [a neighbouring] Ward. [...] It is not possible to cause any problem to me even for a big leader. They have to think about my brother before telling me something wrong." (Foyez, 23.01.2010)

For Foyez, this meant increased protection for his leasehold operation because they were his supporters and others would carefully consider contesting him and his powerful kinship. Furthermore, Foyez also indicated how the family network of relatives protected each other in times of trouble, thus representing a source of power:

> "This is a big family. I have many relatives here and there. As our family is big, a lot of relatives come. We get together if something happened to anyone. If someone comes to know about the problem, then ten other relatives come ahead after hearing about it. [...] That's why nobody comes to create any trouble with us." (Foyez, 06.04.2010)

While Foyez generated power from his kinship relations in addition to other sources, for Rokib, who operated a plastic drying business, powerful kinship was his main source of power. In the Venn diagram (see Figure 23) he mentioned his *bon-jamai*, the husband of his elder sister, as one of his most important contacts in terms of maintaining his (business) position. This is why he put the *bon-jamai* in one of the largest circles. The distance signals the closeness of the relationship in terms of contact, which he did not have as regularly as with the employees of his business or *mohajons*. Rokib explained about the support of his *bon-jamai*:

> "Yes, my *bon-jamai* [is supporting me in my business]. All obey him. He has houses and everything. Everyone knows me through him. [...] No one can tell me anything because of him; people say 'this is the brother-in-law of him'. Everyone knows him. Everyone knows my nephew, too. If I did not have my *bon-jamai* then maybe it would not be possible for me to live here. You have to put him in the largest ball [of the Venn diagram]. I cannot run [my business] without him. [...] My sister got married here. Everyone knows that I am his brother-in-law. That's why no one has the courage to tell us something. [...] I came here [to Manikpara] for him [*bon-jamai*]. In every *mohalla* there is a powerful person. If the powerful person is someone's own relative then no one tells anything, everyone becomes scared. I have in total ten nephews here. Everyone has houses here. That's why I have some power in the *mohalla*. My *bon-jamai* is the main. If he did not stay here then I could not stay and work here, too. Many people tried hard to evict me from here but they could not do it as I am his brother-in-law." (Rokib, 12.04.2010)

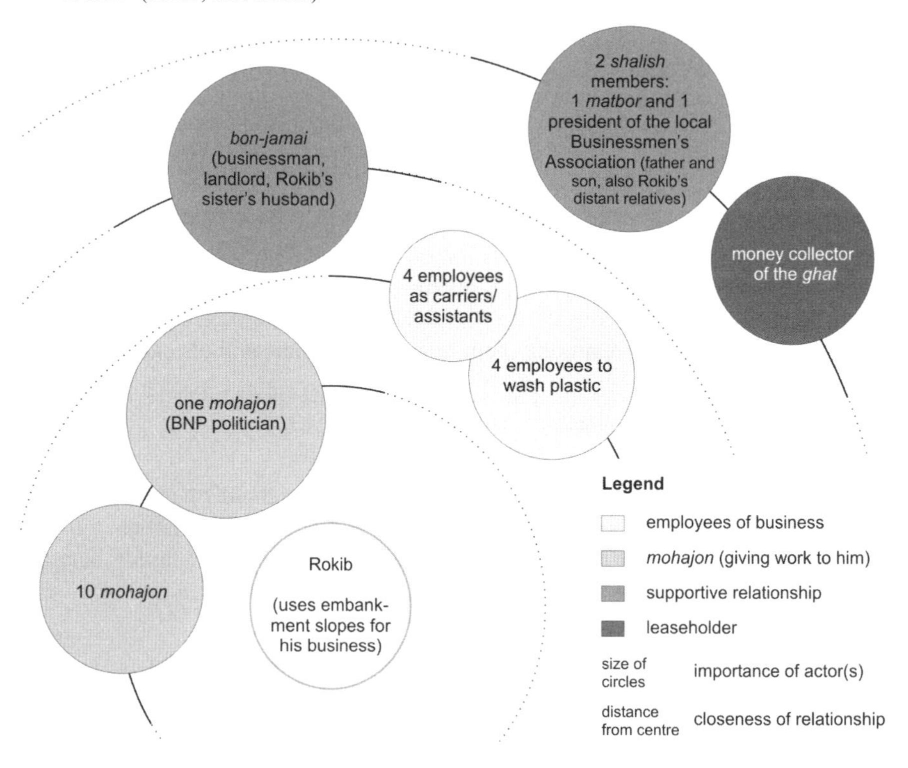

Figure 23: Venn diagram of Rokib, 12.04.2010

Rokib's *bon-jamai* was highly respected in the locality as an economically successful businessman and a member of the local Businessmen's Association. The picture Rokib took of him and his wife during solicited photography shows two very well dressed persons – the *bon-jamai* wears a rather expensive *panjabi* (a male dress of an often knee-long shirt and bulky trousers, similar to the female dress *salwar kamij*) and his wife, Rokib's elder sister, a fine *borka*. Interestingly, Rokib acquired most of his power not via his own family, but via his elder sister's family[10].

In conclusion, among those who have achieved some status within the society and are not among the socio-economically poorest, family relations and the use of these to maintain and secure access to space are of great importance in Manikpara. The two Venn diagrams of Manikpara accordingly show a number of relatives in important positions, while there are no relatives in the Venn diagram of Shahin in Nasimgaon[11].

In Nasimgaon, in contrast, family status and powerful kinship do not so much define one's access to power, and were hardly ever mentioned as a source of power by interviewees (see also Chapter 7.2.2 on rural and urban identities, which found that in Nasimgaon relations to relatives tend to be strongly oriented towards the village rather than within the neighbourhood). Only Rehana managed to get the opportunity to open her (rented) rice shop on Khalabazar with the support of her sister's son, who was involved with the AL committee:

> "When I asked him [shop owner] to give the shop to me [on a rental basis], he did not give it. Then I utilised my sister's son. He told 'My aunt has been living here for many years. But she did not get any land.' [...] Then my sister's son told 'The club [AL Sub-area Committee] gave you the land. Now my aunt wants to take it for rent. Why don't you give it? You have to give it to my aunt.' Then my sister's son helped me to take this place. Now I am sitting here and doing my business. [...] Since he [shop owner] is a friend of my sister's son, he heard what he said and gave the shop to me later on." (Rehana, 23.04.2010)

In both Manikpara and Nasimgaon, it is common to address **non-relatives** using fictive kinship terms (see Chapter 4.4.2). In some instances the relationships assumed point to the generation of power via a protégé-type relation. Foyez maintained such quasi-kin relations with the (official) leaseholder to whom he was like a younger brother, indicating a relationship of caring and respect. Tariq referred to having an uncle-nephew relationship with the (official) leaseholder of Manikghat (2009/2010), indicating that the leaseholder would help/support them in their endeavours and thus represented a source of power. Mesbah Uddin repeatedly referred to his non-kin brother who was the organising secretary of BNP *Chatro Dol* of the area's *Thana*. He mentioned him when he expressed that it would not be prestigious to send his wife to the bazar (see quote in Chapter 6.1.3). This indicates a close relation to this 'brother' as a source of power and respect for Mesbah Uddin. In Nasimgaon, my landlords succeeded once in negotiations with the AL Amlabazar Committee because they achieved the support of the officer in charge

10 In the paternalistic family organisation in Bangladesh, his sister once married belongs to the family of her husband and no longer to Rokib's family in the sense of *gushti*, the patrilineal kinship (see Chapter 5.1.4). Similarly, Rokib's wife has left her family to become a member of Rokib's family. Despite these traditional concepts, there are nonetheless affectionate relationships between maternal relatives, as is the case here.

11 While a single Venn diagram in Nasimgaon and two Venn diagrams in Manikpara do not sufficiently support the argument, these Venn diagram results support the trend already discovered in interviews.

of the *thana* police, who shared their village origin, and thus assumed the role of a more powerful quasi-relative.

The **importance of friendship** for acquisition of power was only mentioned in Manikpara. This also reflects the findings on the importance of spending free time with friends which was most common among (male) inhabitants of Manikpara with higher socio-economic status (see Chapter 6.1.2). Relying on friends as a source of power seems to depend on income, whether one can 'afford' to have friends beyond the family relations. Tariq, who initiated the garden at Dokkin Ghat, relied on friends when the Mosque Committee contested their garden. At that time, they gathered their circle of friends (200–300 people) to make their claims in front of the MP and achieved permission to stay (see Chapter 8.1.3). Similarly, friends were important to Foyez' business operations and he mentioned during the Venn diagram exercise that they were of the same importance as his relatives. Jahangir and his *adda* round friends also kept a close relationship to each other, going on tours together (see Chapter 6.1.2).

The above discussion indicates that family status and kinship relations as sources of power are of great importance in Manikpara. Here the society is closely knitted and has a longer history in the locality, and these traditions place a high level of importance on relatives. This, however, slightly differs for the working population who only came to Manikpara in search of jobs and are not as well integrated into the society. In Nasimgaon, with its more recent history, kinship relations are not of great importance. Here, quasi-kinship, often related to place of origin, and party affiliation (see above) are more important sources of power. Friendship, as already discussed in Chapter 6.1.2, is also of higher importance for power relations in Manikpara than in Nasimgaon.

Bad reputation

While none of the interviewees claimed their own power stemmed from a bad reputation, many narratives mentioned people or groups who acted powerfully with power generated in a non-acceptable or unappreciated way. These people were referred to, dependent on context, as *shontrashi*, *chadabaj*, *gunda*, *rongbaj*, *mastan* and local spoiled/corrupt boys.

Textbox 7:	**Characters of 'bad reputation'**
chadabaj:	extortionist, collecting money (*chada*) unlawfully, forcefully (see Chapter 5.1.1)
gunda:	gangster, used in connection to a political leader, considered a threat to society by causing harm to people
mastan:	muscleman, often local youth working for political leaders (see Chapter 5.1.1)
rongbaj:	gangster, like *gunda* but without necessarily a political connection, rather 'spoiled youth', young men who pass their time aimlessly
shontrashi:	armed gangster, acting against the law and society for his own profit

The ordinary, when talking about the practices of such 'bad characters', usually became very silent and afraid of speaking out loud. This was for example the case when people talked about the conflicts in Nasimgaon with one person and his supporters, who was considered a powerful *mastan*, and was murdered in 2011. For example, Dipa, when talking about the housing developments and resulting conflicts along one of the roads leading out of Nasimgaon, only whispered for fear of being overheard. On the other hand, those who were conscious of their own power talked openly about people they considered to belong to the above mentioned 'characters'. For example, Mokbul Hossain was not afraid to speak his mind during interviews and informal discussions, and thus referred to one high level JP politician as a big *mastan*, accused another one, who had once managed to take *chada* from him, as *chadabaj* and repeatedly talked about the local 'spoiled boys'. The leasehold operator Foyez was also not afraid to refer to those giving him trouble as *mastan*, *gunda* and *rongbaj* (and mentioned them in the Venn diagram accordingly, see Figure 24 in Chapter 9.2.1). The current and previous leaseholders, Foyez and Jahangir, however, indicated using 'bold' expressions that they also had some power and were 'naughty' which could be interpreted as power through bad reputation by others. Jahangir told of how, when he started the leasehold and collected charges, people perceived him as *chadabaj*. As he had legal documents for the leasehold he did not consider himself *chadabaj*, and he treated the people at the *ghats* very well, which can be understood as a means to generate and achieve respect.

The frequency with which interviewees referred to characters of bad reputation showed the prevalence of such practices and the fearful perception of them. The most 'physically violent' characters were mainly mentioned with reference to past times.

Conclusion: Availability of power sources in the study settlements

The overview of the importance of power sources provided in Table 6 shows that in Manikpara there is a larger variety of power sources that actors can draw upon in negotiating spatial claims. An exception is 'doing politics' as a source of power for women. Doing politics and affiliation with the ruling political party is an important power source in both settlements, however, women seem to be less involved in politics in Manikpara. The difference of the importance of social status as a source of power is most pronounced. While in Nasimgaon the *shalish* is important and provides its members with considerable power, the importance of other sources of social status and respect can be considered less important than in Manikpara. The reasons for this can be found in the different structure of society which is a conglomerate of rural-urban migrants rather than being built on a strong belonging as in the case of Manikpara (see also the analysis in Chapter 7.2.2). Furthermore the community is less conservative, which is partly attributed to the socio-economic status of inhabitants, and thus notions of respect and honour have different importance as sources of power than in Manikpara. This directly relates to the difference in religious education, which is less common in Nasimgaon than in Manikpara. With regard to involvement in community work, such involvement in Nasimgaon tends to be understood as being politically guided and discriminatory along the lines of party affiliation and thus not necessarily as a source of respect leading to social status. The social relations that can be drawn upon as a source of power in Nasimgaon furthermore differ due to the socio-economic status of inhabitants (non-affordability of friends, see Chapter 6.1.2) and the more recent migration history.

Power sources	*Manikpara*	*Nasimgaon*
Political system and party affiliation		
- 'doing politics' and affiliation to the ruling party	⬤	⬤
- 'doing politics' as a source of power for women	-	⬤
Social status		
- respect and honour	⬤	•
- involvement in community works	⬤	⬤
- involvement in traditional institutions	⬤	⬤
- 'being a local'	⬤	•
Religion and religious norms		
- religious institutions	⬤	⬤
- religious education, positions	⬤	•
- hosting of religious programmes	⬤	⬤
Economic resources	⬤	•
Social relations		
- family status	⬤	-
- relations to powerful kinship	⬤	•
- relations to quasi-kin	•	•
- relations to friends	•	-
Bad reputation	⬤	⬤

Power sources of ⬤ high importance • low importance - not relevant

Table 6: Overview of sources of power in Manikpara and Nasimgaon

9.2 STRATEGIES TO NEGOTIATE AND CONTEST SPATIAL CLAIMS

In this sub-chapter, I aim to differentiate the strategies applied in spatial negotiation processes. The individual strategies followed by different actors can be understood as reactions to the spatial claims of others or to 'requirements' emanating from the livelihood strategy followed by an actor. The interaction of actors in their use of individual strategies then constitutes the negotiations of access to public space. The strategies will be differentiated based on the aims followed:

- to secure a spatial claim (preemptive),
- to respond to contestations (reactive), and
- to contest the spatial claims of others (offensive).

Furthermore, in the last section I will discuss resistance strategies that do not necessarily have direct spatial implications. Throughout this sub-chapter, I seek to analyse how actors make use of their sources of power in performing specific strategies and how this involves social relations, norms and institutions. At the end of each sub-chapter, the strategies applied are summarised in a table, differentiating the two study settlements and the application by elite groups or the ordinary.

9.2.1 Preemptive strategies to secure spatial claims

Preemptive strategies are applied to secure spatial claims against potential future contestations. This means ensuring that one's spatial claims are recognised and respected by others, and that others will not contest them in the future. Preemptive strategies are closely related to reactive strategies, some of which also serve to secure a claim in the long run (see Chapter 9.2.2).

Maintaining relationships with statutory institutions and actors

In order to secure spatial claims and avoid contestations, especially the elite groups **maintain good relations with the forces of the executive**, such as the police and RAB. This is done through extra payments and provision of 'entertainments'. In the above story of the *ghat* leasehold (see Chapter 8.1), Foyez narrated how he spent money on the police every week. He especially maintained close relations with one police officer whom he sent Flexiload (mobile phone top-ups) of 300 Tk per week, and whom he also mentioned in the Venn diagram (see Figure 24, dark circles for institutions of the executive; these were characterised by him to be of medium importance for his leasehold operation, the contact with these institutions and persons remained irregular and the relationship was not close). In order to increase security and limit contestations at the *ghats*, Foyez extended the 'normal duty area' of the police via additional payments to the police on duty:

> "There are boats till 12pm in the *ghats*. […] No problem occurs. The police are there. We give money to the police. […] We give 500 Tk for the five *ghats* per week. […] They are not supposed to guard only the *ghats*, but they also have the duty to take care of the whole area. But we are paying them money so that they take special care of the *ghats* and no problem can occur here." (Foyez, 26.06.2009)

Similar payments were also made by the previous leaseholder Jahangir to uphold security at the *ghats*. He secured his leasehold further by sending a copy of the leasehold contract to five surrounding *thana* police stations. This can be understood as providing information on and re-confirming his legitimate claim of jurisdiction over Manikghat. But also Kabin Mia tried to maintain a good relationship with the police when he operated his rickshaw garage along Embankment Road before its reconstruction (see Textbox 5 in Chapter 6.3). According to him, the police came regularly to sit at his place and took small amounts of 2–10 Tk from him. Mesbah Uddin told us that the police regularly threatened to evict some small-scale shopkeepers in front of his house, but once the owners served tea and cigarettes, the policemen did not carry out the evictions. In Nasimgaon, the *thana* police are less present in the neighbourhood but do visit the area in case of problems. According to Shahidul, one of the night guards of Amlabazar, the committee then needed to serve food to the police officers of the *thana* – using money generated by the daily collection from the shopkeepers (see Chapter 9.1.2).

Besides the executive, especially the leaseholders in Manikpara secured their spatial claim by **drawing on the support of administrative officers**. While this seems a logical consequence of the statutory leasehold contract between the leaseholder and the BIWTA, the relationship goes beyond the statutory procedures. In the Venn diagram exercise, Foyez mentioned seven officers of BIWTA as being important to his leasehold operation. One of his contacts in BIWTA offered to ensure that he would become the next leaseholder from July 2010. At the beginning of 2010 he completely took over the leasehold from the official leaseholder, who then remained the leaseholder only on paper, while communication about the leasehold went on between BIWTA and Foyez directly and Foyez was also the one obliged to pay the outstanding amount of the 2009/2010 leasehold of 300,000 Tk to BIWTA. The BIWTA officer's offer to receive bribes which, however, was not accepted by Foyez at that time, underlines how maintaining relationships with statutory institutions can be a preemptive strategy to secure a spatial claim:

> "Now the *ghat* is under my authority. Yesterday I gave 25,000 Tk to the officer [BIWTA]. He said to me that, 'Give additional 300,000 with this 300,000 Tk [the outstanding amount of the lease money for 2009/2010]. I will give the *ghat* to you for the next year.' I have a good relationship with that officer [BIWTA]." (Foyez, 06.04.2010)

Jahangir was also convinced that due to his relation to a BIWTA officer, he would not be evicted from the embankment slopes:

> "They [BIWTA] will not be able to do so [evict his shop]. I know an officer at the head office in Motijheel [pronounce Motijhil]. They wanted to evict my house before and that person said 'No. You can stay here. You are taking care of this area'. He also said 'You will inform me if someone comes to evict your house'." (Jahangir, 11.02.2010)

However, during the eviction drive in August 2010 this relationship did not suffice to exempt Jahangir from eviction. The statutory legitimation of the leasehold also caused Rokib to underline the necessity to keep a good contact with the leaseholder to secure his access to space (whom he also included as the 'money collector of the *ghats*' in his Venn diagram, see Figure 23):

> "Everyone needs to do contact. We have to be on good terms with them [the leaseholder/lease operator]. If I am not involved with the *ghat,* then they will not allow us to unload materials in the *ghat*. We have to be on good terms with them. There was bidding from the government, so we have to stay with them mutually." (Rokib, 06.04.2010)

In Nasimgaon, there is no area under the official jurisdiction of a statutory authority like in the case of Manikghat and the embankment slopes. Thus administrative officers are rarely consulted to secure spatial claims.

The main statutory institutions drawn upon to preemptively secure spatial claims were those of the executive, i.e. the *thana* police and RAB, but also administrative officers. As the above discussion indicates, especially elite groups, but also in some cases the ordinary, maintain good relationships with statutory institutions in order to secure their spatial claims against future contestations. However, it also becomes obvious that in many cases the support of statutory institutions is not generated via the statutorily authorised channel. Instead, actors secure the support of statutory institutions by making extra payments and providing 'entertainment' – *de facto* bribes – to employees of statutory institutions. Statutory institutions thus become instrumentalised to contribute to modes of the production of space that do not follow statutory directives but carry with them the 'disguise of legality'. However, as the examples above also indicate, the administrative officers happily get involved in non-statutory modes of production (follow-up discussion in Chapter 10).

Involvement in politics and maintaining relationships with politicians

Involvement in politics and maintaining relationships with politicians and political organisations is another preemptive strategy to secure spatial claims.

The elite groups are commonly **involved in politics** which can be considered a strategy to generate power (see 9.1.3), and consequently this involvement is a means towards securing spatial claims. Accordingly, most of those who make large spatial claims are involved in the ruling political party AL, e.g. the leaseholder Foyez, those who took possession of parts of Nasimgaon Eid Gah Math (the water tank, school and hospital grounds transformed into rickshaw garages and a row of shops) and of the *bhits* at Khalabazar. While some of these claims were nonetheless contested, the spatial claims could only be achieved by 'doing politics'. As such then, 'doing politics' for the ruling political party can be considered a preemptive strategy as it reduces the opportunities for contestation considerably. This strategy is, however, usually limited to one electoral period of normally five years (see also on 'domination of social space' below).

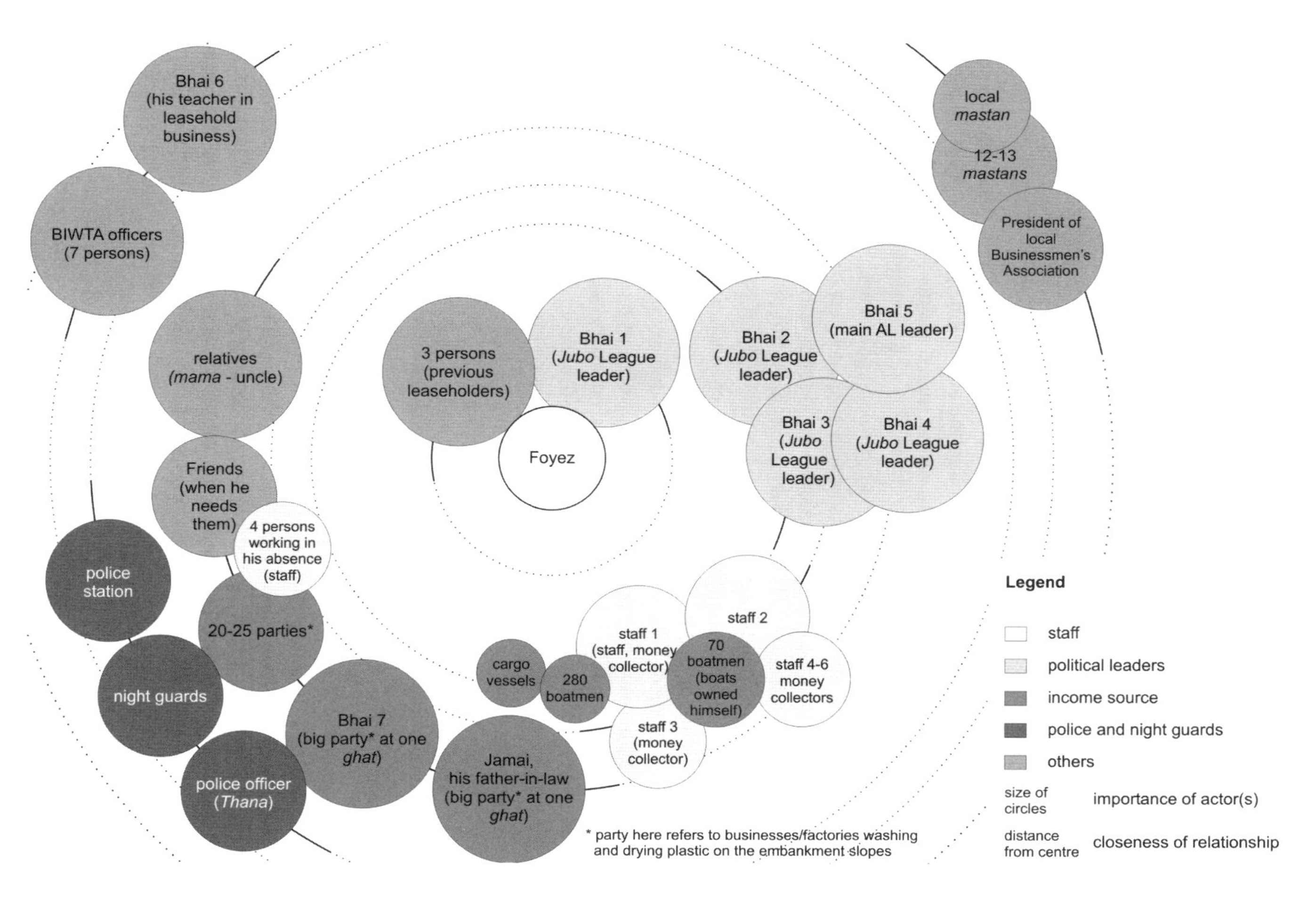

Figure 24: Venn diagram of the lease operator Foyez, 06.04.2010

To secure spatial claims, especially the elite groups **maintain relationships with powerful political actors** – accordingly these are the AL politicians, due to the currently limited power of the BNP – who then support their spatial claims against contestations. These relationships, similar to the relationships with the police, are maintained by providing entertainment, making extra payments, participating in *michils* (political demonstrations) and granting concessions. For example, Foyez reduced or even waived the charges taken from cargo vessels belonging to important AL leaders (see quote in Chapter 8.1.2). In his Venn diagram (see Figure 24) a large number of the people he kept up relations with and considered relevant for his business were AL leaders, both from the main party and its sub-organisation *Jubo* League, the powerful youth sub-organisation of AL whose existence goes back to the Independence War. Foyez dedicated five of the largest circles to these leaders, *Bhai* 1–5, and the closeness of these circles to him indicated that he maintained a very close and regular relationship with these individuals. He described this maintenance of relationships with one of the *Jubo* League leaders as follows:

> "There is another person named [*Bhai* 1 in the Venn diagram] living in this locality. He comes to the *ghat* every day to take a cup of tea and joke with me. He also asks me 'Are you facing any problem here?' I tell him 'No'. He comes to the *ghat* every day." (Foyez, 06.04.2010)

In Nasimgaon the discussion with local leaders revealed how the elite groups maintained relationships with higher level political leaders. The relationships were especially maintained by organising and taking part in political demonstrations and election campaigns in exchange for support during contestations (see also Chapter 9.2.2):

> "As we do politics, the higher level leaders always call us when they need us in meetings or *michils*. We go to do their *michils*. They also come for us. They [big leaders] don't come to *michils*. We always join the *michils*. Yesterday […] a BNP leader barricaded some roads. All of them [who participated in the barricade] were the slum dwellers. The big leaders live in [high-income areas] and they don't come to the road to join in a *michils*. They need us badly for this. They cannot go without us. They definitely respect us. Otherwise they will not get us when they need us." (Sayed, 23.04.2010)

But also the ordinary maintain relationships with politicians to secure their spatial claims. Their participation in *michils* can be regarded as such a strategy, although it often serves rather immediate gains (e.g. the distribution of blankets to the participants, see Chapter 6.2.3). Rokib identified political actors in his Venn diagram whom he could consult in times of contestation (see Figure 23 in Chapter 9.1.3). The two *shalish* members, one of the *mohajons* commissioning work to Rokib and his *bon-jamai* were all involved in politics of the AL or BNP. Furthermore, when in the rainy season he moved his washing and drying operation to an alternative location (see Chapter 6.1.1), he secured this spatial claim by paying *chada* to the 'local boys' (considered *mastans*, with a political connotation).

On Nasimgaon Eid Gah Math, Mokbul Hossain also kept relations with some political actors, some of whom I met when having informal discussions at his place. One of these leaders offered to prevent any future contestations if he would make an extra payment to higher level leaders:

> "I have a very good relationship with a local influential party leader and that leader has suggested me to give about 10,000 Tk to the higher level leaders to prevent the risk of eviction of my shop permanently. But I did not agree. I got hurt in my head one time, I am willing to suffer this again, but I will not give money." (Mokbul Hossain, 25.07.2009)

While this demonstrates how politicians can be supportive in securing spatial claims, Mokbul Hossain's denial points at his perception of such practices as having a 'bad reputation' and being dishonourable. Taslima, in operating her *pitha* stall on Nasimgaon Eid Gah Math, was also able to draw on political leaders' support to secure her spatial claim. However, she did not become active herself in maintaining a good relationship, but narrated that the leaders supported her out of a feeling of social responsibility for her poverty. In contrast, Shahin was not able to secure his spatial claim (he had to start being a tenant instead of an owner of his shop) because he supported the 'wrong' party:

> "I am a supporter of BNP. And they are AL. That's why I did not get the plot. If I was a supporter of AL then I could get the plot. Maybe I would have to spend an amount of 2,000–4,000 Tk to get it. But still I could get it. I am supporter of BNP. That's why I did not get it as the land is in front of AL [office]. So they have a right over this place. It does not belong to us. I am paying rent and doing my business and will continue in this way as long as I am here." (Shahin, 21.04.2010)

In her relationship towards Hatem at Khalabazar in Nasimgaon, Afsana addressed him as *mama*, maternal uncle, and in her narrative he in return also addressed her as *mama*[12]. Here the matter is not one of a close and supportive relationship, but nonetheless the forms of address indicate the power relations between the two. While the two of them are actually in permanent contestation with each other, the reciprocal address as *mama* symbolises how both consider the other to have acquired a respected status in society and thus transport this level of respect into their relationship. By so doing they maintain their own power position, not allowing themselves to make any movements that would result in the loss of their respected status. For example, as Afsana was powerful due to her involvement in politics and her close relation with other (female) politicians, treating her disrespectfully could create problems for Hatem. He thus rather chose a highly respectful way of addressing Afsana. On the other hand, Hatem despite all contestations, was a high level politician and a religiously respected person, so Afsana could not afford to dishonour this status by addressing him without respect.

Several interviewees used their own or relatives' **multi-party involvement** as a strategy to secure space-based livelihoods in the long run. Rokib kept close contacts with two brothers, one of them a BNP politician and the other involved in the AL and the president of a local Businessmen's Association, one of his most important contacts. Furthermore, one of his *bon-jamai's* sons was involved in the

12 In Bangladesh, it is a matter of respect and affection to use a title in a reverse way, i.e. a grandmother can affectionately address her granddaughter as 'grandmother'.

BNP and another one in the AL – which provided Rokib with the means to sustain his space-based livelihoods even when the political government changed. Jahangir, operating his business of plastic sorting from a room on the embankment slopes, also stated that he faced no problem whatever the political party was, as he kept connections with both AL and BNP leaders. Even the leaseholder Foyez indicated that he had contacts to win the leasehold tender procedure during BNP times:

> "I will have no problem if BNP government comes again – because we have a family relation with Nisir Uddin Ahmed Pintu [former MP of BNP, 2001–2006]. Moreover no one understands the *ghat* matter better than me. So I will always be the leaseholder of the *ghat* in future." (Foyez, 25.11.2010)

In Nasimgaon, multi-party involvement did not appear to be a similarly important issue during interviews. While this was not a topic for discussion among the AL political leaders of local committees, it evolved as an issue in the discussion with the AL politician Rizia with regard to NGO committees. Rizia indicated that the members of these committees came from different parties and their aim was to keep a good relationship with all parties, to avoid a situation where a new party coming to power would remove them.

Involvement in political parties is especially a strategy of the elite groups to secure spatial claims; however, involvement in AL demonstrations might also serve as a preemptive strategy of the ordinary, while the involvement of vendors in the BNP is rather counterproductive. Especially elite groups who have to secure large spatial claims do so by maintaining relationships with political leaders through providing entertainment and money. Multi-party involvement of an individual or an individual's powerful (quasi-)kin is furthermore employed as a strategy to secure spatial claims beyond the electoral periods.

Maintaining relationships with other actors

Apart from maintaining relationships explicitly with statutory institutions or political actors, securing spatial claims can also be achieved by drawing on religious institutions or by maintaining relationships with actors beyond statutory and political spheres.

In Manikpara, by making **regular contributions to the local mosque** the leaseholder managed to have his spatial claim perceived as religiously legitimate. For himself, the mosque was not influential for the *ghat* operations and he consciously excluded it from the Venn diagram:

> "There is no influence of the mosque on the *ghat* matter. I give 100 Tk as subscription to the mosque fund. I want to keep them [mosque committee] happy. [...] I used to give 100 Tk every day to the mosque fund and now I give 500 Tk per week as income is less." (Foyez, 06.04.2010)

For Rokib, however, this religious contribution of the leaseholder served as a legitimation of the leaseholder's jurisdiction over the slopes. Rokib understood the payments he made to the leaseholder as a religious contribution (and a government contribution, see above) he generally appreciated. Thus for Rokib, the leaseholder's power over the embankment slopes was religiously legitimated:

> "The mosque is weak; there is no income or improvement. That's why the money is given to the mosque. It means they [leaseholder and money collectors] don't eat the money; they collect the money from us and give it to the mosque by fully calculating it. A receipt is given from the mosque." (Rokib, 12.04.2010)

Foyez' maintenance of relations with the mosque committee through a weekly contribution thus serves as a preemptive strategy to secure his claims (see also 'creating incontestable space' below). Similarly, the vendors on Nasimgaon Eid Gah Math contribute a daily amount to the mosque fund. Only Mokbul Hossain, shop owner at the Math, used the payments as a more active preemptive strategy, as he made the payments[13] despite his knowledge that the money would not be used for the mosque:

> "People from the mosque committee come and take about 50/100/200 Tk. I have to give this money. I know that they will not do anything with this money. But I don't say anything." (Mokbul Hossain, 23.03.2010)

In order to secure their spatial claims, actors also **maintain relationships with kin, quasi-kin, friends** (see social relations in Chapter 9.1.3) **and generally the inhabitants** – although some of these groups may also be related to statutory institutions and politicians and the border remains fluid. The leaseholder Foyez underlined the importance of maintaining connections and contacts in general to secure spatial claims:

> "But you should keep connections, i.e. connection with the inhabitants, connection with the chairman, etc. Otherwise you can't take the *ghat*, or maybe you can take the *ghat* but can't manage it. It is too tough to keep it workable. Suppose we stand here from the morning till the night and anything can happen like anyone can shoot me from any side, any time. That means we have risks. For example, many people come to collect money for many purposes, but we don't give." (Foyez, 06.04.2010)

In Foyez' Venn diagram a number of actors like relatives and friends but also his employees are important for securing his spatial claim and he thus invested in maintaining good relations. Furthermore, Foyez underlined the importance of being sensible about what was supported by the local inhabitants – as the case of the introduction of passenger fees demonstrated (see Chapter 8.1.2). When he started to collect fees from everyone using the *ghats*, the local inhabitants set fire to the stations in protest. Afterwards Foyez realised how important the awareness of local perceptions was as a preemptive strategy to avoid contestations.

Shahin, who together with his wife Rohima operated a fruit and tea stall on Khalabazar, underlined yet another dimension of maintaining good relations by

13 During the interviews, Mokbul Hossain narrated sometimes making payments and sometimes refusing to do so. While this may seem contradictory, it rather has to be understood as his 'strategic choice'.

good behaviour with everyone, which refers to notions of respect and honour, and understood this as a social investment for future contestation:

> "I have good relations with everyone. There is no such person who has a bad impression of me. I also didn't do any bad work by the blessing of Allah. If someone told me any bad word still I gave him a hug rather than having any conflict with him. Maybe it is seen that I have got a loss of 5,000 Tk but still I did not have any conflict with anyone. Today there is money; tomorrow it may not remain; but the word what we say will remain for ever. Suppose we and many other people are living in the *bosti* now. But if the *bosti* is evicted, then people will go to their village tomorrow or some may go to another place. But one will not get a good relation. In this place I walk, sit, talk and laugh with all. That's why I get on with everyone and maintain a good relation. Everyone has a good impression of me and loves me. No one is doing any harm to me. Everyone comes to help me in my danger time. If I have any danger, then all will come running and ask 'What has happened? Tell us'." (Shahin, 21.04.2010)

The awareness of good and bad behaviour expressed by many other vendors points in a similar direction. For women in particular, maintaining a specific dress code that is acceptable to society indicates good behaviour and reputation (see Chapter 7.2.4).

As has already been suggested above (statutory institutions and politics), the **provision of entertainment and/or money** can be considered central to generating support and securing spatial claims. For example, Mokbul Hossain spent about 2,700 Tk per month on entertaining others and his own consumption – especially for bethel nuts and leaves and tea. He explained that people consider him rich because of his large business and thus he had to incur these hospitality costs. Other means of maintaining good relationships – and equally valid for statutory institutions and politicians – is the organisation of functions and the invitation of important actors. This has also been discussed in Chapter 9.1.3 on power sources, for example with regard to invitations to religious functions, such as those hosted regularly by Rokib and Jahangir, but also in Nasimgaon.

The above discussion extended the maintenance of relationships beyond statutory and political actors to include religious institutions, but also general inhabitants, kin, quasi-kin, friends or staff members. Such relationships are often also based on entertainment provision and money, but similarly on spending time and 'good behaviour with others' as the quote of Shahin indicates, or on being sensible about what the local inhabitants would support and when they would protest against specific claims. Maintaining good relationships in the sense of 'good behaviour' is a strategy widely employed by the ordinary, which may be attributed to the fact that this does not necessarily require financial resources.

'Formalisation' of spatial claims

The strategy of formalising spatial claims refers not necessarily to statutory types of formalisation. Ways of formalising spatial claims are rather any strategies that make the spatial claim appear more reasonable/justified or legitimised and impressive.

Spatial claims can be underlined by **constructing more permanent structures** (see also in Chapter 9.2.3, as this at the same time was used as an offensive strategy) which would be more difficult to contest and make a claim appear more justified. This strategy can be assigned to the AL leaders' activities on Khalabazar. From the beginning of 2009 onwards, the leaders who had taken possession of the previously open space justified their charging of rents and deposits by several stepwise improvements. At first, the *paka* platform was established and taken possession of. With each further construction stage, the rent and deposit increased. This strategy of 'informal formalisation' can be understood as a means to increase legitimation and secure the spatial claim, as increases in payments can be justified by improvements in the building structure. This strategy did not work out in all cases – for example Taslima's formalisation of her claim by spreading debris was immediately contested.

The construction of stable rickshaw garages on the south of the Nasimgaon Math can also be understood as a formalisation of spatial claims. Especially the fact that the garages were constructed by the MP served to prevent contestation, as the MP's activities were perceived as being statutorily legitimated (see Chapter 9.2.2). But generally, as soon as structures become relatively stable, as experienced on the Nasimgaon Eid Gah Math, or with the housing compound constructed in place of the fish bazar at Khalabazar, this is understood as a formalisation of a spatial claim that could only be contested with considerable trouble. For example, the structures that had been built on the Nasimgaon Eid Gah Math during BNP times included the three stable shops and community police office, and the rickshaw garages which at the time of the contestations were not of stable/durable materials. The concentration on contesting the garages, while the three shops/office remained uncontested (despite some re-construction to make way for the access road) supports the above observation. Similarly the owners of the stable shop structures of Shaonbazar were not contested, while the mobile to semi-permanent units at Khalabazar were contested in 2009.

The **monthly contributions** made by vendors to the sweepers and night guards (but also to the mosque) were also perceived as a kind of formalisation by Aklima, the semi-mobile vendor on Nasimgaon Math, following the logic of having a right to space through payment of the required fees for its usage. This becomes evident from her story of a contestation she and her husband faced by local *mastans* who wanted to collect *chada*. The leaders to whom they paid the night guard, sweeper and mosque fees supported them against the *mastans*. So her perception was that by paying the fees regularly she also received the support of those she perceived as 'higher level leaders', i.e. the local political leaders.

Another way of formalising spatial claims is to get a **written certificate of ownership** which then can be produced to avoid future contestations. The leasehold contract as a statutory document serves to underline the spatial claim of the leaseholders. Deed documents especially serve to underline spatial claims which is why in Manikpara a group of inhabitants went to court (see Mesbah Uddin in Chapter 7.2.2) and in Nasimgaon Rizia expressed the importance of getting a holding number (here refers to a registered right of occupancy) to secure the right to stay for Nasimgaon's inhabitants in the face of the threat of evictions.

Apart from these statutory certificates, however, the ordinary at Khalabazar also aimed at getting 'local' written agreements or contracts, i.e. not necessarily statutorily verified. In January 2010, after she had experienced a year of continuous increases of rent and deposit payments for her shop, Rohima expressed her wish to get a contract or written agreement from the shop owner, in order to be confirmed as the tenant. However, her endeavour remained unsuccessful. Both she and Shoma (operating a vegetable shop in the same place) furthermore expressed their interest in buying the shops from the current owners (the AL leaders who had claimed ownership in 2009) which would also formalise their claims (see also Chapter 9.2.2).

Afsana's experience of signing for her ownership of a *bhit* at Khalabazar, however, tells a story of scepticism against the usefulness of written documents to secure a spatial claim, as in her case she was tricked into contestations. The local leaders had asked her to sign for *bhit* number 8, telling her that she would receive the *bhit* of the location where she was previously doing her business from. But then she realised that the place in the corner that she had occupied was marked as number 7 and had been taken by another person. When her husband asked her why she had signed for the wrong number she answered: "What they told by mouth will remain, but what was written in the paper will not be valid" (Afsana, 23.03.2010). Eventually, with the political support of Rizia, she managed to get the *bhit* number 7 in the corner position, reinstating her previous location. Afsana's scepticism of the written contract is surprising, giving the value Rohima and Shoma attached to a written document. Her scepticism, however, was also expressed when the RAB asked her to produce a written document in another case of contestations and she told them that if there were one they could have torn it up.

The above discussion has shown how statutory formalisation is mainly a resource the elite of Manikpara can draw upon, especially the leaseholder, but also those who are house owners. For inhabitants of Nasimgaon, 'formalisation' as a strategy to secure spatial claims has a different connotation. The elite groups are able to use the transformation of structures as a means to formalise their spatial claims, as this complicates contestations considerably. But also the ordinary strive for a formalisation of their spatial claims with a view to reducing risks, insecurity and the vulnerability of their position and livelihoods. While Aklima already understood the regular payment of fees as a 'formalisation', the vendors at Khalabazar aimed at getting written contracts to secure spatial claims or even ownership of structures.

Creating incontestable space

The creation of spaces that cannot be contested is a powerful preemptive strategy, and here relates to religious spaces/institutions on the one hand, and spaces that are indispensable due to their public welfare orientation on the other hand.

The special power and respect attached to **religious institutions** has already been discussed (sources of power in Chapter 9.1.3 and above about maintenance of relationships with other actors, the example of contributing to the mosque and thus achieving a religious legitimation). In many cases, spatial claims are related to religious institutions, and thus largely protected from contestations. Mokbul Hossain, normally a person not shy of protesting, did not do so when he was made to leave his business premises on open space during BNP times because of the construction of the *Eid Gah* monument where the *Eid* prayers are held. The political leaders of Nasimgaon also emphasised the importance of religious institutions to underline their spatial claims. Furthermore, the supervision of the Nasimgaon Eid Gah Math by the nearby mosque served to protect the Math, although the principle of protection expressed by local leaders and the practice of encroachments partly carried out by the very same leaders did not correspond. However, in 'theory' the protective function of the mosque was underlined repeatedly by the local leader Harez:

> "The Math management is the responsibility of the Mosque Committee. We left it. We left it because if the responsibility was with us who are the social representatives, then we could sometimes become greedy. But it is vested in the Mosque Committee and it is the Eid Gah Math and no one can interfere in it if there are some religious feelings inside. That's why the Math is under the Mosque Committee." (Harez, 16.03.2010)

While Harez' statement vividly underlines the strategy of designating the Math a religious place in order to be able to draw on the protective function associated with this, the very same leader and his Sub-area Committee were involved in establishing the new rickshaw garages on the Math at the end of 2010 (see Chapter 0).

The Amlabazar Committee also secured its claim of jurisdiction over the bazar area by attachment to the local mosque. The fees collected from the shops were, among other purposes, used for the mosque, *madrasa* and payment of salaries to mosque employees. Here again, religion serves as a legitimation for a spatial claim made by the committee and partly as a legitimation for payments to be taken from shopkeepers.

Furthermore, spatial claims made by religious institutions cannot easily be protested even if they do not have a religious connotation *per se*. This is demonstrated by the example of Tariq's garden which could not expand towards Dokkin Ghat because the mosque operated two rickshaw garages on the site (see Figure 17). Despite the non-religious use of rickshaw garages, the fact that these were managed by the mosque made them incontestable (see religion as a source of power in Chapter 9.1.3). Similarly, the daily *chada* collections on Nasimgaon Eid

Gah Math by the local Mosque Committee are not understood as unjust by the vendors and the religious connotations suffice to achieve acceptance:

> "You can give to the Mosque as you wish. You can give 10 Tk or 5 Tk. It is our own [spiritual] benefit if we give some money here." (Taslima, 18.04.2010)

Only Mokbul Hossain, despite paying the subscription, indicated how these were used for individual benefit rather than for the mosque (see above).

The two gardens in Manikpara are examples of how the spatial claim is maintained by **creating indispensable spaces**. While the Bagan has long been established and would certainly no longer be contested, Tariq's garden could still be. But both the reaction of the leaseholder Foyez and the sparing of the garden during the BIWTA eviction drive indicate how the strategy of Tariq and his friends had worked out:

> Friend 1: "If we plant trees in this place and the trees will grow up, then the government will not come to remove us. Because the government knows that trees are keeping a balance of the environment. Trees are very necessary for the environment of Dhaka City." [...]
>
> Friend 2: "If some people want to develop or improve this place, then it does not make any sense to stop them. The government people should not give veto. We are developing this place though they are not doing it. We are planting the trees. People will get shade and oxygen from the trees. The government also said 'Plant trees here and there'." (two of Tariq's friends, 19.04.2010)

Thus, Tariq and his friends had consciously created a space of value for both the community and the environment in order to prevent the government from actions against them.

A different indispensable space for public welfare was created on Khalabazar. Some of the vendors became members of a saving scheme operated by the AL Sub-area Committee and part of their rent payments went into this saving scheme. Shahin said that in this way he was saving part of his rent payments every month – thus this can also be regarded as a preemptive strategy to secure the spatial claim of the Committee leaders, as contesting could result in Shahin losing his savings. However, the criteria for participation in this scheme were not transparent and not all vendors participated.

The creation of incontestable space as analysed above is a means for elite groups to secure their spatial claims. In the case of the Eid Gah Math, religious protection at the same time serves the welfare of the inhabitants – however, the further analysis reveals that this is mainly a principle, while in practice religious arguments are also used in offensive strategies and to make individual spatial claims (see Chapter 9.2.3). The indispensible spatial claims for public welfare, in contrast, can be understood as serving the community both in principle and in practice.

Withdrawal of spatial claims in anticipation of contestation

The withdrawal of spatial claims in order to avoid contestations is another preemptive strategy (although this is similarly a strategy in reaction to contestations, see Chapter 9.2.2 for further discussion). Here I will discuss the selling of spatial claims and the withdrawal of spatial claims as a means to maintain good relations.

One strategy of elite groups to avoid future contestation is the **selling of spatial claims** made during an electoral period shortly before a change of the ruling party can be expected. This was exercised when the BNP government period ended in 2006. The rickshaw garages established on Nasimgaon Eid Gah Math were let until the end of the BNP period, and then sold to the garage operators shortly before the elections. Similarly, this strategy is anticipated by AL leaders. The owner of Rohima and Shahin's shop at Khalabazar offered to sell the shop to them in 2012/2013. This would be about the time for new elections, and the selling offer seems to be a strategy to avoid contestations of the spatial claims of AL leaders by a future BNP government. If after the election the shops would be in the possession of the ordinary, the AL elite groups could not be contested and would have secured a maximum profit from their claims within five years. Afsana already sold her *bhit* for about 20,000 Tk, because she was convinced that after the WC elections the shops would no longer exist as she planned the eviction of the structures put up by the AL Committee politicians and the re-establishing of the old system of vending (see Chapter 8.2).

The selling of spatial claims and thus securing maximum profit before upcoming elections is a common strategy in Nasimgaon, while in Manikpara this is slightly different. Here, similar spatial encroachments are less common due to the lack of open spaces. Instead, the spatial claims and accordingly the contestations refer to the jurisdiction over specific spaces.

Another, more short-term strategy to avoid contestations is the **withdrawal of spatial claims** to the benefit of or to silence potentially contesting actors. This strategy was employed by the Amlabazar Committee in Nasimgaon. In the beginning of the AL government period, they sat together with the Sub-area Committee to discuss management responsibilities and the jurisdiction over (public) spaces. The Amlabazar Committee, who claimed jurisdiction of the whole bazar area, then withdrew their claim to the comparatively small area of Khalabazar (see Chapter 8.2, narrated by Sayed) upon the recommendation of the Ward level political committee. By conceding part of their spatial claim to the Sub-area Committee, they ensured that the Sub-area Committee would largely keep out of their affairs in the bazar area. Although the Sub-area Committee is the 'umbrella committee' in 'theory' (see the discussion of Figure A-3 below), it is less powerful than the Amlabazar Committee. The following quote of the Sub-area Committee leader Baharul underlines how he would not interfere in the business of the Amlabazar Committee:

> "If I could be in that committee [Amlabazar Committee] then I could do something. Suppose if there is anything good or bad happening in your country, we cannot interfere there. Can we? Thus there are problems in the bazar. Since we are not in the bazar committee, whatever they are doing it is right according to them." (Baharul, 01.03.2010)

He made this statement in relation to improving accessibility of the bazar area by introducing a system of shutters to close shops, so that the extension of shops onto the road would be limited. That he did not suggest this, although it would benefit the whole settlement, shows the differentiated power structure between the two committees (which is reverse to their spatial hierarchy) and points at the Amlabazar Committee achieving its aim by conceding part of the jurisdiction of the bazar.

Similarly, Foyez used the conceding of jurisdiction over space to secure his own claims. He conceded the jurisdiction of the embankment slopes in front of the Bagan to the Bagan Committee:

> "We have given much concession here, e.g. we don't take the money earned from drying of materials in the Bagan. This amount is for the garden. It is used as the expenditure for the garden e.g. there is a security guard in the garden, trees are planted in the garden and food is needed for the birds and animals of the garden. It is right that if I, being the leaseholder, allow them to take that money then they get it, but if I don't give it willingly then they can't get." (Foyez, 26.06.2009)

While this statement seems to indicate that he was free to decide whether to concede or not, in the Venn diagram exercise he referred to one of the persons of the Bagan Committee as giving him much trouble and said "I always keep 100 feet distance" – and accordingly this person, together with a group of *mastans*, is located at maximum distance from himself in the Venn diagram (see Figure 24). In the following quote the conceding of jurisdiction seems to be used more strategically to avoid contestations than the above quote of June 2009 indicated:

> "Now they don't create any problem to me. There is also a *ghat* at Bagan. Do you know it? Do you go to the Bagan? [interviewer: Yes.] I have another *ghat* at that Bagan. I gave it to another person. He gets money from that *ghat*. [Interviewer: Do you not collect the money from that *ghat*?] I don't even touch that money. [Interviewer: What does he do with that money?] He says that he gives this money to people. He expends the money for the welfare of the Bagan." (Foyez, 06.04.2010)

The strategies of withdrawal to avoid future contestations were mainly pursued by elite groups. The selling of spatial claims before a new electoral period was mainly directed towards the ordinary, and the case of the rickshaw garages demonstrated that after selling the claims to ordinary rickshaw garage owners, it became difficult for AL leaders to oppose this spatial claim, as they could not deal with the BNP leaders who had made these claims originally. Conceding spatial claims, especially the jurisdiction over (public) space, was a strategy pursued between elite groups, in order to secure power and avoid interference. This directly relates to the domination of social space, which will be discussed next.

Dominating social space and securing power

Apart from the preemptive strategies discussed above which focussed on using the support of other actors or institutions to secure spatial claims, such a strategy can also build on a person's or actor's own power. Thus here I will discuss how dominating social space and securing (political) power can also serve as a preemptive strategy to secure spatial claims.

As has already been discussed in Chapter 9.1.3 on power sources, a small group of elite actors dominates the social sphere of the settlements, especially in Nasimgaon. This domination can also be a preemptive strategy to avoid contestations. The political leaders in Nasimgaon are members of several committees, and the interlinkages between the various committees and the links established to higher level political leaders ensure and secure this domination (see also the relations to higher level political leaders analysed in Chapter 9.1.3). Figure A-3 exemplifies this for the local committees of one sub-area of Nasimgaon. The different local committees partly overlap in areas of jurisdiction and tasks and, as the orange arrows showing 'close ties' in the figure illustrate, some individuals are involved in several of these committees. The most 'separate' institution in this set-up is the BNP Ward Commissioner who is not consulted by the local organisations. Rather, they maintain relations with (red arrows) and consult in case of problems (green arrows) the political party organisations. Accordingly, the figure illustrates the dominance of AL leaders in conceptualising space and spatial practices by 'making the rules'.

The domination of the social space of Nasimgaon by a few elite members would be even more striking if the different NGO committees were added to Figure A-3, as many leaders are members of several NGO committees (see Chapter 9.1.3). For example, Hatem is a member of almost all the committees and expressed this himself by proudly saying "I am either the president or the advisor of each committee working here" (Hatem, 07.03.2010). Especially women involved in politics are also involved in almost all the locally active NGOs – Rizia proudly said that she was involved in every NGO committee that had been formed until the time of this research. On the other hand she together with the other female members of one NGO committee also criticised the local leader Hatem for becoming involved in all the committees:

> "The women at the meeting expressed that they do not like Hatem. Hatem always wants to put his name in every committee whether people want him or not. The women then discussed that some days ago Rizia said to Hatem in front of many people 'Why do you want to include your name in every committee? Who wants you in any committee? You can ask these people.' Then Hatem became unhappy with Rizia but could not say anything." (field note, NGO meeting, 20.06.2010)

The community police/night guard organisation can also be understood as a means to underline a dominant conceptualisation of space. The night guards in Nasimgaon were organised by the AL-operated Sub-area Committees and the Amlabazar Committee. The night guards started their duty from 11pm. After 1am they allowed only a few tea stalls to remain open for their own supply. By doing so, in accordance with the orders of the political committees, they were actively involved in the production of space and resulting spatial practices at night time, as expressed by the nightguard Shahidul:

> "During night time, there is no selling. Previously no shops remained open after 12 in the night. After AL has come, all people keep their shop open after 12 [but no longer after 1am]. Previously, we did not allow any person to enter after 12 o'clock because if any problem occurs then we have to take the liability. If any shop is broken then we have to explain it in the morning. We were told to ask those people who come to the bazar after 12 [about who they are, where they are coming from]." (Shahidul, 01.03.2010)

The night guards at Amlabazar also changed the tea stall that was to be kept open all night from one located at the entrance of the bazar area to one located further inside, at the border of their area of duty. Their aim was to more easily ensure security by avoiding the gathering of large crowds that was inevitable if the tea stall was conveniently located. Similarly, the night guards at Nasimgaon Eid Gah Math took care to avoid conflicts at night by stopping activities after 1am:

> "When it is 1am at night then we whistle and tell the people 'Don't stay on the road anymore. Go to your room. It is not good to stay outside.' […] For example, people sit and take tea at the tea-stalls. Among those people, someone may be the supporter of BNP. Another one may be the supporter of AL or Jatiyo Party. They may discuss among themselves about politics. This discussion may cause quarrelling, fighting or any bad incident. That's why we tell the shopkeepers to switch off their TVs. There will be no TV. So, nobody will be at the tea stalls and there will be no fighting. After switching off the TVs, they go to their houses." (Kamal, 07.03.2010)

The night guard and community police organisation can accordingly be understood as a means to regulate spatial practices and further underline the domination of AL committees. The objective of securing spatial claims for AL supporters also seemed to guide the selection of employees for the committees. The night guard Kamal reported how he got his night guard job with the Sub-Area Committee because he participated in *michils* of the AL – and another night guard, Shahidul, struggled to keep his job when the ruling party in power changed because he was 'accused' of being a BNP supporter, as he had also done the night guard job during BNP times.

Many of the leaders are involved in traditional judgements, and deciding whether a case should be taken to the *thana* police or not considerably conceptualises the social rules of everyday life in a community, hence manifesting domination. This also becomes obvious in the leaders' own understanding of their role. In Chapter 9.1.3 I analysed how some local leaders perceive themselves as 'intelligent' as opposed to the general inhabitants and thus perceive it as natural and necessary that they shape social space. The traditional concepts of leading and respected community leaders (i.e. the *shalish* members, *shomaj* or *matbors*) being in hierarchical relations to the ordinary justify this social dominance. A similar claim was expressed by the political leader Harez, who underlined the necessity of being active in politics for the welfare of the community:

> "We are involved in politics because sometimes we need some support in case of an emergency. If we want to stop any immoral practice, then we also need political support. The main objective of being involved in politics is to save life and lead a good life in this poor condition. Thus, the emerging party gets the responsibilities to manage the area." (Harez, 16.02.2010)

The above statement contains another indication of domination of the social sphere being achieved through political affiliation. Furthermore, the conflict that evolved about the access road leading to the re-constructed garages can also be understood as being motivated by exercising political dominance. From a neutral perspective, the relocation of the access road and the reduction in the width of the two shops and community police office did not significantly improve accessibility (see Figure 20). This indicates that the conflict did not so much concern the matter as such, but that the spatial claims were rather used as an entry point into conflict with the opposition party. The three units that were reduced to make way for the new access road were also spatial claims made during BNP times, and so were the garages. Thus the spatial claims made by AL leaders have to be understood in the political context of rivalry with the BNP, and as such presented a means for political contestation and the demonstration of AL's dominance.

The domination of the social sphere by the leaders and the limited possibilities for resistance by the ordinary are also vividly expressed by Shihab, operating the blacksmith's workshop on Nasimgaon Eid Gah Math:

> "Who will say that the Math has been filled up [encroached upon]? If someone talks about this, there will be fighting and conflicts. They are the leaders and they are all. We can't do anything." (Shihab, 07.06.2009)

In order to continue domination of the social space, political leaders seek ways to secure power. In Nasimgaon elected politicians sustain their political power by contributing to community development. This was especially done by the MP (see Chapter 6.3, 8.3.1), who came to Nasimgaon quite regularly, bringing various improvements, including the improvements of the mosque, graveyard and the Eid Gah Math. He was thus widely appreciated and voted for by many of the interviewees. But his and other politicians' electoral promises were not always kept in the end, and some interviewees were rather critical of these broken promises. Mokbul Hossain took a long series of photographs of the MP when he came to the Eid Gah Math to discuss the sand levelling, and he would have liked to have a voice recorder so that he could record the words of politicians and have proof of what they did and did not do. When the news spread that a school and a hospital would be established on the Math he also discussed this critically, expecting that the plans would eventually not be implemented:

> Faruk: "There is an idea of building a school and a hospital by breaking one side [of the rickshaw garages in the south of Eid Gah Math]." […]
>
> Mokbul Hossain: "It is only talk."
>
> Faruk: "I also know it will not be built. There is no benefit in telling this thing. If seven or ten years pass, still it will not be built. So there is no benefit of talking about this. These are only words of mouth."

> Mokbul Hossain: "I guess when he [AL candidate for WC elections] will get the position of the commissioner, then he will not do it. He is saying these things to get the votes." (Mokbul Hossain, Faruk, 08.02.2010)

The same person also distributed blankets to those who supported him during a *michil* on Victory Day in 2009 (see Chapter 6.2.3) and he also sponsored the tent put up for *Eid* prayers – both activities can be considered as electoral campaigns.

Another way of showing domination and limiting contestations is the open display of power, by physical or verbal self-presentation and through activities that silence potential protesters. At least during the interviews, Foyez repeatedly used vocabulary presenting himself as powerful and underlining his awareness of his power. This was similarly done by Mokbul Hossain and Afsana, who used rather 'bold' statements in characterising their role, behaviour and resistance towards male political leaders.

In several instances interviewees referred to power as expressed in physical appearance or related to bad reputation – which tended to be displays of power limiting the willingness to contest such people (see also the sub-chapter on bad reputation in 9.1.3). For example, during my observations the political leader Hatem demonstrated power by physical appearance when he arrived at the party office – in the internal area of Nasimgaon with its rather narrow and busy roads – on a showy and noisy motorcycle. A similar expression was made by the plastics businessman Jahangir, who asked his employee Abbas Ali, one of the solicited photographers, to take his picture with an equally showy motorcycle.

After the eviction of vendors from the Nasimgaon Math in March 2009, a pile of firewood was burnt in the evening in one corner of the Math. The young men performing these activities stated that they did so in order to reduce harm as playing children could be hurt by the thorns and branches. However, they also confirmed that the pile of wood they burned was used as firewood by the adjacent housing compounds. Given that this burning happened just a few hours after the Math had been cleared of structures, it seems this was meant as a demonstration of power and seriousness in order to silence those who might want to protest against the spatial claim that had been made by political leaders a few hours earlier.

Furthermore, the payments of bribes to maintain relationships with the executive and political leaders (see above) can also be understood as a means to silence others, who could otherwise start contesting one's spatial claims.

By dominating the social space, as especially the AL political leaders do in Nasimgaon, spatial claims made can become incontestable or at least almost so. Domination power in the social space of the settlement is especially exercised by membership in several committees and community institutions like the *shalish*. Furthermore, the domination power is secured by election campaigning in the case of higher level politicians, by maintaining relationships to these politicians, who provide political legitimation, and by demonstrating power and thus reproducing a dominant appearance.

Adhering to the conditions defined by others

Finally, spatial claims in the sense of access to public space are secured by adhering to the rules and regulations defined for a specific space. This was especially applied as a strategy by the ordinary, who accepted the conditions for access to public space often 'dictated' by the local leaders. The vendors at Khalabazar continuously adapted their livelihood strategy to the increase of payments when the rent and deposit payments were introduced. Only Dipa was not able to do so after a few months and accordingly had to leave the space. The dominating power of the local leaders was mostly interpreted by the vendors as a 'given' force that could not easily be contested. Thus one woman expressed how the local leaders "make the rule of paying rent" (Shoma, 27.04.2010), indicating how she perceived herself to be rather without any resistance power and with very limited agency unless she changed her spatial strategy.

On the Nasimgaon Eid Gah Math, vendors also regularly contribute to the mosque collection and night guards and sweeper fees, and thus adhere to the conditions. Similarly, Rokib paid the charge for plastic drying to Foyez on a daily basis, thus adhering to the access rules formulated by the leaseholder. Adhering to the formulated rules and conditions is a strategy to maintain a spatial position by accepting the dominated social space rather than contesting the rules of the game. As such it leaves little space for resistance, but rather re-enforces the decision-making hierarchies that already exist, as analysed above on 'dominating the social space'. The limited adherence to local conditions as interpreted by Mokbul Hossain, however, points at a resistance strategy:

> "I have to stand up if they push me [symbolic, means he has to do everything according to their words]. When they come I will do according to their order. When they go out I forget their order. I am struggling for my shop. I am fighting for this location. Some days ago I fought for my shop. I brought my materials from outside into the shop. After that I brought out my materials from the shop again. This has become my routine work." (Mokbul Hossain, 18.04.2010)

While his adherence, or making the leaders believe that he adheres to their orders, is a preemptive strategy, his narrative also tells of resistance against the dominating forces.

Adherence to the conditions defined by others was primarily employed by the ordinary in Nasimgaon who are not able to draw on a multiplicity of power sources. Only the adherence of Mokbul Hossain carries with it a notion of resistance and awareness of agency, while the other cases reveal the dependency structures between the elite and the ordinary. In Manikpara, the adherence of Rokib to the rules formulated by the leaseholder has to be understood as a result of his perception of the leasehold as being statutorily and religiously legitimated (see Chapter 9.3).

Subsuming preemptive strategies to secure spatial claims

Table 7 subsumes the discussion on preemptive strategies to secure spatial claims.

Strategies (preemptive)	*Manikpara*	*Nasimgaon*
Maintaining relationships with statutory institutions and actors	-	-
- maintaining relations to police/RAB	elite groups: many cases; the ordinary: few cases	elite groups: few cases
- maintaining relations with administrative officers, statutory authority	elite groups: many cases; the ordinary: few cases	elite groups: few cases
Involvement in politics and maintaining relationships with politicians	-	-
- individual involvement in politics	elite groups: many cases	elite groups: many cases
- maintaining relationships with powerful politicians	elite groups: many cases; the ordinary: few cases	elite groups: many cases; the ordinary: many cases
- multi-party involvement	elite groups: many cases; the ordinary: few cases	elite groups: few cases
Maintaining relationships with other actors	-	-
- maintaining relations with religious institutions	elite groups: many cases; the ordinary: few cases	elite groups: many cases; the ordinary: many cases
- maintaining relationships with a wider range of actors	elite groups: many cases; the ordinary: many cases	elite groups: many cases; the ordinary: many cases
'Formalisation' of spatial claims	-	-
- construction of permanent structures		elite groups: many cases; the ordinary: few cases
- paying contributions (night guards, sweepers) as 'formalisation'		the ordinary: many cases
- getting a written certificate of ownership/tenancy	elite groups: few cases; the ordinary: few cases	the ordinary: many cases
Creating incontestable space	-	-
- religious spaces and institutions as protectors from contestations	elite groups: many cases	elite groups: many cases
- creating indispensable space for public welfare	elite groups: many cases	elite groups: few cases
Withdrawal of spatial claims in anticipation of contestation	-	-
- selling spatial claims made during one electoral period		elite groups: many cases; the ordinary: few cases
- withdrawal to maintain good relations (conceding spatial claims)	elite groups: many cases	elite groups: many cases
Dominating social space and securing power	elite groups: many cases	elite groups: many cases
Adhering to the conditions defined by others	the ordinary: few cases	the ordinary: many cases

Strategy applied by

elite groups: many cases few cases the ordinary: many cases few cases

Table 7: Overview of preemptive strategies to secure spatial claims

The results show that except for adhering to conditions defined by others and the maintenance of relationships outside of statutory spheres, most preemptive strategies are primarily applied by elite groups. With regard to the two study settlements, the table shows the greater importance of regularly maintaining relationships with statutory institutions for elite groups in Manikpara. The difference of patterns for the ordinary may partly be a result of the different condition of public spaces in the study settlements and thus the different spatial claims that can be made. If the ordinary are further subdivided to exclude those who have access to diverse power sources and thus can be understood as being situated between the elite and the ordinary, the pattern emerging in the table would even be less 'favourable' for the ordinary.

9.2.2 Reactive strategies to contestations of spatial claims

Reactive strategies are applied in response to an actual contestation. One reaction can be to secure one's spatial claims against this contestation, with varying outcomes of permanence concerning future contestations of the same spatial claim. Alternatively, the reaction may result in the withdrawal of spatial claims, either temporarily or permanently. Both reactions form part of the following discussion.

Drawing on statutory institutions

The main statutory institutions drawn upon to secure spatial claims as a reaction to contestations were those of the executive, i.e. the *thana* police and RAB, but also the elected government representatives (especially the MPs).

In reaction to contestations of spatial claims (and to avoid further contestations, thus at the same time preemptive), ordinary actors make use of the executive and potentially the judicative system by **filing a General Diary** (GD) at the *thana* police. While of less consequence than filing a case, the person against whom it has been filed nonetheless needs to take care not to attract further attention, as this could prompt the police to take action.

Afsana, after her contestation with the local leaders at Khalabazar, decided to file a GD against one of the local leaders at the *thana* police station. This formed part of her resistance strategy:

> "However, I have a diary in the *thana* against [a local AL leader] because I know his character. If he says something then I will say another thing and if he says for the second time I will beat him since he is such kind of person who tortures a lot in Nasimgaon *bosti*. He does not respect man or woman, speaks very bad languages which cannot be expressed!" (Afsana, 23.03.2010)

My landlords, after the construction materials for the further extension of their compound had been destroyed in March 2009, filed a GD against the offenders in order not to be contested again and to limit the opportunities for action of those who had attacked them. However, in this case the offenders waited for a chance to take revenge, and eventually managed to do so (see also Chapter 4.4.2). After the eviction incident of Nasimgaon Eid Gah Math in March 2009, Mokbul Hossain filed a case in order to avoid future contestations. However, he retracted the case eventually, as he was approached by a local leader whose words he could not disobey, presumably because he was recognised as a religious authority and also a *bishisto murobbi*, a notable elder. However, Mokbul Hossain's telling of the story indicated that he regretted having lost this means of power and security:

> "Where will I get money? It is very difficult to earn money. For this money one person is killing another person. Someone also hit me on my head [referring to the events on 13th March 2009, see Chapter 8.3.1]. I became injured severely. After that I took 5,000 Tk as a fine from that person who hit me on my head. He repented for that matter 'What have I done! I did wrong to him by hitting him on his head.' He became uneasy for 24 hours thinking that he would be punished. That's why they [committee, leaders] told me to sit with them. They told me 'Withdraw the case from *thana*. We are giving you this money [5,000 Tk]. Please don't take any action.' They also went to some big leaders. I also have a good relation with those big leaders. They are very close to me. They [committee, leaders] were not getting any advantage from those leaders. Finally they went to this leader [a religious leader, member of the mosque committee and elderly person]. He is also well known to me. Then he solved the dispute. I could not disobey him. Otherwise I would not have withdrawn the case. They would suffer still now. I would have got a benefit if I didn't withdraw the case. They would have been accused all the time whenever I faced any kind of problem." (Mokbul Hossain, 18.04.2010)

The political leaders interviewed also admitted that many GDs had been filed against them, for example against Harez and eleven other leaders during the conflict of the rickshaw garages on Nasimgaon Math. However, these GDs were removed later on, when the conflict about the access road location had been solved. Similarly, a high number of GDs and cases were filed against the Amlabazar Committee leaders, which made the Sub-area Committee wonder about the subcommittee's activities. Some of the GDs and cases were filed because of land conflicts, others because of the contestations at Shaonbazar (see Chapter 8.2).

The option of filing a GD is mainly chosen by those who are in a comparatively powerful position among the ordinary. In the above examples, my landlords and Mokbul Hossain were in possession of economic resources, and Mokbul Hossain could also build on relationships with religious and political authorities. Afsana acquired power via political affiliation. On the other hand, at the beginning of 2010 Dipa could not draw on any of these power sources when her spatial claim at Amlabazar Road was contested by her neighbour. Accordingly, she was advised and also decided that it was better to not file a GD or a case as she feared this would cause more severe problems. Making use of the executive thus requires a minimum of domination or resistance power recognised by the offenders. In one case, however, despite the powerful position of the actor, filing a case was deliberately avoided. When his *ghat* operations were at first contested by political leaders, Foyez considered filing a case but then preferred to achieve a solution by discussion. Given his aim of keeping the leasehold as long as possible, this may have been a wise solution.

Especially those who maintain a good relationship with the police (see Chapter 9.2.1) also **bring in the police or RAB** in case of problems to secure their spatial claims. Foyez did so when he quarrelled with one of his opponents, who came to the *ghat* with weapons. At this time he called the police officer with whom he maintained the closest relations. In Nasimgaon the police were consulted in several instances by local leaders. Shihab said that the winter fair which was supposed to be held in January 2010 on the Eid Gah Math (see Chapter 6.2.2) was stopped by the police after the conflict between local leaders about the organisation. Similarly, the police came to the Eid Gah Math during other conflicts, called in by elite groups, for example in the conflict with the rickshaw garages.

But also the ordinary were successful in some instances, although only those with some sources of power were able to draw on police/RAB support. Afsana successfully brought in the police and RAB in two instances, after some local leaders had evicted the vendors from Khalabazar. With the police she was able to restore the vendors after they had been removed from the open field for a religious programme and were prevented from returning after the programme (see also Chapter 9.2.3), and with RAB's support she secured her own and others' spatial claims after the vendors had been removed on *Pohela Boishakh* during the caretaker government. At first, RAB spoke in favour of the leaders, but then Afsana accused them of having accepted bribes and brought RAB to the AL office "by pulling their hands" and finally they compelled the main leader involved in the contestations to leave Nasimgaon for good. My landlords solved a conflict about the position of a roof with neighbours by consulting the OC (Officer in Charge) of the *thana* police who then stopped the AL leaders from further repressing the family. Here, the OC supported them because they originated from the same village, but also because he was a BNP supporter. The blacksmith Shihab also explained how when RAB was present in Nasimgaon, he felt safe from the leaders at his (permanent) spatial position on the Eid Gah Math:

> "No [the local leaders cannot evict me now], this is the government's land. But they create many problems. Sometimes they scold. But since there is RAB they are not doing anything now. RAB has given its phone number [to me]." (Shihab, 07.06.2009)

Furthermore, after the eviction incident in March 2009, according to Shihab the police permitted the vendors to return to the Math for selling. The night guards and the community police both in Manikpara and Nasimgaon also consulted the *thana* police in case they could not handle an incident.

Due to the limited power of the current BNP Ward Commissioners (see Chapter 9.1.3), they are hardly ever consulted to secure spatial claims. For example, Foyez did not consider the WC important for his leasehold operation, and thus did not include him in his Venn diagram (see Figure 24). Instead, the MPs, affiliated with AL or its coalition partner JP, are consulted to secure spatial claims as the highest authority of elected government representatives. In Nasimgaon, the rickshaw garage owners in the south of the Eid Gah Math were able to keep their garage position because they could **draw upon the MP's support**. The local AL leaders could not radically contest the MP's directive, as Harez' report of events indicates:

> "Two days before the eviction of these things [garages], [the MP] came and then the garage owners told him 'Sir, we bought these garages. Before that BNP people took possession and got the benefit of it. We are poor people. Those who took possession of the land of the Math

> are not in the Math now. They sold the garages to us and went away with the money. Now, if you remove the garages, then where will we go?' Then he [the MP] thought that they [rickshaw garage owners] live here; if it is demolished then where will these poor people go. Then he told 'All right, I will build rooms for you.' Since he is the Member of Parliament of this area and since he has said so then it has become as law. If you believe in democracy fully then you have to believe that Members of Parliament make the laws. Their mouths' words are like law. When he told the garage owners that he will build rooms for them then they became assured." (Harez, 16.03.2010)

The statement points at the general acceptance of the MP as the highest authority. Harez perceived the MP as being part of the legislative and hence his words as being statutorily legitimated[14]. The AL leaders negotiated with the MP about the location of the road, but it seems that due to the MP's involvement they no longer contested the garages in general – although initially they had wanted to evict them as 'unrightful' claims made during BNP times. Drawing on the MP's support thus became a successful strategy because AL leaders had to recognise him as the highest authority, the 'law' as Harez indicated.

A similar experience of the MP being an authority was made by Foyez at the beginning of 2009, after the AL government had assumed office. Some AL politicians soon caused conflicts and closed the *ghats* for two days. Foyez together with these AL politicians then had a meeting with the MP. The MP in this meeting asked the charge for the passenger boats to be reduced from 20 Tk to 15 Tk per day and thus solved the contestation of spatial claims. Tariq and his friends also went to the MP with a larger circle of friends when they were contested by members of the mosque committee. Again, their consultation of the MP as the highest authority resulted in the committee members stopping their contestations. This was because once the matter had been discussed with a respected authority, the elders of the locality considered the quarrelling with 'young boys' to hamper their social reputation.

Drawing upon a higher statutory authority was employed as a reactive strategy in quite a number of cases. This reaction, however, is primarily carried out by those who are able to draw upon other sources of power and can be considered powerful, whether in a dominant (committees) or resistance (Afsana) sense. The example of Dipa shows that the ordinary, if they cannot draw on other power sources, do not dare to use the executive in contestations. Even for my landlord's family, who especially drew on economic power, the filing of a GD at the *thana* police resulted in a subsequent series of contestations with the committee leaders they had acted against. Remarkable is the local perception of drawing on higher authorities – especially in the case of the MP whose words are considered as law and as incontestable. But both the MP and the police in fact do not necessarily act in accordance with the law, nor can the action of the MP be understood as law *per se* simply because of his position. This perception of 'ultimate power' points at the persistence of patron-client hierarchies in decision making.

14 The MP was often referred to as *saheb* by different interviewees, both political leaders and the ordinary, e.g. the rickshaw garage owners. This honourable address further underlines his position as a highly respected and powerful authority.

Drawing on support of politicians and political organisations

In reaction to contestations, actors also made use of politicians (apart from the MPs, who due to their status as perceived government officials were discussed above) and political organisations. The success in drawing on political support thereby often depends on whether an actor has 'invested' in maintaining relationships with political actors preemptively (see Chapter 9.2.1).

In many instances, to secure spatial claims people **draw on a higher authority**, i.e. a person of considerably more dominant power and influence. This strategy is used both by the ordinary and elite groups. For an overview of the (spatial) hierarchy of political party organisations, see Figure 8 in Chapter 5.1.1.

The political leaders of Nasimgaon generally communicate with higher level leaders whenever they face a problem which they cannot solve by themselves, and which also cannot be solved with police support (see Chapter 9.1.3 on generating political power through organisations of the ruling party, see also Figure A-3). The Amlabazar Committee, confirming the findings already outlined above on the limited power of the Sub-area Committee, reported that they would not consult the Sub-area Committee in any matters, but would directly move to the Ward, *thana* or *ason* level AL committees. Most commonly they approached the Ward level and the WC candidate of AL, as he had the power to take the matter upwards to either the *thana* level or the MP. On Nasimgaon Eid Gah Math, the rickshaw garage owners managed to secure their spatial claim by drawing on the MP (see above), but also on the higher level politicians of JP (Dhaka *Mohanogor*) who then came to Nasimgaon to solve the matter with the AL Ward level leaders. The local AL leaders had approached the AL higher levels (especially Ward) in parallel. However, the rickshaw garage owners involved the MP as the highest authority, also locally perceived as statutorily legitimated (see above), and thus succeeded in maintaining their claim.

In the conflict between Afsana and Hatem, Hatem involved the office secretary of the Ward level AL committee to make her pay the deposit and fees, while she reacted by attempting to get the support of the chairman of the Ward level AL committee, producing a higher power authority. Furthermore, in the opening story to this research (see Chapter 1, introductory note from the field), Afsana was able to call on Rizia as a source of power in her negotiations. The way Rizia talked about Hatem and other AL leaders during NGO meetings and the way she behaved towards Hatem, according to Afsana and during another NGO meeting, indicated that she could be considered more powerful than Hatem.

In Manikpara, Foyez mentioned a number of politicians in his Venn diagram (see Figure 24) whom he consulted if his spatial claim was contested and he faced any problems. In a previous interview, he had referred to the political leaders contesting his spatial claims at the beginning of his leasehold operation, which he managed eventually by maintaining relationships with them:

> "It was not easy for me to manage the *ghat*. I also struggled in the beginning. These party people are hungry for everything. In the beginning they also tried to hold me back by showing weapons to me. But I am doing it for some years and I have some hold in the area. So I am

> slowly resolving everything. I am meeting some respected and powerful people and doing some group meetings and also talking with the party cadres. I also have some power [*khomota,* having the (political) power to do something]. I could file a case against those party people and then the police would arrest them, but I want to achieve a solution through talking." (Foyez, 09.08.2009)

The option to draw on a person of higher political authority did not always involve higher level leaders, as in the above cases. For Aklima, when facing contestations on Nasimgaon Eid Gah Math by a group of *mastans*, it was sufficient to draw upon the local political leaders (see above). When Dipa faced contestations from one of her neighbours at her new location on one of the roads leading out of Nasimgaon, she was supported by her other neighbour. He solved the problem, and according to her he was able to do so because of his affiliation with the AL as well as with a local leader considered a big *mastan* in the area. Shahin mentioned the owner of his shop in the Venn diagram (see Figure 25), who was an AL political leader, as a potential protector for his business if his spatial claim as a tenant of this shop would be contested:

> "Suppose someone attacks this shop [operated by Shahin on a rental basis]. Then he [the owner of the shop, a local leader of AL] will come by running and tell 'What has happened?' I do not have power. I am a person from outside. If he comes and stands, people will give value to it. He also does the local judgements. So if he comes then there will be no more problem. All will leave silently." (Shahin, 21.04.2010)

This political leader and another shop owner at Khalabazar also involved in local AL politics (A. *Bhai* in the Venn diagram) were the most important contacts for his business in terms of securing his spatial claim[15].

In reaction to contestations of spatial claims, a common strategy is to approach politicians of the higher administrative units of the political system. For the ordinary, this means drawing on the support of the local political leaders at the lowest level of the party system, e.g. in the case of Shahin. In contrast, those among the ordinary who can draw on more power sources, e.g. Afsana, and the elite groups seek the support of the higher level politicians, both in Nasimgaon and Manikpara. Approaching the highest political levels is mainly a strategy followed by the elite groups.

15 The other 'circles' of the Venn diagram mainly represent contacts immediately related to his business operation, i.e. the wholesellers from whom he bought the goods for his shop and the cooperatives where he borrowed money or participated in saving schemes.

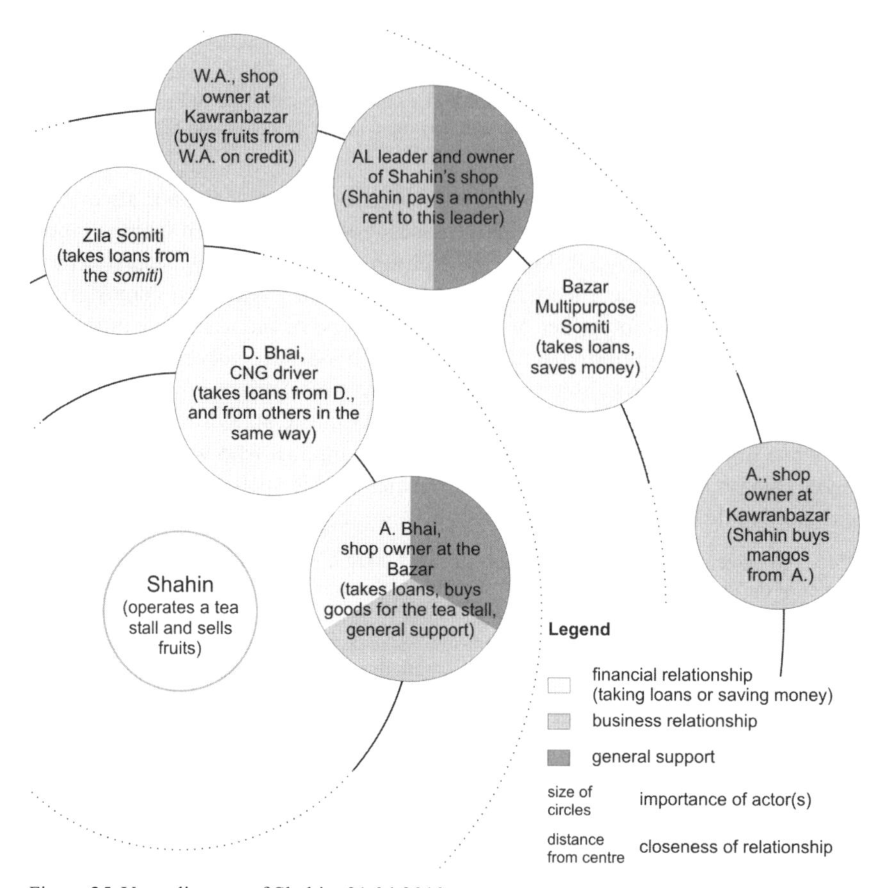

Figure 25: Venn diagram of Shahin, 21.04.2010

Drawing on the support of other actors

Besides statutory and political institutions and actors, there are also other actors and institutions which can be drawn upon to secure spatial claims. Often these may also be related to politics or statutory institutions (as a source of power), however, they are strategically approached because of relationships beyond the state and politics. These may be relatives, respected elders, *shalish* members and the community police. As with the above actors, relationships are often maintained by entertainment provision.

The powerful people a person draws on to secure his/her spatial claims and position can be **relatives**. Rokib referred to his powerful *bon-jamai* as his main supporter in many instances (see also Chapter 9.1.3 on powerful kinship as a source of power and the Venn diagram of Rokib, Figure 23), and repeatedly managed to secure his spatial claims with his support. In the beginning when Rokib started to operate his business, his position on the embankment slopes was contested. As soon as the parties who wanted to take possession of the same place got to know about the relationship between Rokib and his *bon-jamai* they stopped the contestations. According to Rokib's wife Momena, they understood that Rokib possessed power via his *bon-jamai* and could not be contested. Another relative of Rokib was involved in the traditional judgement system (*bichar*) in Manikpara (and was also the president of the local Businessmen's Association). This person was the one Rokib called if anyone caused a problem and if there was a need for a *bichar* by a respected member of society. During the eviction drive by BIWTA in August 2010, his business did not get harmed because his wife's elder brother was a driver of a BIWTA officer and could take a stand for Rokib. Although this elder brother himself was not in a powerful position, the kinship relation nonetheless helped to protect Rokib and his business[16].

In Nasimgaon, however, support of relatives to secure spatial claims tends to be less common than in Manikpara. This may be attributed to the different histories and identities of the settlements (see Chapter 7.2.2; and Chapter 9.1.3 on the low relevance of family status and kinship as sources of power).

In some cases the **support of elders** finally resulted in sustaining the spatial claim. This was the case when Rohima's and Shahin's shop was about to be let to someone else who offered to pay a higher advance deposit of 20,000 Tk. Then an elder person supported Rohima and Shahin's claim and indicated that they had been operating the shop for a long time, and as they were poor it was their only source of income. Upon this respected elder's support, the owner of the shop agreed to continue with Rohima and Shahin as shop-tenants:

> "Then there was an elder person sitting beside us and he told them 'Give them the shop. They are poor people, take 15,000 Tk from them and let them survive.' After that they agreed." (Rohima, 18.01.2010)

Foyez also was supported by elder people of the locality, when he denied the demand of the chairman's son (son of a respected person addressed as chairman) for protection money in January 2010. In most cases, though, those who supported others in their spatial claims may have been respected elders but they were referred to in their functions as political leaders, *shalish* members or relatives rather than stressing their age.

The above discussion confirms the findings outlined in the sources of power on the importance of respected positions in society, especially the *shalish* membership and respected elders (*bishisto murrobi*). Rokib's successful reaction to contestations was also based on a person who acquired power via the

16 This is the version told by Rokib, who believed that the eviction threat was managed via his wife's elder brother. The narrative of Foyez about the same event, however, suggests that Foyez tried to save Rokib via his own contacts. This initiative of Foyez can be attributed to Rokib's powerful kinship, i.e. his *bon-jamai* and the president of the local Businessmen's Association. Foyez characterised the latter as 'someone causing me troubles' (see also this person's distant position in Foyez' Venn diagram), for which reason Foyez may act in favour of Rokib to avoid contestations. Either way, it was powerful kinship that saved Rokib from the BIWTA eviction drive in August 2010.

shalish/bichar system, and was his relative at the same time. Given the differing availability and importance of power sources in Nasimgaon, drawing on the support of relatives in securing spatial claims was not applied in the cases discussed in this research.

Reclusive strategies: Low visibility and 'quiet encroachment'

While the above strategies all draw on third persons'/actors' support, reclusive strategies can be conducted self-dependently. While they are discussed here as reactive strategies, they may also be applied pre-emptively.

One reclusive strategy to keep a claim to space despite contestations is to transform a shop/stall into a **more temporary structure**, lowering visibility. After the eviction drive of BIWTA in August 2010, Jahangir established a new shop on the embankment slopes, but this time he constructed not a stable CI-sheet room but only a temporary CI-sheet roof on bamboo pillars without walls. In November 2010 he still had this temporary structure, although he had 'boasted' how only he and one AL leader would be able to construct stable rooms again after the eviction drive.

The vendors on Nasimgaon Eid Gah Math also resorted to a strategy of lower visibility following conflicts. After the eviction drive on 13th March 2009, they only slowly came back over the following days. Taslima, the woman selling *pitha*, did not put up the permanent structure of a roof and benches but for some days operated her shop only semi-permanently with the three stoves, remaining more flexible and less visible. Similarly, a vegetable seller sold her goods in a semi-mobile fashion by sitting on the ground instead of using the semi-permanent *chouki*. When the conflict about the water tank construction emerged in May 2009, the vendors left the Eid Gah Math and did not come back for some days. It seemed that they remained invisible, waiting for the grounds to be 'safe' again. Only those selling from semi-mobile units were back soon after the conflict. The same happened after the Eid Gah Math was levelled with sand in July 2009 and rumours of eviction and putting up a fence spread – those who had stable shops but usually extended into public space by using it as a storing place, like Mokbul Hossain, kept their goods inside their shops.

In the past, when he did not yet have a stable shop, Mokbul Hossain also adapted his business habits in order to avoid contestations from local leaders who started to demand money from him: he only produced as many concrete products during the daytime as he could sell in a day, because he feared that anything he had to store over night would be broken by the local leaders' supporters. Later on, owning a stable and lockable shop, he always stored his valuable materials inside his shop, reducing his visibility during the night in order to avoid contestations. Finally, Dipa, after she had moved to her new place along Amlabazar Road, also chose a strategy of temporary 'invisibility'. In February 2010 she had plans to extend her house by constructing a second room extending onto the lake to be let to a tenant. However, she did not feel secure enough to do so, and wanted to wait until after the Ward Commissioner elections and for a more politically secure situation. In Bayat's logic of the 'quiet encroachment of the ordinary' (Bayat 2004; see also Chapter 2.2.1) this strategy can be understood as becoming 'more quiet' again, after having become too visible.

Another reclusive strategy is a stepwise procedure of making spatial claims in the fashion of **'quiet encroachment'**, where the next step of the claim-making is only conducted if the previous step has gone uncontested or if the contestation of the previous step has ceased, resulting in an acceptance of this claim. This strategy was applied along one of the roads leading out of Nasimgaon, although in a fairly visible style. However, the spatial claims were made stepwise, and after the occupation of one side of the road in June/July 2009, it took until February 2010 before the other side began to be occupied. Dipa's account of the contestation by RAB in August 2009 confirms this strategy of 'checking out' the acceptability of the spatial claim. She told of the demolishing by RAB of a few constructions that had began to emerge on the opposite side in July/August 2009:

> "The houses on the opposite side were demolished by a RAB officer one day. The RAB officer also wanted to demolish the houses on my side but he didn't manage. The people were putting me in front of the officer and told 'This woman is not having any shelter or any place to go to. So if you evict her she will be totally helpless'. Then the officer didn't do anything to me and the houses next to my house." (Dipa, 08.08.2010)

The 'checking out' strategy is also applicable to the spatial claims made by the *Sromik* League in the area between the tea stall and Taslima's *pitha* stall (see Chapter 8.3.3). The space in-between was only slowly and quietly transformed into a building. Similarly, the water tank on the Eid Gah Math was under construction for more than a year, including many contestations, until finally it was transformed into a room for a religious group. These strategies of 'quiet encroachment' have to be understood as both preemptive and reactive. Preemptive in the sense that they were used to avoid contestations, but as thc above examples show they were also employed as reactive strategies after an initial contestation.

Reclusive strategies as reactions to contestations with the aim of maintaining one's spatial claim in the long run are often strategies employed by the ordinary, independent of their access to power sources. At the same time they are employed by elite groups, especially those subordinate to the main party organisations, but similarly in visible spheres, such as on one of the roads leading out of Nasimgaon. The above examples of the ordinary's application of reclusive strategies, however, also carry with them a notion of 'permanent temporariness', as spatial claims are hardly ever secured statutorily. For example, the vendors on the Eid Gah Math resorted to temporary strategies repeatedly, whenever a new conflict arose. Thus these strategies also translate into a need of constant re-negotiation of spatial claims.

Withdrawal of spatial claims

Another set of strategies to avoid further/future contestations and at the same time secure public space-based livelihoods is to withdraw the spatial claims either temporarily or permanently.

To secure space-based livelihoods even when the previous location is no longer available, vendors **resort to temporary locations**, sometimes resulting in permanent alternative locations (see below). When the AL leaders evicted Khalabazar in order to construct the *paka* platform in January/February 2009, many vendors made arrangements with neighbouring shop owners who allowed them to temporarily sell their goods in front of their shops. In doing so they did not contest the local leaders' claim. At the same time, the AL leaders suggested Rohima and Shahin should resort to more mobile activities by buying a rickshaw van and selling from the van – but they were not able to do so because of Shahin's heart problems. Dipa, after her displacement from Khalabazar, for some days sold vegetables from a shop in Tejkunipara while the owner of this shop was absent:

> "After the election of the AL government, the leaders of AL took our shops from us. I became helpless when they took my shop [in June 2009]. After becoming helpless, one day I went to Kawranbazar to manage to do something for my living. I bought a basket from there. Then I took my measuring instrument from here and the basket to Tejkunipara. I told them [the people at the market] 'Please give me some space, I want to do some business for my living, and if you don't give any space then I have to live without any food. Then they told 'This shopkeeper will not stay here for ten days and you can do your business here for these ten days'." (Dipa, 10.02.2010)

In December 2009, while the shop of Rohima and Shahin was under reconstruction by the committee and negotiations about the payment of advance money on deposit and rent were going on, Rohima and Shahin sold vegetables from the open space in front of their shop. These stories point at a general solidarity among shopkeepers and vendors – however, these may also have entailed some monetary transactions.

While some temporary strategies have been outlined above, others had to **permanently shift to alternative locations** due to contestation of their spatial practices. Jahangir, after the BIWTA eviction drive in August 2010, permanently shifted his store room for plastic goods to a place along Embankment Road as a reaction to the contestation. However, he decided to keep his *adda* place on the slopes. Kabin Mia had to shift his rickshaw garage and repair shop to another location permanently after the reconstruction of Embankment Road and the increased speed of traffic resulted in his previous location being no longer accessible (see Textbox 5 in Chapter 6.3). In Nasimgaon, Mokbul Hossain's story is one of repeatedly choosing alternative locations due to contestations. However, he remained in each location for a considerable time (years) and thus his strategy cannot be considered temporary but fits this context of seeking permanent alternative locations:

> "At that time I had my shop on the Math, close to the water tank and toilets on the Math. At that time the water tank wasn't built yet and I used the Math for my work. I was putting my

> materials in a rickshaw garage – I didn't have a closed shop like now. Then people were always pressing me to take my materials away from the Math, the *mastans* and committee people. Then I got a contract to construct toilets from an NGO. For some days no one bothered me as I was constructing them. Then people started pressing me again, afterwards, so some people advised me to take a proper shop. I was searching for some shop close to the Math and found this place and bought it for 50,000 Tk." (Mokbul Hossain, 26.05.2009)

While he shifted several times (see also Chapter 9.2.3, for an offensive contestation of his spatial claim with religious reasoning), he finally resorted to a stable shop which could not easily be contested. At Khalabazar, Dipa finally also reacted to the contestations of her spatial claim by searching for an alternative location. When the local party leaders took possession of the market space and introduced rent and deposit payments, her spatial claim was contested. For some time she tried to adapt to the new access rules, but eventually she was not able to pay the new fees. Accordingly, she agreed with the offer to set-up a new place along the then unoccupied Amlabazar Road. Her withdrawal from Khalabazar and change of location was painful:

> "I have to move, I never wanted to change the place, because I was staying in that place for a long time, but I have to move because I have no other option. It remains as a big scar in my heart that I have to move from the place where I stayed for such a long time." (Dipa, 17.06.2009)

Her withdrawal can thus not be understood as completely voluntary, and she perceived it as a dislocation. The process of adapting to the alternative location, given her low business turnover and additionally her long illness, was all but smooth (see above for contestations by RAB, see Chapter 6.1.4 for 'spiritual contestations' by someone she understood was wishing her 'bad').

Another reaction to contestations is **apparent withdrawal by sub-contracting** to another person. This delegation of jurisdiction only applies to the leasehold system of Manikghat. The new leaseholder, who lived outside of Manikpara, was not able to collect fees from Manikghat after winning the tender procedure in July 2009. His jurisdiction of space, despite being legitimised by a statutory contract, was contested by the local inhabitants, but especially by political actors. Accordingly, his only choice was to sub-contract the leasehold operation to Foyez, who had already operated Manikghat for a year. By sub-contracting to Foyez, the leaseholder was able to generate an income from his legitimate spatial claim, as he and Foyez negotiated a daily fee for the sub-contracting. On the other hand, by delegating the everyday operations to Foyez, who was experienced in operating the *ghat* and able to draw on a higher level of local acceptance (see Chapters 9.1.3, 9.2.1), the leaseholder avoided having to deal with the everyday contestations.

The temporary or permanent withdrawal of spatial claims can especially be considered a strategy of the ordinary in reaction to contestations by more powerful groups, mostly belonging to the elite. The consequences of the withdrawal of spatial claims can thus considerably influence a person's livelihood strategy and livelihood vulnerability. Especially for Dipa, withdrawal and relocation to a place which at that time was highly insecure meant a tremendous increase in her vulnerability and the insecurity of her space-based livelihood. The example of sub-contracting only applied to the case of the leasehold in Manikpara as a statutory procedure.

Demonstration of dominant power

The strategy of demonstrating dominant power was used by Foyez one day, when the boatmen at the *ghats* he supervised went on strike. His account of the events is as follows:

> "Two days ago the boatmen at one of the ghats called a strike for 15 to 20 minutes. I was at another ghat. I was informed and came to the [respective] ghat. There had been a problem between a boatman and the person who collects the money from the boats [Foyez' staff]. They had been quarrelling with each other and in the end the money collector had slapped the boatman. After that, all boatmen decided not to pull boats. There were lots of people waiting at both sides of the river to cross the river. The public would beat me, not them. So I discussed with them [the boatmen] and told them to launch the boats but they refused. Then I said to them 'Ok. You can stop your boats and I will bring some boats from other *ghats*. I have lots of boats at different *ghats*. I can manage this *ghat* by bringing 30 boats from three other *ghats*'. Then they said that they were going to start operating the boats again. After that I suggested to the money collector to say 'sorry' to the boatman. At last we could solve the dispute without making each other worse off." (Foyez, 23.01.2010)

The conflict was solved by Foyez using his dominant power. He threatened to bring other boats, and by doing so indicated to the boatmen on strike that he had considerable power which he could use to threaten their livelihoods. Accordingly, the boatmen gave in and resumed normal activities. Foyez in this case drew on his 'ultimate power authority', while being very aware of his power. This is the only example of demonstrating dominant power in reaction to a contestation of a spatial claim. This strategy of power demonstration was rather in other cases employed pre-emptively (see Chapter 9.2.1, especially on the domination of the social sphere) or as a strategy for offensive contestation (see below).

Subsuming reactive strategies to secure spatial claims

Table 8 subsumes the discussion on reactive strategies to secure spatial claims. The results show that especially the strategies which draw on statutory and political institutions are strategies employed by elite groups. Only those of the ordinary who can draw upon other power sources could also use these strategies in some cases. Generally, the reaction of the ordinary to the contestation of spatial claims especially in Nasimgaon is more defensive. The most common strategies are to resort to a reclusive strategy of lowering visibility for a certain time period or even to withdraw the spatial claims either temporarily or permanently.

Strategies (reactive)	*Manikpara*	*Nasimgaon*
Drawing on statutory institutions	-	-
- filing a GD or case at the *thana* police		● elite groups: many cases; ● the ordinary: many cases
- bringing in the police/RAB to support individual spatial claims	● elite groups: few cases	● elite groups: many cases; ● the ordinary: many cases
- drawing on the MP as a perceived government official	● elite groups: many cases; ● the ordinary: few cases	● elite groups: many cases
Drawing on support of politicians and political organisations	● elite groups: many cases	● elite groups: many cases; ● the ordinary: many cases
Drawing on the support of other actors	-	-
- securing spatial claims with the help of relatives	● elite groups: many cases; ● the ordinary: many cases	
- securing spatial claims with the help of respected elders	● elite groups: few cases; ● the ordinary: few cases	● the ordinary: few cases
Reclusive strategies: low visibility, 'quiet encroachment'	-	-
- temporary structure and lower visibility	● elite groups: few cases	● the ordinary: many cases
- 'quiet encroachment': stepwise procedure		● elite groups: many cases; ● the ordinary: few cases
Withdrawal of spatial claims	-	-
- temporary location		● the ordinary: many cases
- permanent alternative location	● elite groups: few cases	● the ordinary: many cases
- (apparent) withdrawal by sub-contracting	● elite groups: many cases	
Demonstration of dominant power	● elite groups: many cases	

Strategy applied by

elite groups: ● many cases ● few cases the ordinary: ● many cases ● few cases

Table 8: Overview of reactive strategies for securing spatial claims

9.2.3 Offensive strategies to contest others' spatial claims

Offensive strategies can be applied to take possession of public spaces claimed by others or to protest against the spatial claims of others, and thus lead to the dislocation of another person. But the contestation of others' spatial claims can also be motivated to reduce the power of the other person holding the spatial claim.

Establishing physical structures

The establishment of physical structures with a view to contesting the spatial claims of others is used by elite groups as an offensive strategy.

The **establishment of permanent structures** in public space can be understood as an expression of dominant power. This was the case when the AL leaders of the Sub-area Committee at Khalabazar established the *paka* platforms and claimed ownership of the *bhits* – contrary to the long-term occupation of the space by semi-mobile and semi-permanent vendors. The physical spatial claim by constructing the *paka* platforms was a very dominant expression of power. Accordingly, two of the vendors, namely Dipa and Afsana, protested against the transformation of their vending places, although with varying outcomes (see Chapter 8.2). Despite this resistance the AL leaders, able to draw on the power acquired through being members of the ruling political party, succeeded in their offensive strategy. From that moment onwards, they dominated the public space and manifested their spatial claim by continuously extending the physical structures (see the discussion of their strategy of 'informal formalisation' in Chapter 9.2.1). Similar strategies were followed on Nasimgaon Math with the construction of the water tank in May 2009 and the construction of the foundations of six *paka* market stalls in April 2010 by AL leaders (see Chapter 8.3) – although neither succeeded in their offensive claim.

Another strategy to implement a certain 'conceptualisation of space' against the claims of others is to **establish a physical boundary**. Plans for and attempts at fencing were repeatedly a source of conflict on Nasimgaon Eid Gah Math. On the one hand the fencing was to serve to protect the Math from encroachment in order to keep it as a safe space in case of fire outbreaks, as a playing field for children and for the *Eid* prayers – a conceptualisation that was generally supported by many of the ordinary. On the other hand, the fencing plans discussed by the Eid Gah Math Committee from May 2009 onwards alerted many semi-mobile and semi-permanent vendors and shopkeepers because the implementation would threaten their space-based livelihoods. The offensive strategy of making a spatial claim by installing a fence was thus originally not an attempt to claim space for personal benefits but for the long-term benefit of the community. Accordingly, Harez indicated that sacrifices had to be made for the aim of protecting the Math from encroachments. But when he continued to talk about the fence as a physical boundary, he saw it as a means to avoid further encroachments, while permitting the (mobile to semi-permanent) vendors to continue activities:

> "Those who operate small shops beside the road, they will be able to operate their shop even if a barbed-wire fence is put up. The meaning of providing the barbed-wire fence is: so no one can build rooms within the Math. And if someone sells *chanachur* [Bengali snack], nuts etc. dispersedly we have no problem so we do not prohibit it." (Harez, 16.03.2010)

Despite the plans, the fence intended to protect the Math was never finalised (see Chapter 8.3.1, Figure 19). The reasons for this may be manifold, while the spatial claims made in 2010 indicate how such a fence would have made the spatial politics of the AL leaders visible. Instead of the 'protective fence', another fence was erected in response to the reconstruction of the rickshaw garages in the south of the Math. This fenced area, erected by the same political leaders who had expressed the need to protect the Math from encroachment, was then transformed into rickshaw garages. The fence was thus used as an offensive strategy for personal gains, no longer serving the community. These claims would not have been possible if a fence prohibiting any further encroachments of the Eid Gah Math had been established earlier.

Similarly, a physical fence was used in the conflict about future access to the rickshaw garages. Here the fence acted as a blockade to underline the offensive spatial claim which was then solved by negotiating a new access route.

Establishing physical structures/boundaries was found in this research to be an offensive strategy adopted by elite groups who can draw on dominant power, especially that acquired through politics. However, the examples show that this strategy does not always go uncontested. Especially on Nasimgaon Math, resistance was successful in the case of the water tank and the access to the rickshaw garages, but in both cases the resistance movements also drew on political power and did not consist of the ordinary alone.

Drawing on religious norms to dislocate others

While in the above sub-chapters I pointed out how religious norms are used to secure spatial claims, religious functions and institutions can also be (mis)used to contest the spatial claims of others.

The organisation of religious events is generally respected by most members of society and thus a religious use of public space would not easily be contested. Thus on Khalabazar during the caretaker government, the vendors considered it only natural to give way to the hosting of a religious programme:

> "One day [four political leaders of AL] came and told that they will arrange a programme of *Bhandarsharif*. Then they removed all the shops by telling 'You can operate the shops after the completion of the programme'. So everyone gave them the chance since it was a religious programme. But after the following day they did not allow us to set up our shops. And they created a boundary all around by putting bamboos so that we could not do shop-keeping anymore." (Afsana, 23.03.2010)

With due respect to the religious practices, the vendors had not protested their one-day displacement from public space. However, this power of religion was then abused by the local leaders to make a permanent spatial claim against the vendors' claims. Despite underlining their claim with the erection of a physical fence, they eventually did not succeed (because Afsana brought in the police as a reactive strategy) and soon the vendors were back.

The importance attached to religious places and buildings (see Chapter 9.1.3), here the Eid Gah Math, mosque and *madrasa*, is also used as a means to contest the spatial claims of others. When the vendors were evicted from Nasimgaon Eid Gah Math in March 2009, the reason provided was the Math's religious purpose under the jurisdiction of the mosque – a reason that tends to be accepted because of the religious connotation. However, this is inconsistent, as the mosque also generated part of its funding from the vendors, and the mosque committee did not interfere when the Math was encroached upon later on. On Khalabazar shortly after the new AL government came to power, the vendors had been told that a *madrasa* for poor children would be built. Given the respected connotation of this plan, the vendors left the place without resistance, only to eventually realise that the construction of *paka* platforms instead of a religious building had been started.

The examples above, and the findings illustrating the use of religion in preemptive strategies, underline how religious norms and institutions – a religious legitimation – can be strategically used in making and underlining spatial claims against the claims of others. In the above cases this strategy was employed by elite groups, while the ordinary were not able to resist, underlining the respect and power of religious institutions, as analysed in Chapter 9.1.3.

Removing/evicting physical structures or jurisdiction over space

Another option to contest the spatial claim of others is the physical dislocation of vending/production units and other structures, or the revoking of jurisdiction over space.

The offensive strategy of eviction was used by the statutory authority BIWTA as a means to re-instate their authority along the embankment. During an interview in February 2010, the BIWTA officer stated that BIWTA was planning to evict all informal/illegal activities along the embankment slopes. The eviction drive carried out in August 2010 removed various structures (see Chapter 8.1) and resulted in the statutory authority re-establishing its jurisdiction over the slopes which are not part of the *ghat* leasehold contract. While this can be understood as a contestation of other's spatial claims, it was not consequential as activities deviant from the statutory directive soon resumed and remained uncontested. The example can be understood as a demonstration of state power and as a reminder that the state authority is still aware of its formal obligations (see Chapter 10.2 for the discussion of the state's involvement in informal space). The enforcement of state power cannot be understood as a permanent claim for jurisdiction. The option to cancel the leasehold in case of misconduct provided a further option for the state to offensively contest the spatial claims of the leaseholder by drawing on a statutorily legitimised process. This was, for example, made use of in February 2010, when the leaseholder of Manikghat received a letter informing him that "proper legal measures will be taken against him including the cancellation of the leasehold" if he did not pay the remaining lease money that was due in December 2009. While the examples of the BIWTA are based on an initiative taken by the statutory authority itself, in other examples the statutory authority was called by the person/group contesting a spatial claim, e.g. when the police was brought in to react to contestations (see reactive strategies in Chapter 9.2.2) or support contestations.

Physical dislocation was also employed as a strategy by the political committees in Nasimgaon, both at Khalabazar and on the Eid Gah Math. These cases have already been outlined above, as they involved the establishment of physical structures/boundaries coupled with a community-welfare legitimation and/or the drawing on religious norms and institutions with a religious legitimation. In the case of the AL *Shechhashebok* office that had been constructed on the Nasimgaon

Eid Gah Math (see Chapter 8.3.3), the political leader Harez expressed his discontent about the spatial claim of these 'undisciplined' people, and underlined that he planned to evict them. However, he did not consider this offensive contestation possible without the support of the police as he considered the AL sub-organisation's politicians 'bad type people', and as of the end of 2010 their spatial claim continued uncontested.

While in the above examples strategies of domination were employed, Afsana's plan to re-establish the old system of vending at Khalabazar by removing all the structures erected since the election of the AL in 2009 (see Chapter 8.2) has to be considered an offensive strategy of resistance. Her plan, for which she wanted to draw on her political power was, however, never implemented and the further developments at Khalabazar (especially the conversion of shop units into a housing compound) probably mean that it will never be. It is rather unlikely that any political leader who generally supports Afsana would support the demolition of houses, a rather strong contestation of the spatial claims of other leaders.

Removal of physical structures/spatial claims is generally an offensive strategy employed by those who can draw on effective power sources, statutory power and politics, and who are part of elite groups. This is also the case for Afsana, as her plans to remove physical structures were part of her political activities and not carried out from the position of an ordinary vegetable vendor. Except for Afsana's plan, the strategies are acts of domination, either to demonstrate power and the territory of jurisdiction as in the case of the BIWTA, or to actively take over the spatial claims of others, as in the case of the committees' actions in Nasimgaon. Afsana's case, in contrast, symbolises an act of resistance power, although it has to be questioned whether the outcome of the removal would have guaranteed more inclusive access to space.

Subtle strategies to contest spatial claims

Furthermore, a range of 'subtle' and enigmatic strategies is applied to contest the spatial claims of others. These are different from the reclusive strategies (reactive) discussed in Chapter 9.2.1, because they aim at contesting a spatial claim of others rather than reacting to contestations.

Given the importance of social respect and status in the local society, a strategy attacking a person's respect, prestige and honour can be applied to contest a spatial claim. Such strategies, however, are difficult to verify if only told from one perspective. In the interviews this applied to only one incident. Tariq and his friends told of how the local elders of the mosque committee tried to create a bad reputation for them by referring to the 'young boys' involved in unsocial activities:

> "Our issue is that some people are spreading bad rumours about this place and us. [...] As an example, they are saying that mugging is being done at this place. They are spreading a lot of foul words about us. They are trying to relate many foul activities from the outside to us. Actually, all of us do our business more or less. We just pass our leisure like one hour or two

> hours altogether at this place. Tariq has come with you from his shop, hasn't he? We have called him after we came here." (Tariq's friend, 19.04.2010)

However, eventually the local elders did not succeed in this 'campaign' and became quiet after Tariq and his friends had consulted the MP.

Another subtle strategy to contest others' spatial claims was referred to as 'tricks', indicating cunning moves to successfully contest parallel spatial claims. This was repeatedly used by the lease operator/leaseholder Foyez to refer to the way in which he finally managed to become the official leaseholder in 2010. His playing of tricks was meant to annoy the leaseholder who had sub-contracted the operations to him. Foyez continuously called him about small issues until the leaseholder had had enough (see Chapter 8.1.1). Foyez' consciousness of his own power presumably supported the application of this strategy.

Such subtle offensive strategies could only be uncovered in a few instances. In both cases above, elite groups applied such strategies. However, in some instances the ordinary also resorted to subtle strategies in resisting more powerful actors. These will be discussed in Chapter 9.2.4.

Open protest against spatial claims

Finally, a range of other strategies of open protest can be considered offensive reactions to the spatial claims of others.

Some activities of protest have a physical expression. This was especially the case when local inhabitants in Manikpara protested against the leaseholder's attempt to introduce a toll collection system at Manikghat and start collecting from every passenger using the *ghats*:

> "We thought about charging a toll. But because of the inhabitants it didn't happen. Even everything was already complete and we had announced the toll. We spent around 50–60,000 Tk to set up all the arrangements everywhere, and planned to charge the toll from the next day. But the public put fire during the night on everything." (Foyez, 26.06.2009)

The physical expression of protest here was to burn the toll stations, which finally resulted in the leaseholder removing the toll collection stations. As of the end of 2010 he had not again attempted to collect such passenger fees. The 'public' he mentioned as protesting most probably consisted mainly of the 'local youth' who are related to political parties and rather not the 'ordinary'. In Nasimgaon during various conflicts, expressions of contestations also included physical fighting, finally resolved with the help of the police. In the conflict at Shaonbazar (see Chapter 8.2), the political leader Sayed uprooted the sign put up by the Sub-area Committee to establish a club for the Jatiyo Party, and subsequently a violent conflict started.

In another incident, related by Mokbul Hossain, one of the night guards at Nasimgaon Math protested against his non-payments by vandalising the rickshaw van Mokbul Hossain usually parked outside his shop overnight:

> "The rickshaw van is mine; it stays on the Math and is safe because of the night guard. During BNP time, I paid 60 Tk each for the two night guards on the Math. Now I am only paying the guard on my side of the Math 120 Tk per month. The other night guard is angry about it and also wants some money from me. So sometimes he is taking the air from the tires of my rickshaw van. One night the night guard on my side saw this and told me. I haven't said anything yet but if he is doing it again I will take revenge." (Mokbul Hossain, 26.05.2009)

At the same time, Mokbul Hossain's promise of revenge indicates his self-perceived power and ability to resist.

Another less physical expression of protest is open contestation by not behaving as requested. Both Afsana and Mokbul Hossain, who can be considered powerful among the ordinary, did not accept the dominance of leaders and protested against this by not paying contributions as requested or by openly expressing their discontent.

Afsana protested against having to pay contributions for shop improvements she did not ask for (see Chapter 8.2). Before, when during the caretaker government the AL leaders had cleared the Khalabazar for a *Pohela Boishakh* programme, she had refused to remove her *chouki* and thus protested. Mokbul Hossain did not contribute *chada* regularly to the mosque committee, saying that "they should also work for their money", indirectly indicating that they did not only use the money for the mosque but also for their personal benefit. Furthermore he threatened to resist the local *mastans* if they bothered him again, saying "if they cross my limits, I will see them" (26.05.2009).

The political leader and Imam of the mosque, Rabiul, said that the rickshaw garages (which had been contested at the beginning of March 2010) did not pay *chada* to the mosque, asking why they should contribute as they were supporters of JP.

Furthermore, the BNP-dominated *somiti* at Amlabazar (see Figure A-3, *Somiti I*) used an offensive strategy of open contest to protest against the dominance of the AL-operated Amlabazar Committee. This protest was especially expressed in the social sphere, as a large number of cases had been filed against the AL-operated Amlabazar Committee, according to one of its leaders:

> "Now there is a committee [Amlabazar Committee] beside the *somiti*. So, a conflict exists between the *somiti* people and the committee people. The *somiti* is very powerful. They [the *somiti* people] are supporters of BNP. They show their influence in the bazar though their party is not in government power. They want to show their power in the bazar. As a result, about two to four conflicts have happened so far in this bazar between the *somiti* people and the committee people. About two to four cases have been filed in the *thana*." (Sayed, 23.04.2010)

The strategies of open protest outlined above especially include those among the ordinary who can draw on power sources, or they involve elite groups backed by party power. Many of the above cases can be considered resistance strategies to the dominant practices performed by elite groups. However, in not all cases do these resistance strategies carry with them a notion of permanent change to the social order. Especially in the last case of the contestation between AL and BNP leaders in the Nasimgaon Amlabazar area, it can be anticipated that after a change in government BNP leaders would establish a similarly dominant social order as the AL leaders have, changing only the name of the dominant party, but not the general practices.

Subsuming offensive strategies to secure spatial claims

Table 9 subsumes the discussion on offensive strategies to contest the spatial claims of others. The results show that offensive strategies are less commonly employed in Manikpara, while the elite groups in Nasimgaon draw on a range of offensive strategies more frequently. With regard to the appearance of public space, the offensive strategies of establishing physical structures are most notable as in some cases these lead to a decrease of public space. The table of results shows almost no application of offensive strategies by the ordinary. The two exceptions in Nasimgaon, again, were conducted by those of the ordinary who can draw on larger power sources than others.

Strategies (offensive)	*Manikpara*	*Nasimgaon*
Establishing physical structures	-	-
- permanent structures in public space		elite groups: many cases
- physical boundary		elite groups: many cases
Drawing on religious norms to dislocate others		elite groups: many cases
Removing/evicting physical structures or jurisdiction over space	-	-
- enforcement of state power	elite groups: many cases	elite groups: few cases
- enforcement of committee's power		elite groups: many cases
- removing physical structures as a resistance strategy		the ordinary: few cases
Subtle strategies to contest spatial claims	-	-
- creating bad reputations	elite groups: few cases	
- playing 'tricks'	elite groups: few cases	
Open protest against spatial claims	-	-
- 'physical' expression of protest	the ordinary: few cases	elite groups: many cases
- not behaving as requested		the ordinary: few cases
- contesting dominance in the social sphere		elite groups: few cases

Strategy applied by

elite groups: many cases few cases the ordinary: many cases few cases

Table 9: Overview of offensive strategies for securing spatial claims

9.2.4 Resistance strategies

The strategies outlined above already include some strategies of resistance. However, as the discussion focussed on avoiding, securing and challenging spatial claims, some more subtle strategies of resistance that do not openly contest spatial claims, but that are often mere expressions of non-agreement, are not covered by the analysis. Accordingly, these will be discussed in the following.

In a number of cases I witnessed how the ordinary **made fun of those in dominant positions**, both in their presence and absence. When Foyez appeared during Rokib's Venn diagram exercise to collect the charge for the day, his wife Momena addressed him as 'thief' in a mocking way, while this carried an awareness of the power constellations between them and maybe even a warning that Foyez should take care not to overdo it. During an NGO meeting, the women participating started to make fun of the (male) local political leaders. One of the women told a story about how one of them was caught in a 'non-acceptable position' with a woman who was not his wife. Afsana reacted by laughing and saying "Rain stopped due to such works of *mama-chacha*". Rain is generally considered a blessing and she meant that this blessing would stop due to such unacceptable behaviour by the elders. The address *mama-chacha* – translating as 'maternal uncle-paternal uncle' – indicates how the leader was still considered a respected person, but at the same time his position was at least being questioned. In another incident, the night guard Shahidul came to collect the daily contribution for the sweepers from the shop of my landlords. My landlord in a joking way told him to go away. Shahidul went on collecting from the other shops, and my landlord, when I asked why he didn't pay, just said 'just so'. Later on Shahidul came back with tea, which was shared among those present in the shop. All the above stories of 'making fun' carry some subtle moments of resistance, although they do not have spatial consequences.

A number of interviewees expressed **resistance by verbally challenging other actors** during the interviews. Thus these were expressions of resistance made in the absence of the 'accused', however, they indicate much of the interviewees' opinion and possibly even desire to resist, even if implementation was impossible. For example, Mokbul Hossain took a picture of a person and explained that this Ward level politician once managed to take *chada* from him, and the way he expressed this was rather in an 'angry mood' that he had had to give in. Furthermore, during other interviews and informal discussions Mokbul Hossain threatened to challenge the *mastans* if they ever bothered him again. He also expressed his anger about the committee leaders who 'eat money' and exclaimed that he would not give money to them again but would beat them.

Especially Afsana underlined her potential to resist those in power time and again. Furthermore, her resistance also presented a challenge to male domains, as she frequently trespassed over the 'border' drawn by gender norms. The following examples of 'bold' expressions are taken from the interview with her (23.03.2010) and show how she was aware of her power and ability to resist and contest the powerful:

- In a case when a local leader told her to remove her shop she answered: "I have been facing losses for the last three to four days. Instead of removing my materials [doing it myself], you can take my goods and eat them by dividing them among you leaders."
- Still referring to the same contestation she continued to talk about how she built on the support of RAB: "Then I told [to RAB] 'If you do not believe the mouth's word and a written document is given then you may tear it up. You have to come with me right now.' Then I brought RAB people to the club [the AL office] by pulling their hands by myself."
- About a leader whom she had filed a GD against she said: "[…] I know his character. If he says something then I will say another thing and if he says something for the second time I will beat him since he is the kind of person who tortures a lot in Nasimgaon *bosti*. He does not respect man or woman, speaks very bad languages which cannot be expressed!"
- When Hatem wanted to collect money from her for providing a roof for her shop she described her reaction as follows: "I told 'No problem *mama*, do the work'. I made him do the work by telling bla bla [*onare hawvaw diya kajta koraisi* – indicating that she made him to do the work by giving a positive response to his demand without planning to ever give him the money]."

Afsana's behaviour could be considered as deviating from accepted female behaviour (see Chapter 7.1.1) but in her case, as she had effective power relations, it was not openly contested. Furthermore, her husband was not involved in politics, leaving her to be the dominant person of the household also towards the public, again challenging traditional gender norms. A similar expression of political power was made by Rizia when AL leaders contested her in 1998 on the matter of Nasimgaon eviction (see Chapter 9.1.3). Both Afsana and Rizia spoke out very strongly against male politicians on various occasions and seemed to move self-consciously in a male domain.

Ramisa, selling rice from Khalabazar as a mobile vendor amidst the beggars, expressed her discontent about the AL Sub-area Committee leaders who wanted to make her pay 30 Tk per day for sitting in the area – due to protests by rice vendors who had semi-permanent or permanent vending structures. According to her, she protested this payment and since then was not always allowed to use the space – but she expressed that she had made her spatial claims 'forcefully' against the political leaders' directives.

The resistance strategies discussed above build on 'subtle moments of resistance' which do not necessarily lead to direct changes in dominant patterns in the social sphere and space. Nonetheless, they are important for understanding the expression of the ordinary's agency. Table 10 subsumes the discussion on resistance strategies. The results show that these are strategies of the ordinary. However, most of the examples referred to above involve the more powerful ordinary, while the ordinary who are not able to draw on a multiplicity of power sources tend to also remain quieter about such subtle resistance strategies.

Strategies (resistance)	*Manikpara*	*Nasimgaon*
- making fun of the dominant	●	⬤
- verbally challenging others in their absence, self-conscious presentation of resistance power		⬤

Strategy applied by

elite groups: ⬤ many cases ● few cases the ordinary: ⬤ many cases ● few cases

Table 10: Overview of resistance strategies for securing spatial claims

9.3 LEGITIMATION OF ACTORS' SPATIAL CLAIMS

The above discussion of power sources and strategies to secure and contest spatial claims carried with it different notions of the legitimacy of spatial claims, of how actors legitimate their spatial claims. Five categories of legitimising spatial claims could be identified from the interviews (see overview in Figure 26). Legitimation here refers not only to a democratically or statutorily/legally accepted path of legitimation, but also to the self-perceived legitimation of actors.

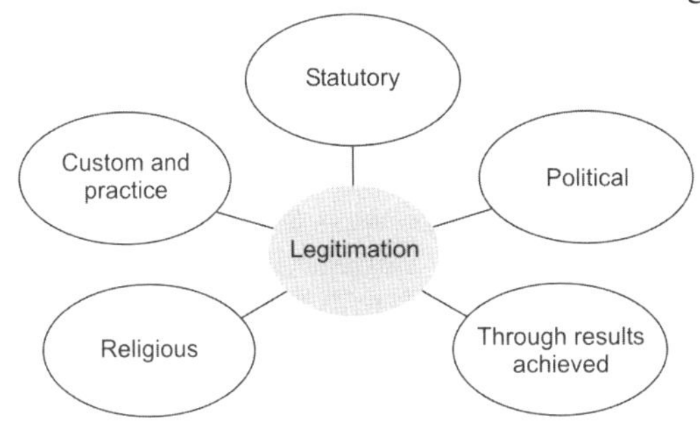

Figure 26: Overview of actors' legitimation

Statutory legitimation

Spatial claims can be legitimised via statutory/legal institutions. Below I will discuss two cases – the 'legitimate' legitimation via statutory institutions and the claim to such a legitimation which is *de facto* not in accordance with these institutions.

Legitimation via statutory institutions especially occurs in Manikpara which might be related to the fact that the legal status of Manikpara with its official property deeds differs from that of Nasimgaon which is located on land owned by the government (while local leaders aim to get a holding number to legalise the right to stay). The BIWTA and the leaseholder drew on statutory contracts to legitimise their actions – although as will be analysed below, the leaseholder also used this legitimation as a backbone that enabled him to make additional claims which were not statutorily legitimised and that were illegal according to the contract. The BIWTA officer, however, insisted that with a formal lease contract in place there were no such abuses of statutory legitimation. Whether acting according to the contract or not, the leaseholder's power was accepted as being statutorily legitimated, which is evident from Rokib's perception that the charges for plastic drying were a government charge (although this charge was not legitimised by the contract) and his understanding that he followed the laws of the government:

> "But we are working here following the laws because I have to give the money according to the government rule." (Rokib, 12.04.2010)

In order to uphold his statutory legitimation, Foyez paid bribes to BIWTA to legalise the increased charges for the *ghat* usage (of course not including the illegal charge for plastic drying) – which again allowed him to circumvent contestations as he could refer to his statutory legitimation.

A second instance of statutory legitimation is the possession of deeds. On the one hand, those who have deeds say that they have a legitimate right to a space. On the other hand, Jahangir also told of how he doubted that the process of land distribution in Manikpara had been conducted correctly. Although everyone had deeds, he thought people had seized the land and managed to find ways to get deeds. He also said that he had deeds of the land on the slopes where his shop was – but the slopes are government land, so this deed cannot really be legitimate in a legal sense. In other cases, however, Jahangir referred to his right of space because he could show a deed. The claims to space in West Manikpara were perceived as legitimised by the residents (especially by Mesbah Uddin), because they had deeds, land filling had been supervised by the DCC Commissioner, the government serviced the land with gas, water, and electricity, and they paid the bills and income taxes (see also Chapter 7.2.2).

The claim to a statutory legitimation which *de facto* does not exist is a legitimation practice mainly found in Nasimgaon. The political leader Baharul of the Sub-Area Committee in Nasimgaon understood the Committee to be statutorily legitimised as the sub-area system had been established by government – however, there are no explicit regulations about how the sub-area committees are to be set-up and operated. This tendency to mix-up government institutions, committees and members of the ruling political party was expressed repeatedly and indicates how deeply rooted and accepted the dominance of the ruling party and its takeover of all statutory institutions are (see below).

Similarly, many of the ordinary perceived the activities carried out by the politically operated committees in Nasimgaon as 'government activities'. Some, however, also realised the fine distinction between using the government as a legitimation and *de facto* not being legitimised accordingly. This, for example, became obvious from the discussion between Mokbul Hossain and Faruk, active in the *Chatro* League at the Ward level, and similarly between Mokbul Hossain and Wahab, a political leader of the AL at the Ward level. In both conversations the discussants analysed how, in response to the reconstruction of the rickshaw garages, the political leaders declared the claim made to the Nasimgaon Math to be a government activity involving construction of a school, a hospital and a community centre in order to achieve general acceptance and avoid contestations.

Another claim to statutory legitimation was made by Mokbul Hossain. In the past, when his shop was located in a different part of the Nasimgaon Math, he was appointed to construct toilet slabs for a toilet building on the Math. He perceived this as a government job and considered himself protected from evictions due to this official function.

Political legitimation

As indicated above, statutory and political legitimations tend to be mixed-up especially in Nasimgaon, given the common perception of the absolute dominance of the supporters of the party that has won the national elections.

The statement of Nabin, an AL leader of Manikpara, on the right to claim public spaces as part of and as a reward for doing politics (see Chapter 9.1.3) indicates such a political legitimation. In Nasimgaon, Baharul's emphasis on the importance of maintaining contact with higher level political leaders can be interpreted as a way of legitimising the committee's activities in the locality via the recognised political structures (see Figure A-3). In many of the instances of spatial claim-making and contestations analysed in Chapter 9.2, political organisations served as legitimation for action.

Doing politics is widely perceived to legitimise spatial claims. Rehana explained that the husband of her daughter was involved in politics so he had a 'right to space', while Shahin expressed his understanding that AL people now should 'eat money' at Khalabazar so as to benefit from politics making up for their previous sufferings:

> "This space is in front of their [AL Sub-area Committee leaders'] club. They had lots of sufferings. Police and RAB made them run. Now they have come to power. If they don't eat 2–4 Tk now then wouldn't it be a problem?" (Shahin, 21.04.2010)

This exemplifies how deeply engraved and apparently legitimated the practice of party politics and clientelism is.

In the local perception and the self-perception of political leaders, the differentiation between statutory legitimation and political legitimation remains blurred. In the local perception, the ruling party's political leaders often have a quasi-government status even if they are not in any official government position. Baharul, when explaining how his committee related to the higher level party organisations, referred to the government and at the same time to the AL *thana* level office which gave its signature to establish the Sub-area Committee. The establishment of the Sub-area Committee through signature by the *thana* level is a solely political procedure, but politics and the state seem to be so closely interwoven in the minds of people that it *de facto* becomes the same. So political organisations like the Sub-area Committees are perceived as being legitimised via the statutory system, although they are not acting according to statutory rules.

Legitimation through results achieved

Dominance over public spaces, spatial claims and practices are furthermore legitimised by the achievements, improvements and mutual benefits they result in, according to the perception of the dominating, claim-making actors. This understanding of legitimacy is beyond statutory involvement and rather rooted in the contributions to local conceptualisations of space. Foyez, beyond the statutory legitimation he had via the leasehold contract, justified his position with improvements made and security gained in the locality, which was a mutual benefit for him and the local population. Tariq and his friends emphasised how the garden had improved the local environment and thus benefited the local population. They also perceived their activities as being government responsibilities so that with their garden they fulfilled what originally were statutory tasks, thus contributing to their perceived legitimation. Similarly, the AL leaders of the Sub-area Committees in Nasimgaon legitimised their existence by bringing improvements to the area's security and environment – which indeed they did by organising the night guards and sweepers in the area. In the same way, their claims on Nasimgaon Eid Gah Math during the conflict with the rickshaw garages were legitimised by contributing to the community through building a school, a hospital and a community centre. That these were not built even after the conflict was solved indicates how such a legitimation may be misused to pursue political aims.

In making spatial claims political leaders often legitimise these by a redistributive effect serving community welfare. Baharul narrated how the *matbors* of the area had invested in the road network. While they recovered part of the costs from the community, they also added their own funds if the collection was not enough. Harez underlined his selflessness in catering for the community and the wellbeing of residents, without even considering party affiliation:

> "Those who are with me, we all always work for the social development. We work for the welfare of the people. We have been doing this from the beginning of the slum. We do it irrespective of party and opinion. We go forward when people are in danger or need medical treatment." (Harez, 16.03.2010)

Furthermore, in the case of Khalabazar, the political leader Hatem told of how he and the Sub-area Committee had formed a *somiti* and asked for contributions from vendors to make the place *paka,* while the vendors became the owners of the shops. In practice, however, those who paid and became owners were the political leaders themselves and not any of the previous vendors (except Afsana). This indicates the distortion between theoretical talking about a welfare-based claim to legitimacy and the implemented reality where leaders appropriated public space. Shahin, operating the fruit shop and tea stall with his wife Rohima, however, indicated on the one hand his frustration about having to pay rent and a deposit for a space he himself had prepared for usage, but on the other hand he understood his payment as being an investment as some of it went into a saving scheme and he would eventually profit from this. While not all vendors contributed part of the shop rent to such a saving scheme, Shahin's perception indicates how welfare orientation and political leaders' personal gains are closely interlinked.

Serving community interests can also generate power from an individual's perspective. Rokib felt that by cleaning up part of the slopes which had been covered in garbage, he also worked in the interest of the community thus giving him the backing and power to stay and perform his business. Ripon similarly stressed how the community police initiative protected the inhabitants and served the community. All of them thus claim legitimacy for their actions by referring to their serving of community interests, independent of whether these initiatives are recognised by the community or not.

Another strategy of avoiding contestations can also be considered as a claim to legitimacy: the strategy of 'formalising' spatial claims, for example providing improved and more stable structures on Khalabazar. The increase of rent and deposit for the vendors was legitimated by the gradual improvement of the vending units. The pillars constructed to demarcate the extent of Eid Gah Math similarly provided legitimacy, and Mokbul Hossain felt here was improved security due to this 'formalisation'.

Religious legitimation

Religious institutions, especially mosques, have very strong backing in society. Thus claiming space for a religious purpose or serving a religious institution is a strong legitimation of such activities which is not easily contested. This legitimation is not formal in the statutory sense, but is formalised by religious institutions that are widely accepted and respected in society. But as the examples show, such religious legitimation, similar to statutory, can also be abused.

For example, the temporary mosque at Dokkin Ghat in Manikpara was accepted on government land, although normally the slopes have to be kept free from buildings. But the necessity of re-constructing the original mosque led to the erection of a temporary mosque on the embankment. This was perceived as legitimate because it was a religious building. Although Tariq and friends faced some problems due to the existence of the (temporary) mosque, they would never openly contest this religious institution. They had to accept it as a legitimate claim to public space. Other examples have been discussed in detail in the Chapters 9.1.3, 9.2.1 and 9.2.3.

Legitimation through custom and practice

Many of the ordinary based their legitimation on custom and practice, but one of the local leaders in Nasimgaon also used such an argument to legitimate his spatial claims.

A widely accepted legitimation to spatial claims is the initiative to create a usable public space. Rokib had cleaned up the embankment slopes where he operated his business by himself and thus could claim the right to use this space for his income generation. He also emphasised how the environment had improved due to his initiative and that he had thus contributed to the community (see above). At Nasimgaon Khalabazar, Rohima and Shahin's right to space, as perceived by themselves and many of the original vendors, was also built on their initiative to clean-up and fill the low lying waste depot:

> "Shahin came here at least 20 years ago and at that time this Math was a big hole like a lake. Shahin filled it up by putting lots of sacks of rubbish and all the garbage of the slum. That Shahin is now operating the shop and has to pay rent. He is paying a rent of 2,000 or 2,200 Tk. Isn't that it was his right? He, who is the inhabitant of this place for the last 20 years, also does not get a shop here. Why didn't he get one?" (Afsana, 23.03.2010)

Similarly, Dipa also considered that she had a right to the place she finally had to leave in June 2009 (see her quote in Chapter 8.2, extract: "I used to do my business over there for 7–8 years. It was a low place before and I repaired it by spending money and did my business. So I have a right.").

Against the political leaders' claims these legitimations, based on custom and practice and a mutual consensus among the mobile vendors, did not suffice, and Rohima's family and Dipa had to start paying rent to the AL Sub-area Committee to maintain their space-based livelihoods in the same location.

Another legitimation via 'custom and practice' was to refer to others who had made a similar spatial claim. Tariq's friend referred to this when explaining that they were not afraid of any BIWTA eviction:

> "If they [BIWTA] come to evict, then all of these houses [indicating the houses beside the road] should be removed first, because all of these houses have been constructed on illegal land. All people have been living here for a long time. It will be tough to remove all of them. They can demolish all of these structures but it will return to its original situation after some days of demolition." (Tariq's friend, 19.04.2010)

In Nasimgaon, the local leader Baharul also referred to the 'everyone is doing it' legitimation when talking about the contestations after he had built a veranda on the second floor of his house. He argued that everyone was doing so and thus his veranda should not be contested, or all verandas should be contested. At the same time, he narrated that he was among the first to build *paka* walls and afterwards many people imitated this. Afsana told of how in the early days of Khalabazar there were only a few vendors, but after she had started her shop and had been operating peacefully for some time, other vendors followed her example and set up shops.

Ownership of structures – even if ownership of land could not be achieved, as in the case of Nasimgaon – was commonly perceived as a strong legitimation of a spatial claim. This is especially obvious from the low level of contestations owners of structures, especially of stable shops, experienced. Neither Mokbul Hossain nor the rickshaw garage owners in the south of Nasimgaon Math faced any considerable opposition to their spatial claims after initial conflicts. The construction of housing compounds at Khalabazar will also not be easily contested. At the same time, many interviewees expressed awareness of the fact that they did not own the land, as Nasimgaon is erected on government land. For example, Azad, one of the rickshaw garage owners, stated "it is government land, not someone's grandfather's land" (16.03.2010).

Furthermore, being from the locality served as a legitimation for spatial claims. When discussing the contestation they experienced by the members of the mosque committee, Tariq and his friends underlined that they were 'local boys'. This emphasis of their local background can be understood as a legitimation for having a right to occupy the space of the garden. Similarly, when Rohima and Shahin's continuation of their business was endangered due to another person offering a higher deposit, they were supported by others who stressed that they lived close by and had been operating the shop for a long time.

Often related to the custom and practices identified above, residents also referred to their citizenship as a legitimation for spatial claims. In the following quote of Shihab, he expressed a claim to political citizenship as a legitimation for staying in public space:

> "[...] nobody can evict me, but sometimes they scold me! But we don't move! I am a voter of Nasimgaon." (Shihab, 06.07.2009)

Similarly, the rickshaw garage owners when they went to the MP to avoid being evicted made a claim to citizenship, as Azad said:

> "At first [the MP] also wanted eviction but later we made him understand that we knew this was government land but we bought this land during the BNP period. We said 'Now you [the government] want to evict us because it is illegal, but we are also citizens of this country and people of this government'. [...] Then [the MP] understood our problem and told 'I will do nothing that will lead to any loss of your money and I will also rehabilitate you'." (field note, 06.03.2010)

These claims to political and civic citizenship made by the ordinary speak of an awareness of citizen's rights, presenting a potential for resistance and agency against dominating power authorities. Most of these claims to citizenships were, however, expressed by the political leaders in relation to eviction and achieving resettlement and rehabilitation of Nasimgaon inhabitants. While here the notion of citizenship is elemental, it is at the same time expressed only by those dominating the social sphere, questioning the outcomes to be expected if Nasimgaon were evicted and its inhabitants resettled and compensated (see Chapter 5.4.1 for the common rumours that a few members of the elite will get most of the compensation while the ordinary might go empty-handed).

Legitimacy of legitimations?

The above discussion has indicated a large variety of 'legitimations' that the different actors draw upon to underline their spatial claims. Legitimation in the understanding of actors is not only a statutory process, but rather the organisation of society allows for manifold ways of legitimising an actor's spatial claims. Thus it is not only the elite groups who find arguments for the legitimacy of their actions, but also the ordinary who especially claim legitimacy with reference to customary practices and notions of citizenship. The multiplicity of legitimations indicates a pluralistic understanding of modes of the production of space, and close interconnections between the legal and extra-legal sphere. This discussion will be followed up especially in Chapter 10, considering the notion of urban informality.

9.4 RESULTS OF THE NEGOTIATIONS OF ACCESS TO SPACE

While above I have discussed the power sources, strategies and legitimation in negotiation processes, here I will discuss the results of these negotiation processes (for a critical assessment of the term 'negotiations' in the context of this research, see Chapter 9.5). The discussion focuses on the results for the condition of public space in the urban fabric, the ordinary and the elite groups.

9.4.1 Results for the condition of public space in the urban fabric

The processes of spatial claim-making do not only influence space-based livelihoods of the ordinary and elite groups, but also manifest themselves in physical space. They considerably influence the urban fabric within the study settlements, especially so in Nasimgaon. In Manikpara the road network is fixed in its appearance and within the settlement no spaces which could be encroached upon or protected from encroachment exist. Instead, densification concentrates on the built-up plots. The embankment slopes are not suitable for permanent occupation. Only the use of the slopes by the gardens (Bagan and Tariq's garden) can be understood as an explicitly spatial result (in the sense of physical space) of negotiations and will be included in the discussion below.

Encroachment of public space

During the course of this research and the several negotiation processes of access to space and of spatial claim-making that I observed, the public space available within Nasimgaon was reduced considerably.

The last two remaining public spaces within the internal settlement structure available in 2009 have decreased considerably during the two year period of this research. The main activities leading to the decrease of openly accessible public space were triggered by the appropriations of previously open spaces by local political leaders. Both in Khalabazar and on Nasimgaon Eid Gah Math (see Figure 27 for a black and white plan of the changing urban fabric of Khalabazar) all the encroachments of a stable and permanent nature were conducted by politically-backed elite actors. Encroachments planned by the ordinary, such as Taslima who wanted to raise a platform of debris at the place where she sold *pitha*, were in contrast prohibited by the very same leaders who gradually took possession of open spaces.

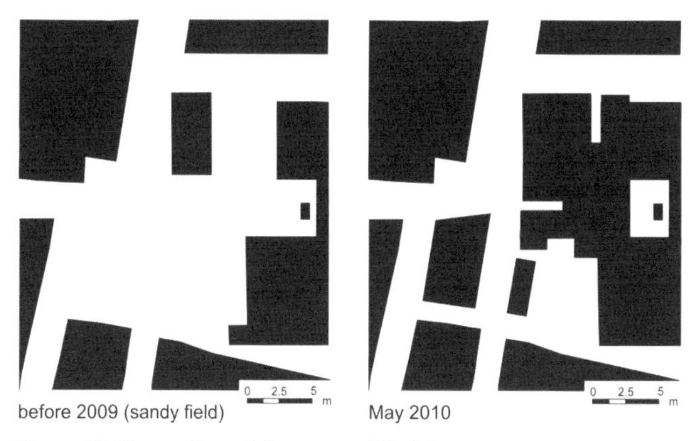

Figure 27: Decreasing public space at Khalabazar

The contrast between the actual encroachments and the simultaneous commitment of the 'encroachers' to the protection of public space shows the high pressure on public space as a resource[17]. Furthermore, it shows how the institutions at work – both the political committees and the social institutions of protection – are not sufficient to protect this resource against the high profit-motivated development pressure. A public space at the fringe of Nasimgaon was protected for a considerable time, but in summer 2009 a local *mastan* tried to take possession. Although he was defeated by the people of the area for the time being, in the following months the space was gradually encroached upon and the protective fence dismantled, making it only a question of time before the space was occupied by housing compounds on the initiative of the same *mastan* or leaders with similar power. While the political actors often emphasised the importance of public spaces, the following quote also illustrates the importance of 'development' as manifested in stable structures:

> "This bazar will be developed if the government doesn't demolish the slum. Now the shops are located on the open space. These shops will be developed more. But we will have nothing to do if the government demolishes this slum. If the government will not do it then we will arrange the bazar following a plan. [...] This market will be developed and well organised like the Amlabazar market [referring to the indoor structure of Amlabazar with a number of stable shops under one large roof structure]." (Hatem, 07.03.2010)

17 The high pressure is illustrated by the high values of structures which are sold – based on ownership of structures rather than land due to Nasimgaon being on government land. For example, the stable *bhits* at Khalabazar that had not yet been transformed into full shops with shutters at the front were sold for 20,000 Tk – for a unit of less than 4 m^2. In one of the less integrated locations where house rents were almost half the price of the bazar area, one political leader had sold 11 rooms of his housing compound for 200,000 Tk.

The reference to government at the same time indicates the insecure tenure status of Nasimgaon, and thus the dependency of the elite and the ordinary on the ruling government's support. The following quote of Rabiul, Imam of the mosque and AL supporter, also indicates the practice of possession-taking and the decrease of space with reference to the Eid Gah Math:

> "In different times the area [Eid Gah Math] reduced in different ways e.g. personally, politically and socially." (Rabiul, 23.03.2010)

The differentiation between 'personally', 'politically' and 'socially' refers to the reasons why spatial claims were made – for individual, political or community-oriented benefits.

Many inhabitants complained about the decreasing public spaces and increasing densities in their neighbourhoods. During the solicited photography Meher took a picture of a public space at the fringe explaining that this was one of the few spaces where children could come to play. Other references were made to open spaces as protection in emergency situations (see below). In Manikpara, the high density was observed as problematic by many interviewees. The high density was among the factors that caused Mesbah Uddin to refer to the area as 'industrial' instead of 'residential' (*abashik elaka*) (see Chapter 7.2.1).

Hortem also complained about the 'face-to-face condition' and the lack of spaces for leisure time activities and expressed his preference for a previous development stage when there was more attachment among people, more space and fewer people. The working women in the focus group discussed the crowded living conditions with hot rooms and a lack of windows as unfavourable, preferring to live in houses with windows, daylight and a veranda. Similarly they complained about the narrow lanes, and during the focus groups Fahima's colleagues referred to the urban area of Mirpur as a place where there was enough space for roads so that both walking and rickshaw driving were possible at the same time. However, the perceptions of high density were also contradictory. Fahima at the same time disliked the low density in the neighbourhood where she lived before coming to Manikpara: it frightened her, especially as she feared that someone would harm her daughter who was approaching marriage age at that time. Mashrufa, although she did not go to the bazar herself, also referred to a convenience of density, as it meant that everything was close by and there were no long trips to the bazar.

The processes of decreasing public space and densification result in a diminishing urban functionality within these neighbourhoods by making them more prone to various public-health related consequences of dense living conditions (see also Baumgart et al. 2011), and especially to fire hazards. The compromising of urban functionality can be seen as a consequence of high development pressures and the simultaneous effects of ever-present political contestations whereby 'doing politics' is understood as a means to make immediate gains during one electoral period (see Chapter 9.1.3, where the AL leader Nabin discusses the need of rewards for political activities, and Chapter 9.2.1, 'Withdrawal of claims in anticipation of contestation', for the practice of selling spatial claims before the end of an electoral period).

Protection of public space and improvements

A number of the initiatives analysed above served to protect public spaces from encroachment. However, the analysis has also indicated that there is a gap between the wish and desire to protect public space in principle and its encroachment in reality.

Both the ordinary and the elite groups repeatedly underlined the importance of protecting public space from encroachment as it was a community asset. The most prominent argument was protection from fire outbreaks. The fear of fire is ever-present in the minds of Nasimgaon's inhabitants. At the time when the first solicited photographers had received cameras in April 2009, a fire broke out in the northern part of Nasimgaon, destroying several housing compounds. Even those solicited photographers living in the southern part of Nasimgaon went to the place to take pictures. Rohima explained about one of the pictures that only the existence of the graveyard as a nearby open space prevented the fire from spreading, thus acknowledging the protective function of public spaces while seeing the dangers of dense living conditions. This was similarly referred to by the political leader Harez:

> "Slum dwellers and all expect that this Math should be kept as open. It is because there is the chance of a fire outbreak in the month of *Choitro* [Bengali month, end of dry season, mid-March to mid-April]. Any time an accident can take place. If there is an outbreak of fire then the situation becomes extremely bad. This type of incident happened several times. Once there was an outbreak of fire and seven people died. However, everyone expects that the Math should be clean and open for all and the slum dwellers." (Harez, 16.03.2010)

He continued to complain about the encroachments, especially the rickshaw garages which had been the source of a large conflict (see Chapter 0). Again, his statement underlines the ambiguity of local leaders between commitment to protection and active encroachment.

The plan to fence the Nasimgaon Eid Gah Math and provide a road surrounding it was the only initiative taken to protect public space from encroachment (although its designation as an Eid Gah Math was also meant to protect it from encroachment, but without success). The planned approach, i.e. putting up concrete pillars and a barbed wire fence, indicated a conceptualisation of the protected space that did not take into account the perspective of the ordinary who wanted to access the public space for their space-based livelihoods without erecting permanent structures (see discussion in Chapter 9.4.2). Whether the approach would have been the most compatible one remains questionable. However, the fencing remained a plan, as only the concrete pillars were erected. Instead of protecting the Math, what happened later on was exactly the opposite. Furthermore, the commitment expressed by local leaders to even enlarge the Math by removing the structures constructed during BNP times can in retrospect only be understood as the 'mouthing of words', as they were also the ones supporting further encroachments later on.

During one interview the Sub-area Committee leader Baharul proposed a way to improve accessibility to the settlement, and he especially underlined how this would improve access for ambulances. His proposal was to change the system of opening the shops from putting the front CI-sheet up on bamboo pillars into a system of shutters (even though this solution would be more expensive), thus reducing the extension of shops onto the road space. However, due to his limited power in interfering in the market area under the Amlabazar Committee (see 9.2.1) he did not succeed, although he had been convinced that an implementation would be possible. The only successful initiatives were the NGO-supported improvements of the drainage system and road coverage (brick surfacing). Furthermore, in some small areas, local shopkeepers jointly contributed money to improve the roads in front of their shops, e.g. surface levelling to avoid water accumulating in the rainy season (see Chapter 6.3).

In Manikpara, improvements of accessibility were achieved especially by the reconstruction of Embankment Road, although this also resulted in some negative side-effects for those operating shops along the road (see Chapter 6.3). This was carried out by DCC and circumstances differ considerably from those in Nasimgaon because the road carries much of the traffic traversing between Old Dhaka and the wholesale market and bus station area of Gabtoli, thus being of considerable importance on a city level – in contrast to the road network of Nasimgaon which is only for internal traffic.

While in Nasimgaon public spaces disappeared, along the embankment slopes in Manikpara spaces for public welfare were created or improved. The Bagan, already in existence at the beginning of this research, was improved considerably by constructing a roof structure including a stage, by extending the 'zoo' area and by surrounding the garden with *paka* walls (although the latter improvement at the same time means increased control of access and a shielding from the wider public that may adversely affect access for women). Tariq and his friends started to develop a smaller garden at Dokkin Ghat from 2009 onwards, with a view to providing another space for recreation, although during the course of this research the garden was used only by those who had developed it, which might also be due to the lack of 'convenient' access when the temporary mosque was located at Dokkin Ghat.

9.4.2 Results for access arrangements regarding the space-based livelihoods of the ordinary

The access arrangements for the space-based livelihoods of the ordinary tend to be shaped by insecurity and a condition of 'permanent temporariness'. Only in a few cases was security of access achieved during the time of this research.

The insecurity of access is manifested in access arrangements that are not fixed but that change depending on the decisions of dominant power holders. This indicates how vulnerable those with space-based livelihoods can be and how access remains temporary, subject to sudden or gradual changes. The gradual change of access arrangements at Khalabazar from a daily contribution to night guards and sweepers of about 5–8 Tk to a monthly shop rent of between 1,000 and 2,000 Tk and an additional deposit payment impressively illustrates this insecurity. If vendors were not able to adapt to these new access arrangements, they could not maintain their position in the bazar, as for example Dipa who had to leave the place. While Shoma managed to rent a shop later on, she also explained that mobile vendors nowadays made a better profit because they did not have to pay rent and contribute to night guards and sweepers.

The vendors on the Eid Gah Math experienced repeated disruptions of their livelihoods due to the conceptualisations of the space by local leaders and conflicting situations. Similarly, the rickshaw garage owners in the south of the Math and also Mokbul Hossain experienced insecurity of their positions for a long time, however, their stable structures finally appeared fairly safe from evictions or further contestations. Concerning the use of the Nasimgaon Math to store materials, Mokbul Hossain continued to experience insecurity and accordingly shifted his materials inside from time to time. This dependency on others' decisions about how public space should be used produces vulnerabilities and the need for continuous re-negotiations of access rights to public space.

The temporality of access arrangements is inherent in many of the stories and strategies discussed above. Many of the space users changed locations due to the non-permanence of access arrangements. The history of Mokbul Hossain's shop is one of continuously moving to alternative locations due to contestations of his present locations. For years he was always in a temporary and insecure condition regarding access to space, until he finally bought a stable shop. Nonetheless, he seemed to still fear contestations, as although he had planned to make his shop two-storied in summer 2009 he did not do so until the end of 2010. Furthermore, in April 2010, he again referred to new rumours that his shop was designated for removal:

> "Nowadays I am hearing something about this Math, like 'You don't pay any money for this shop. So the decision has been taken that your shop will be taken from you.' They [leaders] think about this matter. I said 'OK. Tell them to come to me and discuss with me about this matter. I do not agree to discuss with them deliberately.'" (Mokbul Hossain, 18.04.2010)

At the time that this research was completed nothing had happened to Mokbul Hossain's shop, however, his story very vividly shows the creation of insecurity and security as a continuous process. Even he, who was able to draw on a range of power sources and had shown considerable ability to resist, seemed to be repeatedly irritated by new plans for the Nasimgaon Math and his shop that evolved from discussions among local leaders.

The repeated disruptions of vendors' livelihoods exemplify how access to public space is always temporary and time and again access is newly negotiated, often silently, with a strategy that rather resembles the quiet encroachment of the ordinary, meaning that vendors are testing the grounds and only slowly coming back (see Chapter 9.2.2).

On the embankment slopes in Manikpara, the eviction by BIWTA also show how many arrangements of access are temporary – while those with a community welfare orientation, i.e. the gardens, were not contested. The temporality was here not created by the leaseholder, who took care to keep good relations with the users of the space as they were also his source of revenue. It was rather created by the statutory authority in a move to demonstrate its ultimate power and jurisdiction over space. Similarly, the owners of the rickshaw garages established by the Mosque Committee at Dokkin Ghat after the temporary mosque was removed underlined that they had no guarantee of security in this space, but faced similar conditions of temporariness, as they could also become subject to government evictions.

Despite these feelings of insecurity of access to space, some users of space also expressed that they felt secure in their right to stay. Mokbul Hossain, after discussions about evicting him from his shop location had come up between June and August 2009, expressed how since the concrete pillars had been put to demarcate the boundary of the Nasimgaon Math he felt secure again about his right to stay. The concrete pillars manifested the extent of the Math in a way that no longer necessitated the partial eviction of his shop (see Figure 19 in Chapter 8.3.1 for the original re-design and fencing plan that would have involved partial removal of shops in the north and east of the Math, including Mokbul Hossain's – although even this 'permanent' security was not long-term as rumours about eviction of his shop repeatedly came up). As already discussed above, Aklima understood the payments she made to sweepers, night guards and the mosque as a guarantee of security against unpredictable demands from *mastans*. The rickshaw garage owner Azad also confirmed feeling that the continued existence of his garage in its location was more secure after the matter had been solved with the support of the MP. Rokib in his business on the embankment slopes also felt secure in his position as he had powerful supporters (especially relatives, but also the leaseholder Foyez himself).

The conditions of security can thus be summarised as emanating from ownership (at least of the structures), from powerful supporters and from the ability to draw on a wider range of power sources than the other ordinary. In the absence of these conditions, the insecure and temporary conditions analysed above persist.

Those considered powerful in the society also claimed to have increased the feeling of security for others by acting as 'role models'. The leader Baharul was the first who built *paka* walls and then a second storey. When others saw him doing so they considered this an indication that the slum would not be evicted and imitated his behaviour. After Afsana had fought back the political leaders during the caretaker government and started operating her semi-mobile shop again, this seemed to provide a feeling of security for other vendors who then also returned.

9.4.3 Results for elite groups

The outcomes for the elite groups tend to be more secure, and are the results of negotiations among different elite groups that claim jurisdiction over public space. A special case is the Manikpara embankment, where elite groups instead negotiated directly with statutory authorities.

The dominance of elite groups in the social sphere, analysed above as a strategy, but also based on the availability of sources of power that tend to reproduce already existing structures of domination, translated into increased spatial claim-making. Claiming the jurisdiction over space is often closely tied to personal gains of income generation, although the leaders and committees at the same time commit themselves to contributing to social welfare and redistributing resources. However, reality and principles here differ considerably, as the examples have illustrated (see Chapter 9.4.1). Those who are able to dominate in the social sphere tend to experience a higher security of their spatial claims, as they at the same time are involved in defining the 'rules of the game'. Insecurity is mainly experienced in cases where the political situation remains fuzzy and the spatialities of power are ever-shifting. This is the case with the Ward Commissioner elections which have not been held despite being long overdue. This caused even the lease operator to express a feeling of insecurity. Secondly, the power of BNP supporters in the Amlabazar area in Nasimgaon, as explained by Sayed, also led to a higher degree of insecurity of jurisdiction over (public) space.

In addition to what Bayat framed as the 'quiet encroachment of the ordinary', which I also identified as one of the reactive strategies (at the same time preemptive) applied by the ordinary as well as by elite groups (see Chapter 9.2.2), the spatial claim-making of the elite groups in many instances more resembles an 'organised encroachment of the powerful'. This idea, developed in close cooperation within the DFG research project with reference to both public space and supply of infrastructural services (Hackenbroch, Hossain 2012), could be put forward based on the finding that the permanent and stable spatial encroachments have predominantly been made by the elite groups. The use of political power in particular supports their dominance in social space, which translates into the possibility to make encroachments in physical space (see Chapter 9.4.1). Furthermore, the example of the spatial claim-making process along the Amlabazar Road illustrates how the local power holders strategically use the urban poor to access opportunities. Here, the strategy of starting the construction process by handing the land to the poor served to test and limit contestations (see the case when Dipa was brought in front of RAB to underline the poverty of the new house owners).

9.5 CONCEPTUALISATIONS OF (PUBLIC) SPACE

Based on the preceding sub-chapters and the narratives in Chapter 8, in the following I will analyse the conceptualisation of (public) space inherent in the negotiations of access to space. Specifically I will differentiate between modes of the production of space that build on dominant conceptualisations as opposed to resistance conceptualisations, and in so doing will relate the findings to the theoretical departures outlined in Chapter 2.

Public space within the study settlements has to be understood as a contested resource, as a multiplicity of actors with diverging interests seek to access public space and make spatial claims. Public spaces – even within the study settlements which do not represent the most central places in the city context – thus present a very visible arena of socio-political contradictions (see Chapter 2.2.1) on all spatial levels. As Lefebvre put it, the "socio-political contradictions are realized spatially" (Lefebvre 1991 [1974]: 365) which in the cases analysed above becomes visible time and again with contestations between dominant groups and between dominant groups and the ordinary. Occasional acts of resistance of the ordinary against the decisions and 'dictates' of the leaders bring to the forefront the dominance experienced on the ground in everyday life. The contested space and the negotiation of access to these spaces make visible the spatialities of power, manifested in dominant and resistant modes of the production of space and the entanglement of different actors.

The analysis of the negotiation processes of access to public space has pointed out the prevalence of a mode of the production of space shaped by actors holding dominant power. Referring back to Lefebvre's dimensions of the production of space, the access to public space is predominantly negotiated in the dimension of the representations of space, the conceptualised space (see Textbox 1 in Chapter 2.1.1; Lefebvre 1991 [1974]: 38-39). In this conceptualised space or social sphere, an entanglement of statutory, political and religion-based actors define the access rules to public space, no matter whether explicitly formulated or rather existing as behavioural rules and norms. The composition of actors determining the social space as the space produced by society over time makes explicit the persistence of strong patron-client based hierarchies of society, leading to a dominant mode of the production of space where the outcomes, the spatialities of power, are largely determined by a dominant social group. Thus the power structure at work in negotiating access to public space has to be understood as being differentiated along the divisions of political affiliation and social leadership claims, and the livelihoods strategies of the ordinary.

Those exercising dominant power in the social sphere are primarily legitimated politically, religiously and 'socially' in the context of 'social leadership' in Bangladesh, but they operate in constellations of close entanglements with statutory institutions (see also the further discussion in Chapter 10). The actions and self-perception of the political organisations in particular underline the existence of a *de facto* state (Roy, AlSayyad 2004b) that dominates especially the social sphere of conceptualised space even in the relative absence of statutory institutions. Outside of the state, these institutions create a conceptualisation of space Lefebvre would refer to as the production of abstract space, experienced as a repressive space (Lefebvre 1991 [1974]: 352; see Chapter 2.2.1). Use rights of public space are to a large degree defined and determined by dominating institutions, be it the state or other actors acting as *de facto* state but with similar logics of authority, which shape the negotiation processes considerably (see Chapter 9.4.3 on the 'organised encroachment of the powerful'). The spatialities of power lead to an increased vulnerability of the ordinary, while the dominant are in a position to extend their spatial claims and dictate rather than negotiate social space and thus access to physical public space. As the analysis in Chapter 7 has indicated, the institutions of dominance are not only defined in the political field and as rules of access, but social institutions and norms especially concerning social gender relations similarly define levels of access to public space.

Resistance power – the ability to resist (Sharp et al. 2000: 3) – tends to be low in this environment, where social space is dominated largely by local leaders who secure and legitimise their role via the party system, religion and traditions of patronage relations. Nonetheless approaches challenging the dominant conceptualisation of space exist, albeit often rather in 'subtle moments of resistance' than as open challenges and counter spaces. It becomes apparent how resistance power and its successful application are highly linked to individual's political power or ability to draw on power sources in general. Someone with political connections risks less if resisting, while these strategies are hardly employed by the ordinary who will rather accept a dictate than increase their exposure to risk by contestations. Only a few of the ordinary interviewed in this research made such visible attempts to claim a counter space, the majority of the ordinary's actions rather focus on a strategy of invisibility – be it the 'quiet encroachment of the ordinary' (Bayat 2004; see Chapter 9.2.2), the resorting to temporary strategies and alternative locations, or the adhering to rules negotiated in social space among the dominant – and thus do not constitute a movement "potentially destabilizing the centre" (Simone 2010: 40, quoted in Roy 2011: 232). This includes the notion that these acts help to stabilise the state or the activities of the dominant in general, in the sense of strategies to compensate the failure of state provision ('nonmovement', see Bayat 2004 in Chapter 2.2.1), rather than challenge them. In Lefebvre's terms, the adaptation strategies of the ordinary would have to be understood as a reproduction of existing spatialities of power, rather than as a new mode of the production of space challenging the dominant mode.

Pronounced forms of resistance tend to be 'political' and rather do not entail the production of a counter space. In the cases of resistance in Manikpara, those who established the garden at Dokkin Ghat, for example, did not generally contest the dominant modes of the production of space. They resisted against a particular move by the mosque committee to create a bad reputation for them. Being part of the economically successful middle class in Manikpara, they are rather tomorrow's power holders than creators of counter spaces on a large and challenging scale. Similarly, when expressing resistance to some groups disturbing his leasehold operations the leaseholder did not employ resistance as a way of challenging but rather re-instated his domination of the social sphere against the claims of others.

The above results suggest a differentiated power structure and a mainly dominant conceptualisation of (public) space and indicate similar divisions with regard to agency. The claims to resources by the elite groups and the ordinary are made in different spheres, underlining the division of society, as for example analysed by Chatterjee (2004) who differentiates 'political and civil society' or by Baud and Nainan (2008) who refer to different platforms of 'invited space'. For the ordinary, the negotiations of access to public space are characterised by temporary outcomes, translating into a 'temporality of achievements' (Chatterjee 2004: 60–62) and the negotiation of 'seemingly endless liminalities' (Abourahme 2011: 459). For example, the continuous re-negotiations of access to the stalls newly constructed at Khalabazar are a vivid example of how achievements in terms of negotiated access rules and regulations do not provide vendors with a secure and guaranteed right to stay. The restriction to spaces that are characterised by temporary arrangements and achievements at the same time translates into insecurity, risks, vulnerabilities and uncertainty for those not able to participate in dominant modes of the production of space.

The spatialities of power produced by dominant forces translate into continuous contestations and a lack of fixed regulations about the access arrangements to public space for the ordinary. Accordingly, the risk of being dislocated and exclusionary practices are ever-present in the space-based livelihood strategies of the ordinary. The vendors in Nasimgaon especially are thus permanently located in a 'gray space' (Yiftachel 2009) produced by those who hold dominant power and do not wish to 'formalise' the spatial claims of the ordinary. Negotiations of access to space are therefore subject to a permanent 'unmapping of space' (Roy 2009b) and "constant negotiability of values" (Roy, AlSayyad 2004b: 5). The examples of the study settlements furthermore underline the importance of seeing space as a product and a process where social relations are inscribed and continuously negotiated. The same applies to risk and uncertainty which can also be seen as socially constructed, and thus as a product of power relations in social space translated into the spatialities of power.

For the negotiations of access to public space, a highly contested resource as the analysis has shown, the dominant mode of the conceptualisation of public space and a differentiated power structure translate into 'differentiated negotiations'. The actors negotiating access to public space do not have access to the same spaces of negotiations and do not have the same scope to act. This notion of unbalanced power relations has been discussed in Chapter 2.1.1 with reference to Bose (1998) and Habermas (1981) and in Chapter 2.2 with reference to the different 'spaces' and 'spheres' of society where actors can make spatial claims (especially in literature referring to Indian examples such as Chatterjee 2004 and Baud, Nainan 2008). Habermas' assumption of an equal access to resources and power in communication and negotiation processes cannot be transferred to Bangladesh, as this research has demonstrated. While some actors can shape and conceptualise public space, others mainly have to adapt their strategies accordingly, and thus can hardly take part in an interactive process of negotiations. This leads to the question of whether the term 'negotiation' is appropriate to frame the processes leading to the production of space. Negotiations here are often not a process of open discussion and meetings among actors, but many of the processes go almost unnoticed as they are guided by the rules of society and thus are deeply inscribed into everyday life and actions. They are informed by the strategy an actor follows and can be understood as a process of action and reaction, rather than a mutual decision-taking process. In the context of access to public space in Dhaka the term negotiations needs to be understood as a result of actors' power sources and relations, and their scope and agency to apply certain strategies and draw on 'accepted' legitimations.

CONCLUSION: THE NEGOTIATIONS OF ACCESS TO PUBLIC SPACE

In this short concluding chapter, I aim to summarise the main findings of Part III before I move on to discuss the implications of the empirical findings for framing the concept of urban informality, for spatialities of (in)justice and for urban planning approaches in the last part (IV) of this research.

- Successful negotiations of access to public space (securing and maintaining spatial claims) are only possible for those who are able to draw upon a multiplicity of power sources. The ways of acquiring power tend to reproduce existing structures of hierarchies and patron-client relationships.

To successfully negotiate access to public space, actors draw upon a multiplicity of power sources as analysed in Chapter 9.1.3. The most important power source is political power, i.e. affiliation with the ruling party or close contacts to politicians of the ruling party, as the followers of the ruling party tend to dominate the social sphere and conceptualised space and thus define the rules of the game. At the same time, those drawing on political power also derive dominant power from their social and religious status within the neighbourhood. This is especially the case in Manikpara with its more differentiated society, local identity and tradition of social institutions with a community welfare orientation.

- The strategies applied to pre-emptively or reactively secure spatial claims include drawing on statutory institutions in an entanglement of legal and extra-legal activities. This entanglement with statutory authorities, however, tends to be available primarily to the dominant and elite power holders and only in a few cases to those of the ordinary who are able to draw on sources of power.

In Manikpara access to the embankment slopes is closely negotiated with statutory institutions. For example, the statutory authority (BIWTA) legitimised the leaseholder's spatial claims via the statutory contract. However, the negotiation of jurisdiction over public space also involves extra-legal activities such as bribery, which are actively made use of by the leaseholder to secure his spatial claim of jurisdiction. Similarly, while performing their official tasks the police are at the same time drawn in to support the leaseholder by an extra-legal arrangement of extra-payments for extra-security. In Nasimgaon the use of statutory institutions focuses on the police and RAB as part of the executive and the filing of General Diaries or cases at the *thana* police station. Drawing on statutory authorities remains an option predominantly available to elite groups. The ordinary are able to access this option when they can generate power for example through the political system or economic resources. The example of Dipa shows that without sufficient power sources, the statutory institutions tend to be inaccessible for the securing of spatial claims.

- The ordinary draw largely on less openly pronounced and rather reclusive strategies, characterised by adaptation to dictated rules of the game, or quiet, invisible and stepwise processes of making spatial claims.

The analysis has indicated that only those of the ordinary who can claim a range of power sources, and in some respect resemble elite groups, actively resist contestations and offensively articulate their spatial claims. This was done especially by drawing on political and economic power but also on social relations. For the ordinary who cannot draw upon such sources of power, reclusive strategies present an opportunity for spatial claim-making that is less easily contested.

- The access arrangements to public space for the space-based livelihoods of the ordinary are characterised by conditions of permanent insecurity and temporariness, as a result of the dominance of social space by elite groups.

As already indicated above, the ordinary, especially if they are not able to draw on a wider range of power sources, have to perform their space-based livelihoods according to the rules and arrangements of the dominant groups. These rules tend to be unfixed outside of the legal framework, and thus access to space is not secure. This limited agency in making permanent spatial claims means a continuous need for re-negotiations of only temporary achievements and arrangements which can constantly be contested.

- The analysis of negotiations of access to public space reveals a dominant conceptualisation of space manifested in the social sphere. Resistance conceptualisations, in contrast, are embodied mostly as 'subtle moments' or plans which cannot be implemented, and do not translate into larger movements to claim a different conceptualisation of public space.

The social sphere is dominated by elite groups, mainly backed by the party system and affiliation, but often also in entanglements with statutory institutions and religious norms. In this entanglement of dominant power forces, the representations of space are largely conceptualised, making the dominant mode appear as a 'powerful encroachment of the ordinary'. Resistance movements remain limited and seldom challenge the dominant conceptualisation of space. While some small achievements were made depending on the availability of power sources, larger movements to create counter spaces or differential space could not be identified.

- The political domination of access arrangements to public space coupled with high development pressures leads towards a decrease of public space, despite commitments in principle to protect public space as a community resource.

Especially in the two narratives of public spaces in Nasimgaon, the transformation of and encroachment on public space became very obvious during the course of research. This underlines the high pressure of development and densification within the city and on a local neighbourhood scale. The local socio-political organisations are not able to prevent the decrease of public space despite their verbal commitment to do so. Instead, they are even involved in actively encroaching on the last resources of public space within Nasimgaon. Institutions preventing encroachments and safeguarding public space are yet to be found in an environment characterised by confrontational party politics and maximisation of profits on the part of local political leaders. Safeguarding public spaces thus seems to become an issue of working democracy and involves overcoming traditional party politics (see Chapter 11.2 for a continuation of this discussion).

PART IV – RECONNECTING THE RESEARCH TO THE THEORY DEBATE AND URBAN PLANNING

Based on the empirical findings presented and analysed in Part II and Part III, I here seek to reflect upon these from a theoretical perspective as well as from a more normative urban planning based perspective. Thus in Chapter 10 I will relate the findings to the wider debate on urban informality with a view to adding to the discussion of the role of the state in urban informality and to critically assess the usefulness of the concept. In Chapter 11 I aim to bring the discussion towards the last research topic of 'spatial justice', reflect the findings against the more normative concept of the right to the city, and outline potential entry points and consequences for urban planning.

10 FRAMING THE CONCEPT OF URBAN INFORMALITY IN THE NEGOTIATIONS OF PUBLIC SPACE

In this chapter I aim to discuss how the findings presented in the previous chapters can contribute to framing the term 'urban informality'. In Chapter 2 on theoretical departures and specifically in Chapter 2.3.3, I outlined possible relations between the concepts of 'urban informality' and the production of space. According to the objectives and research questions set in the research framework, I focus the discussion on conceptualising informality within the negotiation processes of access to public space by first exploring which modes of the production of space are produced by urban informality (10.1). The subsequent sub-chapter (10.2) analyses the entanglements of actors by framing a concept of negotiated space that combines statutory and informal space. Finally, I aim to conclude and critically assess the usefulness of the concept of urban informality in framing the negotiations of access to public space, or more generally the production of public space, in the study settlements and for further research (10.3).

10.1 URBAN INFORMALITY AS A MODE OF THE PRODUCTION OF SPACE

In Chapter 2.3.3, I have discussed how informality as a mode of the production of space can take on two styles, namely as an expression of resistance or an expression of domination. The results of this research clearly underline how informality cannot only be understood as a mode of the ordinary as it has equally revealed an elite informality carried out often in complicity with statutory institutions. The following discussion takes up some of the key points and findings outlined in Chapter 1 and the concluding chapter of Part III, and reviews these with reference to the concept of urban informality.

With regard to urban informality as an expression of resistance, including informality both 'as a way of life' (Bayat 2004; Altvater, Mahnkopf 2003; Chatterjee 2004; Roy 2012; see discussion in Chapter 2.3.3) and as forms of insurgency (Holston 2009), the results of this empirical research point out the pre-dominance of informality 'as a way of life'. In Chapter 9.4.2 and the conclusion of Part III, I have outlined how the spatial claims of the ordinary are rarely expressions of active resistance challenging the institutions of dominant orders, but rather tend to be quiet and unassuming claims to urban livelihoods. The expressions of urban informality of the ordinary thus consist of reclusive strategies characterised by adaptation and the aim to be as invisible as possible. While not absent, the agency to produce their own modes of the production of space is at least severely limited among the ordinary. Despite the attempted invisibility, these spatial claims necessitate negotiations with those using urban informality as a dominant mode of the production of space.

Urban informality as an expression of domination has shown a variety of 'faces' in the empirical research. It operates by primarily shaping the conceptualised space (the 'representations of space' according to Lefebvre). This is carried out in an entanglement of state and non-state actors, who can be considered *de facto* state in many instances as they considerably shape the 'rules of the game'. This seems to be the most visible expression of urban informality as a mode of the production of space – an informality which is shaped by elite groups and 'state complicity' (Meagher 1995) rather than by the ordinary in their claims for access to resources.

The spatial patterns emerging from these styles of urban informality have already been outlined in the previous chapters. Urban informality, if not primarily driven by 'deep democracy' (Appadurai 2001) but through appropriation and dominance, translates into patterns of exclusion. Thus there are those who benefit, those who maintain their status-quo, and those whose position deteriorates, in some cases resulting even in displacement. The informal logics lead to increased gaps between winners and losers, as the latter do not have sufficient power and institutional backing to advocate for themselves. The effects in terms of temporariness of achievements and continuous re-negotiations have been discussed in detail in Chapter 9.5, and the specific spatial patterns have been discussed in Chapter 9.4.

Urban informality as a mode of the production of space is confirmed in this research to be both an expression of resistance, albeit not challenging dominant orders, and an expression of domination, of 'state complicity' and elite groups. The findings thus correspond with the perspectives discussed in the relevant literature, as outlined in the theoretical departures (Chapter 2.3). Subsequently, I will discuss negotiated spaces and urban informality by especially revealing the entanglements and relevant categories emerging from the two concepts.

10.2 NEGOTIATED SPACE AND URBAN INFORMALITY

The analysis of the negotiations of access to public space has indicated a rich variety of sources of power, legitimations actors draw upon and strategies employed to make and secure spatial claims. What has not yet been discussed is how the concept of 'urban informality' could help to understand these processes and structures. With the aim of moving beyond the notion of a continuum towards a more hybrid understanding, I here propose to conceptualise informality as part of a triad of three interwoven 'spaces' which shape the everyday and extra-everyday life experienced in a specific place: statutory space, informal space and negotiated space[1]. 'Statutory space' is used here to indicate any arrangements that are laid

1 Parts of this chapter have been published in disP (see Hackenbroch 2011).

down in statutory rules and regulations, while other institutions, norms and social relations are considered part of 'informal space', although it is acknowledged that these may be similarly binding as 'statutory' rules.

Negotiated space as the third perspective determines a mode of the production of space, as it enables statutory and informal spaces to be interlinked and coexist without eliminating each other's existence. In this sense all space has to be understood as negotiated, and the context-dependency of 'styles' has to be acknowledged. Negotiated space cannot be seen as a 'result' but as a temporary condition that is continuously contested and re-produced in relation with the other two dimensions. By introducing these three 'spaces', informality is understood as a hybrid concept, one that is characterised by diverse transactions, where statutory and informal spheres permeate each other (Kreibich 2012). Roy (2009b: 82) also refers to such a hybrid mode arguing that "legal norms and forms of regulation are in and of themselves permeated by the logic of informality". Such a mode of the production of space, between statutory, informal and negotiated space, will be discussed in the following, while the main points of discussion are summarised in Figure 28.

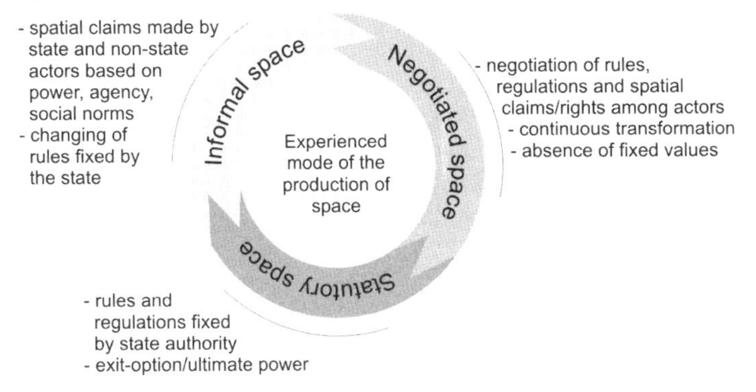

Figure 28: Negotiated space and urban informality (Source: Hackenbroch 2011: 67)

Statutory space

In a first step, statutory space can be understood as being shaped by statutory institutions. In the empirical analysis this included the police and RAB as statutory actors of the executive, local and central government representatives, administrative organisations and authorities, and institutions defining spatial use rights such as planning documents and property rights. While the majority of spatial claims investigated in this research were not made solely in statutory space, in a number of cases the access to certain statutory spaces was used as a source to legitimise and secure spatial claims (e.g. the statutory leasehold contract as a statutory process of deregulation and Foyez' interest in legalising the *ghat* fees, or the filing of General Diaries in many cases of spatial contestations in Nasimgaon).

However, statutory actors and institutions do not remain within the 'limits' defined by statutory space but instead extend to what Roy framed as 'calculated informality' (Roy 2009b). For the statutory actors involved in negotiation processes of access to public space, the findings point to an entanglement of the state and its active involvement in processes of spatial claim-making deviant from what can be understood as statutory space. In terms of the discussion on urban informality, this could be understood as 'state complicity' (Meagher 1995) becoming normality. For example, the negotiation process evolving around the management of the embankment indicates how a procedure of statutory deregulation is highly permeated by the logics of informality.

Statutory space for state actors, however, means both a statutory legitimation of their activities and the availability of an 'exit-option' or 'ultimate power-authority', which could bring back any negotiated mode of the production of space to a statutory position through the execution of state power. This was most explicit in the empirical analysis of the leasehold in Manikpara, where the state reserved an option to terminate the lease contract at any time. The eviction carried out on the embankment slopes can also be understood as exercising such ultimate power, where the state achieves – even though just for a moment – the dominance to conceptualise space as repressive (via the representations of space) and to determine the dominant mode of the production of space experienced in everyday life. These disruptions are what Kudva (2009: 1618), in the context of her analysis of the production of urban space in Delhi and Ahmedabad, refers to as an element of the politics of informality as "an everyday politics of resistance punctuated by the threat and reality of systematic yet episodic expulsion and displacement".

Informal space

Informal space is the sphere of spatial claim-making accessed most frequently in the two study settlements in everyday and extra-everyday practices – outside of the statutory or legal framework. Here it is important to note that I do not adopt a legalist understanding of urban informality in the sense of De Soto (see Chapter 2.3.1). Although outside of the statutory framework and thus extra-legal, activities in 'informal space' are not necessarily illegal or illegitimate, especially when considering social relations and institutions such as religious norms. Spatial claims here depend especially on the sources of power and the dynamics of power relations between actors. This broad delineation already suggests the limitations of the concept of urban informality, which will be discussed further in Chapter 10.3.

This informal space, analogue to the discussion in Chapter 10.1, contains informality both as an expression of resistance and of a dominant urban order. Who makes achievements in claiming access to public space then depends on the power sources, power relations and legitimations an actor can draw upon. The actors who access public space for their livelihoods or claim jurisdiction over public spaces legitimise their actions in various ways (see Chapter 9.3 and especially the conclusion on 'legitimate legitimations'). Some legitimations are built on statutory institutions, but many are built on other sources emanating from societal organisation, i.e. informal institutions and social relations. Via position in the local society, for example as a member of traditional institutions contributing to local community development or of religious organisations, spatial claims and use rights are 'informally legitimised' or 'informally formalised'.

Negotiated space

The statutory and informal spaces outlined above, and their respective actors, are interlinked in what can be termed 'negotiated space'. Here statutory and informal spaces may unite in a hybrid relationship to produce a consensus and a relatively fixed social order shaping everyday life; or they may oppose each other leading to the production of spaces of continuous resistance and contestation. In these negotiations often the different actors' positions are not clearly distinguishable, as their profession (e.g. state employees) may differ significantly from their behaviour (e.g. involvement in local politics and in 'protection activities' that are not according to statutory regulations).

The examples of 'negotiated spaces' (analysed in Chapter 9) in an environment with a richness of 'informal space' indicate an absence of 'fixed values' (Roy 2009b), thus requiring continuous re-negotiation and balancing of powers. The whole system is not very stable due to being highly dependent on personal relationships, especially with political leaders but also with powerful kinship. Thus it is not a system that will maintain a status quo for a long period, rather "this informal political system functions in the here and now, not for the sake of a hypothetical tomorrow. Its legitimacy rests with its immediate achievements, not with long term ambitions" (Anyamba 2006: 221). Given the rather limited voice of the ordinary it further shows a dominance of 'elite informality'. While the government does not directly exercise power over public space, the deregulation process provides elitist individuals and groups with the power to extend their control over public space, while the state and its officials can take their share, both officially and unofficially.

Subsuming the discussion on negotiated space and urban informality

All space has to be considered as negotiated between actors, and above I have outlined a possible analytical model which takes note of the entanglements of actors and their potential operation in different spheres or spaces. Rather than being binary categories, statutory and informal spheres meet in a complex negotiation process and become closely interlocked with each other, producing the very specific space that is experienced today – but that might be completely different tomorrow, due to the continuous nature of the negotiation processes and ever-shifting power structures.

10.3 THE USEFULNESS OF THE CONCEPT OF URBAN INFORMALITY

Above I have outlined how the concept of urban informality can help to link the empirical findings of this research back to theory, and how the empirical findings can contribute towards extending the theoretical considerations of urban informality as a mode of the production of space. However, it also seems necessary to reflect on the usefulness of the concept and whether it adds specificity or rather blurs boundaries and borders. The discussion here is motivated by the critical view on the concept reflected in the session to be hosted at the International Geographic Congress (IGC) in Cologne in 2012 on "The non-/viability of "informality" – Taking critical stock of a contested concept" by Keck and Sterly.

After reflecting the empirical analysis and results of this research thoroughly, I also tend to be critical about the usefulness of the concept of urban informality. While the recent research on urban informality (see Chapters 2.3.2, 2.3.3) has contributed and highlighted important aspects of the entanglements of different institutions in the production of space, the concept seems as yet too broad to clearly categorise and operationalise the processes at work. The notion of informality as a strategy consciously employed by the state is important to understand state actions, yet my own empirical analysis indicates other categories that could be employed to differentiate modes of the production of space (see below).

Despite this critical view of the usefulness of urban informality, the concept also offers a new perspective. When understood as an expression of the claiming of access to resources by marginalised groups as much as by elite groups and the state, the concept of urban informality could be useful in de-criminalising the activities of the urban poor. The differential notion of citizenship inherent, for example, in Yiftachel's (2009) analysis of 'gray spaces' and the 'blackening' and 'whitening' processes employed by the state, could be analytically overcome by understanding informality as a mode of the production of space that encompasses all strata of society.

Accordingly, I propose to use the concept of urban informality as an analytical category to subsume the actions of all strata of society and the state under a concept that uncovers extra-legal entanglements – not used here in the legalist perspective as for example by De Soto (2000) – without criminalising or celebrating them. This understanding of urban informality as a mode of the production of space still needs concretisation by investigating the central categories, which this research suggests can be patterns of exclusion and inclusion (spatialities of (in)justice), power structures and relations (spatialities of power) and actor's entanglements (including state involvement both in legal and extra-legal ways). These explanatory variables then would help to understand whether an informal mode of the production of space is beneficial for a community and/or for the development of the city, or whether it is based on temporary benefits and patterns of exclusion. The analysis of patterns can then be used to inform, for example, urban planning in a strategic way (see discussion of entry points in Chapter 11.2).

11 HOW TO PRODUCE 'SPATIAL JUSTICE'? UNJUST SPATIALITIES AND ENTRY POINTS FOR URBAN PLANNING

In this chapter, I would like to resume the discussion of spatial justice and urban planning of Chapter 2.4 and review the empirical results regarding exclusive spatial patterns and the resulting 'spatialities of (in)justice'. I aim to discuss entry points for achieving a higher degree of spatial justice and a functional city. In sub-chapter 11.1, I thus will summarise and conclude on the production and reproduction of spatialities of (in)justice in the empirical research, relating the results to the theoretical departures. In the second sub-chapter (11.2), I will discuss potential entry points for urban planning with a view to creating spatial justice but also a functional city.

11.1 SPATIALITIES OF (IN)JUSTICE: EXCLUSION AND INCLUSION

The spatialities of (in)justice and the resulting patterns of exclusion and inclusion will be discussed considering both the city level and the local (settlement, neighbourhood) level. For the city level discussion, I especially refer back to the discussion on spatial processes and differentiated citizenship in Dhaka in Chapter 5.2, but also to the discussion of the city perceptions and perceived status of belonging of inhabitants of the study settlements in Chapter 7.2. The local level discussion is based on the empirical results of this research, especially on Part III.

Spatialities of (in)justice at the city scale

In terms of planning and planners' approaches, the 'multiple readings' of the city postulated as a precondition for strategic urban governance by Healey (2002; see Chapter 2.4) are not explored adequately in Dhaka. The dominant conceptualisations of the city by urban planners and elite groups do not assign equal citizenship to the urban poor, who instead tend to be regarded as temporary residents (see Chapter 5.2.2). As has been suggested in Chapter 2.1.1, the social relations between different groups of inhabitants in Dhaka are inscribed into space by the development processes occurring at the city level. In consequence, the urban poor are in a constant struggle to claim their share of urban space to live in the city. Their share is permanently contested, starting from eviction rumours producing feelings of insecurity to real evictions carried out to implement development projects. The 'whitening' and 'blackening' of 'gray' living spaces (Yiftachel 2009) seems to be internalised in the dominant production of the cityscape, as encroachments of elite groups go largely unchallenged (see Chapter 5.2.2). The effects of a global dimension of urban restructuring and real estate developments might further aggravate these contrasts. The housing schemes that have been planned for the urban poor will lead to new unjust spatialities instead of moving towards spatial justice. None of the existing resettlement projects truly served the target groups (see Chapter 5.2.3). The plans for relocation to places far from the city centre will, if implemented, result in new spatialities of injustice concerning mobility and increased opportunity costs. Exclusionary practices and the creation of unjust spatialities continue on a city scale, and especially affect settlements like Nasimgaon whose inhabitants cannot draw on any legal documentation of ownership or occupation rights. In the case of settlements like Manikpara, which are statutorily acknowledged and legitimated, the exclusionary practices and unjust spatialities are less visible on a city scale and depend more on the socio-economic groups living in the area than on location *per se*. The current framework of an inadequate statutory planning system and a highly politicised informal mode of the production of space (see Chapter 10.1) does not enable the urban poor to claim their 'right to the city'. With reference to the spatial justice that residents of settlements like Manikpara experience, one more note is necessary. If considering notions of environmental justice, such as access to a healthy living environment, then Manikpara can also be considered as being off the map of spatial justice.

The spatiality of injustice created by the dominant conceptualisations of the city is confirmed by the feelings of belonging and rural/urban identities of the inhabitants of the study settlements (see Chapter 7.2.1). Manikpara, as a place integrated into the city fabric and serviced by urban amenities, is also a place where inhabitants feel that they belong to the city. This is also manifested in their active claims to citizenship, e.g. in the initiatives to pay land tax as a way of recognition of their ownership and thus citizenship rights. In Nasimgaon, on the other hand, the exclusionary practices and differentiated citizenship established on a city scale can be understood as one factor encouraging the persistence of a rural identity and socio-cultural identification with rural society. The unjust spatialities on the city level, including the differentiated notions of citizenship, are thus deeply inscribed into the everyday spatial practices and the expressions of resistance and claims to citizenship.

The spatialities of injustice on the city scale directly relate to the local scale. Exclusionary practices, especially the denial of full citizenship to sections of urban inhabitants and the neglect of their living areas in terms of services, can be understood to open up room for dependency relationships at a local scale.

Spatialities of (in)justice at the local scale

The non-recognition of some settlements as an integral part of the city – in this context Nasimgaon – and the keeping of these in a 'gray space' or even their 'blackening' results not only in exclusion on a city scale, but also 'invites' the spatialities of injustice on a local scale. In the relative absence of state provision (planning and services), other institutions of regulation emerge and dominate the social space, often in close entanglements with state actors (see Chapters 9.5, 10.1). The mode of the production of space created by this entanglement of actors and institutions rarely increases spatial justice, but builds on differentiated power structures and establishes unjust local spatialities. To understand these local processes as only emerging because of a relative absence of the state would, however, fail to represent the situation adequately. The case of Manikpara has shown how the partial presence of the state in the form of state actors also does not mean the absence of other institutions and ways of regulation. Furthermore, both cases have revealed 'state complicity' in urban informality as a mode of the production of space (see Chapter 10.2).

This local institutional setting, as the analysis in this research has shown, produces local patterns of exclusion rather than spatialities of justice at the local scale. The domination of elite groups mainly re-enforces exclusionary patterns expressed socially and spatially, and the welfare orientation does not extend far beyond beneficiary networks based on political affiliation (see Chapter 9.5). The strategies of spatial claim-making applied by the elite groups result in unjust spatialities:

- The process of 'informal formalisation', identified as a strategy of local claim-making that operates especially by transforming temporary structures into more stable units, builds on exclusionary practices. Although in this case formalisation takes place within the 'informal sphere' and not within the statutory system (see Chapter 10.1, 10.2), Roy's observation fits the context: "If informality is a differentiated structure, then formalization can be a moment when inequality is deepened" (Roy 2005: 153). The changing access arrangements due to informal formalisation lead to a differentiation of the ordinary. Inequality is not only manifested between the elite and the ordinary, but also between those of the ordinary who can adapt and those who have to withdraw from a location.
- The domination of the social sphere by a political elite during one electoral period results in an increasing gap between the small elite group that is able to claim jurisdiction over public space and the ordinary who in many cases in Nasimgaon 'lose' their spaces to the elite groups. The re-production of differentiated power structures and socio-economic positions is expressed spatially in a person's or household's access to resources, including public space.
- The security to access certain spaces does not necessarily translate into spatial justice. For example, the perception of a relative security of use rights on the embankment slopes does not mean an equal access to resources and opportunities. Furthermore, the access arrangements still reflect differentiated power structures and dependency relationships. Public space here is not accessible on equal terms for everyone, leading to Madanipour's question 'Whose Public Space?' (Madanipour 2010), asked in relation to the inclusiveness of processes of creating and having access to public space.

Despite these unjust local spatialities evolving along the lines of differentiated power structures, it has to be noted that there is always an in-between and a fluidity of the border between the ordinary and the elite, and thus between who is included and excluded and who can claim access to resources effectively. This fluidity and dynamic of power structures has been discussed concerning traditional institutions in Chapter 5.1.4 and it opens up the "room for manoeuvre" (Lewis, Hossain 2007: 295) that is then utilised by those aiming at resistance conceptualisations of space (see Chapter 9.5).

Success stories of urban insurgency, such as the success of the Bus Riders Union in Los Angeles (Soja 2010, see Chapter 2.4), against unjust spatialities are rare in the examples discussed in this research. While Soja's example can be regarded as one of a high level of organisation and institutionalisation, success stories at lower levels of organisation, for example the claims made via the legal system by urban poor in Mumbai (Eckert 2006) or narratives of insurgency as suggested by Holston (2009) are also hardly found in the study settlements.

In conclusion, the conceptualisations and negotiations of access to public space are far from resulting in fair and equitable distribution of the access to public space (Soja 2010). Furthermore, the processes leading to unjust spatialities tend to be inaccessible for the ordinary, limiting the readings of the city to members of dominant interest groups (Healey 2002).

11.2 ENTRY POINTS FOR URBAN PLANNING

The entry points for urban planning presented here are based on the above discussion of the spatialities of (in)justice, but they also relate to the functionality of the city. The functionality of the city and its neighbourhoods and the spatialities of (in)justice however seem deeply interrelated. In Chapter 9.4.1, I have analysed how the negotiations of access to public space and the spatial claims made by different actors have compromised the availability of public space in Nasimgaon. The densification processes at work, not halted by effective institutions, have reduced the functionality of the settlement especially in terms of accessibility and safety. These processes have their roots in a spatiality of injustice on a city scale – where the urban poor have only 'squeezed' access to a small percentage of city space – but similarly so on a local scale – where the differentiated power structure between elite actors and the ordinary enables elite actors to make permanent and stable spatial claims compromising public space.

The findings of the research have indicated a number of criteria that need to be considered and attended to in any planning process, whether local and decentralised or on a city scale:

- recognition of differentiated power structures and consideration of the implications of these structures in any planning approaches,

- recognition of the diversity of urban actors and orientation of urban planning on city to neighbourhood scales towards the specific demands of user groups of, for example, different socio-economic status or gender,
- recognition of the political processes at work that by distributing resources along the lines of party affiliations rather than to defined beneficiaries and target groups might hinder implementation of planning intended to create spatialities of justice, and
- understanding of all neighbourhoods and inhabitants as integral parts of the city which have to be included in planning approaches.

Given the current state of planning in Dhaka, the question is how these considerations could become integral elements of future planning approaches. A first step could be explicit integration into the education and training of spatial planners in Bangladesh. Secondly, the necessary awareness could be created by recognising these considerations in the relevant policy documents, especially in the National Urban Policy which to date only exists as a draft, and a re-consideration of planning laws and instruments. Furthermore, capacity building and awareness campaigns for urban planning practitioners and the streamlining of donor contributions to support such activities could provide the means for a long-term integration of these considerations. However the integration of these considerations into planning practice was not the focus of this research, so the discussion will not be further pursued at this point. Similarly beyond the scope of this research is suggesting changes to create a more responsive practice of democracy (as opposed to the current 'illiberal democracy', see Chapter 5.1.1) and a less confrontational environment for the politics that to a large degree influence the processes discussed here.

In the following, I will discuss four specific entry points for urban planning, understood as a complex and collaborative process of delivering spatial justice. The entry points are based on the following key topics for a just and functional city:

- creation of public spaces,
- requirements and quality of (public) spaces according to the diversity of users,
- protection and preservation of public spaces, and
- equal citizenship and inclusionary rights to the city.

Creation of public spaces

The first step to ensure a functional city in relation to public spaces is the creation of such spaces in a city and neighbourhood context. This can be understood as the original task of urban planning. Such an approach should start at a city level with an identification of public spaces considered valuable for the functionality of the city. While the definition and the confirmation of these in statutory plans is at the heart of urban planning, the protection and preservation of these spaces cannot be carried out by the planning system, especially when considering the ineffectiveness of the planning administration. Accordingly, below I will discuss options for an institutional set-up to protect and preserve public spaces.

Requirements and quality of (public) spaces according to the diversity of users

A second entry point is the consideration of the needs of a diversity of actors in the creation of urban spaces and public spaces in particular. This research has uncovered a large array of spatial requirements pertaining to different population groups. The analysis in Part II has revealed spatial practices to be dependent on gender and socio-economic status. These specific spatial practices need to be taken up by urban planning to create spaces that serve the needs of different population groups, or following a more normative approach to, for example, create spaces that enable a more explicit female production of space. This research has revealed how women access space differently depending on socio-cultural values that are related to economic and social status, but also depending on the perceived familiarity of a neighbourhood (see Chapter 7.2, especially 7.2.4). These are important inputs to planning city neighbourhoods while acknowledging difference, as in a polyrational approach to planning (Davy 2008) and a collaborative planning approach informed by multiple readings (Healey 2002; see also Chapter 2.4).

The value of neighbourhood or niche spaces, as inherent in the concepts of familiar and stranger's publicness analysed in Chapter 7.2, also provides an input for the current discussions of urban resettlement schemes. The spatiality of livelihoods and the findings on gendered space reveal the value of a finely grained urban fabric in contrast to the multi-storey resettlement schemes discussed in Bangladesh (see Chapter 5.2.3) and India (e.g. the resettlement plans for Dharavi). The Urban Density Project, a research initiative of the London-based International Institute for Environment and Development (IIED) in Karachi (Hasan 2010; Website Urban Density Project), seeks to preserve finely grained structures which offer niche spaces instead of combining large tower blocks surrounded by an open space that is easily perceived as 'strange'. In remodelling four low-income settlements in Karachi, Hasan proposes an incremental approach to urban transformation and densification, where new rooms can be added on top of existing housing units, largely preserving existing relations of private, semi-private and public spaces. A redevelopment of Nasimgaon where the inhabitants were resettled on part of the land in high-rise buildings would mean an increase of 'strangers' publicness' and the disappearing of the in-between sphere of familiarity. The consequences of this increased permeability especially for female spatial mobility but also for the internal economic and social livelihood activities in everyday and extra-everyday life have to be considered in any planning approach that aims to create a different representation of the city. The socio-spatial requirements and preferences of low-income urban households have also been discussed by Ghafur (Ghafur 2005) in the context of Dhaka and his suggestions as to housing approaches could be further pursued in a participative environment with the affected inhabitants.

Protection and preservation of public spaces

The discussion has indicated the importance of public space as a resource for space-based livelihood strategies, but also for recreation, festivals and safety. At the same time, however, this research has shown that the institutions in place are not able to preserve public spaces against the high pressure for development and common practices and rewards of 'doing politics'. None of the institutions, not even the religiously legitimated actors, were able to prevent encroachment and the decreasing of public spaces. Only very small-scale initiatives that surfaced road space among neighbours were successful, while as soon as public space existed in the form of a square, and thus allowed for different uses, gradual encroachments were unavoidable. The research of Kombe and Kreibich in Tanzania (Kombe, Kreibich 2006; Kreibich 2012) on institutions of social regulation in settings shaped by a fragile ineffective state has revealed a similar dilemma. Although effective in an infancy stage of settlement development, institutions of social regulation collapse with excessive densification. In the study settlements in Dhaka the effect is the permanently decreasing public space in Nasimgaon, despite its important function, and the almost complete absence of public spaces except for the (very limited) road network in Manikpara (see Chapter 9.4.1). As a first step, a visualisation of existing spaces in the form of an urban or neighbourhood inventory could provide an overview of existing public spaces. If carried out as community mapping (see for example the community-commissioned surveys conducted in Mumbai referred to by Appadurai 2001) such mapping could draw attention to the local needs and the perceived deficits and potentials of public spaces. Furthermore, community mapping of this kind could provide a way to actively make a claim that cannot be overheard by politicians or administrative planning officers.

It seems unrealistic to consider the current urban planning institutions as a means to preserve or restore the last public spaces in settlements like Nasimgaon and Manikpara. Such neighbourhood approaches are neither the core concern nor interest of urban planning and politics at the city level, especially not those in areas that developed rather 'informally', nor is the planning system capable of providing effective solutions. The only entry point to maintain a functional urban fabric is thus a decentralised approach. Given the failure of local institutions analysed in this research, the central question is which institutions could establish a local consensus about the non-encroachment of public spaces while maintaining the important functions these spaces have for economic livelihoods, recreation, the hosting of religious and cultural festivals and other events and safety. In Manikpara, the Eid Gah Math has been fenced off by the mosque committee and is thus protected. However, after fencing it cannot be used for other activities any more. Fencing off Nasimgaon Math in a similar style thus does not seem to fulfil the needs of its inhabitants to access public space for their everyday livelihoods.

The system of the leasehold of Manikghat can be understood as a statutory process of deregulation. The analysis has shown that the leaseholder as a statutorily legitimated actor is able to exert considerable influence on the local spatial practices. Thus the question is whether this could be a model to protect and preserve public spaces. Such a deregulation process entails the potential of effective management by a local institution while the state remains in the background but provides an 'exit option' in cases of misconduct. However, sufficient care has to be taken not to produce new patterns of exclusion and spatial injustice.

Accordingly, if considering a leasehold system a number of points have to be considered and weighed against the concepts of spatial justice and functionality of the city. Such a leasehold system of public spaces would have to be based on income opportunities, and thus it entails a commercialisation of public space and access arrangements. A commercialisation of public space, however, often establishes new spatialities of exclusion – for example, how could the rice-selling beggars at Nasimgaon Khalabazar be provided with a niche in a commercialised public space which would entail monetary costs. The way the leasehold of the *ghats* is organised, deviant at the local level from the statutorily prescribed practices, does not create spatial justice but re-produces differentiated and discriminatory power structures. Accordingly, any new system aiming to protect public spaces from further encroachment at the same time needs to ensure spatial justice in access to public space and preserve the character of public space as a common property resource against exclusive use rights (see delineation of public space in Chapter 2.1.1). This includes considering the spatial requirements of diverse user groups (see above) and ensuring that one actor's power does not dis-empower others from claiming space, what Doderer (2003: 266) based on Fraser (1993) framed as a scenario of 'differential difference' (see Chapter 2.2.1). Another potential institution for a decentralised management of public spaces is the traditional *shalish*. The research has shown how the *shalish* can be employed to solve small-scale conflicts between neighbours. However, the research also pointed out the continuation of patterns of patron-client relationships in these institutions. In the current set-up the *shalish* is thus similarly unable to provide an institution ensuring equal and fair access to public space. Nonetheless, both the leasehold system and the *shalish* provide potential entry points in the search for institutions able to establish regulation mechanisms that ensure the preservation of public spaces and their access for livelihoods.

The division of the Dhaka City Corporation into two municipalities decided on 29th November 2011 could offer an opportunity to re-organise institutions of local government and planning administration to make them more responsive and integrated at a neighbourhood level, and thus able to contribute to the protection of public space. However, it seems unrealistic to think that much will change in reality, given that the division has been brought up as an issue of political contestations between the two main parties rather than as a strategy to improve urban governance.

Equal citizenship and inclusionary rights to the city

Finally, the existence of a differentiated citizenship in a hierarchically organised society leads to the question of how to consider or implement the interests of minority groups and of groups who do not have access to power in a way that allows them to claim (social and physical) spaces.

The spatial claims made by the ordinary as part of their everyday practices are not recognised by the statutory planning system as they are not based on legal property rights or other provisions regulating access to space. This is also the case for the spatial claims of elite groups; however, the elite groups have other mechanisms, based on dominating power, to secure their spatial claims. The non-applicability of planning leaves negotiations of spatial claims in a 'gray space', where outcomes depend primarily on access to local institutions of power and domination. Thus urban planning has to find ways of dealing with spatial claims made outside of 'formal' property regimes.

Furthermore, the absence of comprehensive narratives of resistance in the study settlements leads to the question of how spaces for successful urban insurgency and the claiming of citizenship rights can be created. The current framework of an inadequate statutory planning system and a highly politicised local sphere leaves little room for manoeuvre within the dominant conceptualisations of space. A "legalism from below" as discussed by Eckert (2006) would be a way of making spatial claims by using the statutory institutions, but whether this would effect considerable change in the patterns of domination and injustice must at least be questioned. Coming back to the necessity to recognise difference and diversity, this implies the creation of an enabling environment where the urban poor can access relevant institutions and claim agency by stepping out of dependency relationships in political society towards participation in civil society, with civic instead of only political citizenship (Chatterjee 2004). A precondition here is to make visible the spaces and people concerned instead of keeping them in conditions of invisibility, permanent temporariness and continuously unmapped territories and spaces (Roy 2004).

Concluding remarks on entry points for urban planning

The above discussion of potential entry points for urban planning has indicated a number of challenges which need to be overcome to effectively change existing spatial practices, both for the benefit of urban dwellers as well as for the functionality of the city. The recognition and understanding of the processes of power and domination at work at the neighbourhood level is a first step towards producing spatial justice and minimising instead the enlarging of differentiated power structures and injustice by (planning) interventions.

The entry points for planning discussed above focus on the public space of the city. However, the research has also revealed the transformation and densification processes at work in the built-up areas of the study settlements. Especially the increase of densities with their adverse effects on the functionality of the city has been discussed (see also the discussion in Baumgart et al. 2011). Apart from the preservation and restoring of public spaces in a neighbourhood, the question of how to regulate excessive densification processes on private plots (irrespective of formal land ownership patterns, which are non-existent in Nasimgaon) and how to avoid the 'face-to-face' living conditions that consolidate in Manikpara emerges. These excessive densification processes also create spatial injustice as the 'freedom to build' translates into a 'first come first served' approach to urban transformations. Regulation with a view to creating a more just and a more functional city could be based on similar preconditions and entry points as those identified above in relation to public space.

12 CONCLUSION OF THE RESEARCH

This research has explored a wide range of issues related to public space in urban life. In this conclusion, I will summarise the main findings including a short reflection of results, before giving an outlook to potential further research.

Summary of the main findings

The following summary is organised according to the four main research topics and questions. I will also provide a cautionary note concerning the implications this research can and might have as opposed to what I intend.

In Part II of this research I have discussed the spatiality of livelihoods. Public space constitutes an arena for a multiplicity of livelihood activities related to everyday and extra-everyday life. It has a high importance and value for urban life, including economic activities but similarly recreational and domestic ones. It furthermore is an arena where religious beliefs are expressed and religious, cultural and political celebrations and activities can take place, interrupting the everyday but also continuing its spatial practices. At the same time, given the high pressure on public space to be used for livelihood activities in an environment characterised by excessive densities, development dynamics and scarcity of public spaces, the spatiality of livelihoods often carries with it a notion of conflicting uses and interests. Thus public space is a contested resource and access depends on the negotiations of access to public space between a diversity of actors.

The spatiality of livelihoods is furthermore divided by social gender relations, producing gendered space or spatialities of gender difference. This research found that with regard to gender, the hierarchy of spaces as private, semi-private and public does not suffice to explain the hierarchies of urban space in Dhaka, as public spaces, for example, are not always equally accessible. The addition of the notion of 'publicness' being perceived as familiar or strange helps to explain many of the everyday mobility patterns of residents, determined by the urban fabric on a city and neighbourhood scale, the perceived identities, the self-perceived notions of citizenship and those ascribed by mainstream society, and the norms of gender relations. It is important to note here that the borders between nuances of publicness can never be considered fixed, as they are flexible and fluid in relation to events of everyday and extra-everyday life. An analysis of public spaces in Dhaka thus always has to consider what determines shifting borders and with which effects. The shifting borders can be related to specific events, but also to the location and environment of public spaces within the urban fabric of a neighbourhood.

In Part III of this research, I then discussed the negotiations of access to public space, based on the findings about the spatiality of livelihoods. The negotiations commonly taking place between the ordinary and elite groups, but also in entanglements with statutory actors, are primarily based on the spatialities of power. The availability of power to dominate or to resist is largely characterised by persisting patronage structures in society. Especially the affiliation with and involvement in political networks determine an actor's power to make and secure spatial claims. Accordingly, the strategies applied in negotiation processes of access to public space to a large degree depend on membership of and affiliation with such networks, and the degree to which an actor is able to successfully utilise these connections.

The resulting spatial patterns suggest exclusionary processes, where elite groups are able to sustain their spatial claims while the ordinary experience their achievement of access arrangements to be temporary and to require continuous re-negotiations. Resistance against dominant practices is largely absent and only those of the ordinary who can draw on power sources similar to those of dominant elite groups are able to resist. The analysis of negotiation processes has furthermore revealed that negotiations in the context of this research cannot be understood as taking place among actors with equal access to power and other resources. Instead, negotiations have to be understood rather as a process of interaction characterised by differentiation and patron-client related networks.

With regard to public space, the outcomes in the study settlements show a further disappearance of public spaces, despite the values attached to public space which have been addressed in the spatiality of livelihoods. Although often committed to safeguarding public space in principle, the dominant actors are not able to effectively do so in practice. Instead, they themselves are predominantly those making the most permanent and durable spatial claims that endanger the functionality of the city. The high development pressures on land in the densely populated city and the power of 'doing politics' largely trigger these developments.

Urban informality in this research operates mainly as an expression of a dominant mode of the production of space. As a mode of the production of space guided by resistance, it mainly exists in quiet and subtle moments of resistance and claim-making in the absence of other modes of provision, rather than as a challenging of the dominant mode. Informality thus enables the state and elite groups to make spatial claims and to dominate the conceptualised space which is then experienced as 'repressive' on the ground. For the ordinary, this entanglement of dominant power means a limited agency and a temporality of achievements in spatial claim-making in the urban sphere.

Under the assumption that all space can be considered as negotiated, I have outlined a potential analytical model which takes note of the entanglements of actors and their potential operation in different spheres or spaces. Statutory and informal spheres/spaces meet in a complex negotiation process that produces the space experienced at a given moment. This includes an understanding of negotiation processes as being continuous and power structures as ever-shifting. In the dominant mode of the production of space identified in this research, 'negotiated space' as the sphere of entanglements between 'statutory space' and 'informal space' is shaped primarily by the elite groups of society.

Besides this applicability of the concept of urban informality in the context of this research's results, I would like to also add a critical note. While urban informality can be used to frame the entanglements of actors involved in a mode of the production of space, it seems useful to draw attention to the explanatory variables more concisely. The modes of the production of space as discussed in this research can be understood by the patterns of exclusion and inclusion (spatialities of (in)justice), power structures and relations (spatialities of power) and actors' entanglements (including state involvement in both legal and extra-legal ways). An analysis using these categories would enable an understanding of whether a mode of the production of space contributes to spatial justice or is based on exclusionary practices.

The research results demonstrate the reproduction of prevailing spatialities of injustice as effects of the negotiations of access to public space and dominant modes of the production of space. On a city scale, the low-income urban settlements and their inhabitants tend to be neglected in terms of urban planning, servicing and recognition of citizenship rights. These exclusionary processes are also reflected in the inhabitants' self-perception of inclusion and exclusion in the city context. These differential practices and spatialities of injustice experienced at a city scale then find their continuation at a local neighbourhood scale. The local institutional set-up dominated by the local leaders of the ruling party re-enforces exclusionary patterns both socially and spatially. Access to public space cannot be considered to be fair and equitable. Narratives of inclusion and spatial justice or insurgency against the production of unjust spatialities are rare.

In order to create an urban environment of spatial justice, but also in order to safeguard public spaces as important livelihood resources, this research concluded in formulating entry points for spatial planning. Four general considerations evolve that are to be recognised in any planning processes, independent of the scale. These are the recognition of differentiated power structures and their implications in planning; the recognition of a diversity of urban actors and their specific interests and needs; the recognition of the political processes at work and the effects these might have on resource distribution; and the understanding of all neighbourhoods and inhabitants as integral parts of the city. Additionally, four entry points for spatial planning were formulated with specific regard to public space as an asset for a just and functional city. Besides the creation of public spaces (not the focus of this research), these are the orientation of requirements and qualities of public spaces towards the demands and needs of users; the protection and preservation of public spaces existing in the city; and the recognition of equal citizenship and inclusionary rights to the city. Specifically the protection and preservation of public spaces presents a challenge for urban planning as the results of this research have revealed that local organisations and institutions are not able to safeguard public spaces. Yet to be found is an effective decentralised system that is not easily subject to practices of party politics and personal gain, that can cope with high development pressures and that at the same time is inclusionary in providing especially the urban poor with a space for their livelihoods. To aid the translation of both the general considerations and the entry points for urban planning into practice, the recently launched Bangladesh Urban Forum as a forum of planning practitioners and scientists could be one potential stakeholder to discuss the implications for planning emerging from this research.

My research, presenting findings on exclusionary processes within low-income areas could – if not read carefully and interpreted properly – provide city governments with arguments for a drive against 'slums' in order to achieve the 'city without slums'. My arguments, however, are far from this aim, and my findings do not at all suggest 'doing away with slums'. In laying open internal processes in low-income settlements, I seek to point out the degree of exclusion the urban poor have to negotiate in everyday life, but my findings do not identify the existence of 'slums' as the cause of this injustice nor the abolishing of 'slums' as a solution. The exclusionary practices triggered by the prevalence of a mode of the production of space characterised by urban informality as an expression of the dominant elites can be presumed to occur not only in low-income settlements. The problems to be tackled are not the low-income settlements *per se* but the causes behind them. The problems that produce spatialities of injustice lie in society and the socio-political entanglements continuously recreated by dominant spatial practices.

Reflections on the methodology

In this research I have combined a grounded theory approach and an ethnographic approach in analysing a topic related to urban planning. Especially the ethnographic component – rather uncommon in planning related research – enabled me to arrive at findings that point out the deeply rooted social concepts and traditions that always affect planning approaches but too often go unnoticed or are not tackled, with the results analysed as spatialities of injustice. A long-term and embedded planning research can thus in my view considerably contribute to understanding the mechanisms at work and the ways in which insufficiently thought-out planning, despite being committed to spatial justice in principle, can have adverse effects when implemented on the ground.

The empirical research furthermore enabled me to reflect on the existing theories especially with regard to the concept of urban informality. In openly analysing my materials I was able to critically reflect on the theories and categories and to suggest the expansion of common notions and concepts. Especially the use of visual methods, here the solicited photography, enabled a perspective that included the research participants to a large extent and in the course of data analysis proved to reveal much more than originally anticipated.

Further research

This research revealed the diversity of factors at work in producing and reproducing spatial practices with varying outcomes based on largely differentiated power structures. While entry points for urban planning with a view to increasing spatial justice have been defined, this is also the area where further research could begin: Why are the institutions at work ineffective in protecting public spaces, and why does access to public space remain exclusionary instead of just and equal? An analysis of best practices and potential existing institutions could reveal which institutional setting best ensures the protection and management of public space as a public resource. Secondly, the research pointed out the importance of public spaces accessible as neighbourhood spaces. Common approaches to resettlement still focus on spatial compositions that do not offer adequate 'familiar' space from a socio-cultural perspective. Here a combination of research on 'familiar and stranger's publicness' combined with approaches such as the Urban Density Project of Arif Hasan (Website Urban Density Project) could provide additional input for urban planning.

This research has focused on settlements of low-income groups, although Manikpara has a considerable middle class. However, the research has not to a larger extent included the emerging middle class whose spatial claims shape the city space considerably. Their demands trigger both housing developments and the commercialisation and privatisation of public spaces in shopping malls with regulated access, if not explicitly so then by way of 'invisible' social norms. The analysis of clothing habits and nuances of publicness has indicated the prevalence or even re-emergence of conservative gender and social norms among the middle class in Manikpara. In this context, scaling up and enlarging this research to include the spatial practices of the emerging middle class could provide a promising field for further research.

The last topic suggested for further research is at the same time a critical reflection of the research conducted here. During the course of my research, I read a large diversity of papers producing representations of Dhaka from various viewpoints and angles. Furthermore, a large array of especially foreign research focuses on low-income settlements of Dhaka – and so did this research in parts. The image of the 'slum' is thus produced and reproduced by these research projects and publications, leading me to ask how this focus of (foreign) scientific and donor-related interest is perceived locally, and what effects this image production has on the inhabitants of a particular settlement and on the views of political leaders and statutory authorities. Such a focus 'because it is so obvious and illustrative' can also be understood as part of Brandt's (Brandt 2010) critique elaborated in a book review titled 'Bangladesh as a footnote in social science discourse'. I am aware that my research has contributed yet another dimension to what I am criticising here – a production and positioning of a settlement in Dhaka as a 'world slum', even if I do refrain from using the term. I therefore propose to take this issue further by conducting follow-up research about the production of 'world slums' by the international scientific community and its local implications, both on a statutory level and among local inhabitants.

GLOSSARY OF BENGALI TERMS

The following glossary contains mainly Bengali terms, although some of the terms are actually borrowed from English, Persian or Arab and commonly used in everyday Bengali. Where necessary, the direct English translation is supplemented by further explanations and comments concerning the cultural meanings the term carries. For the spelling of Bengali terms, see the preliminary notes on p. 15.

Bengali [1]	**English**	**Explanation, further comments**
abashik elaka	residential area	Refers to an urban residential area that is well equipped, with 'good-looking', wide roads, and generally a calm and quiet place; examples in Dhaka are Banani, Gulshan or Dhanmondi – thus the areas of high-income groups.
adda	conversation, chat	Lively discussion among friends, often about politics; "the practise of friends getting together for long, informal, and unrigorous conversations"; understood as something quintessentially Bengali (Chakrabarty 2001: 124).
apa	sister	Used to address either an elder sister or a non-kin person of similar age and/or status.
ason	constituency	Defined area from which each Member of Parliament (MP) is elected.
Awami League (AL)		Currently the ruling party (won the December 2008 elections), translating as 'People's League'.
badi	complainant	-
Bangla shomoy	Bengali time	Refers to the 'normal time' as opposed to the daylight saving time introduced in June 2009.
Bangladesh National Party (BNP)		Currently in opposition, ruled the country from 1991–1996 and 2001–2006, in Bengali: *Bangladesh Jatiyotabadi Dol.*
bari	(village) home	The village home which is usually owned by the family.
basha	house	Rented house, town house (even if this is owned, people might refer to it as *basha* as opposed to their village home, *bari*).
Baul	member of a devotional community	A community relating their spirituality to both Hindu and Muslim traditions, related to Sufism.

1 If there are commonly used English transliterations, which do not exactly match the Bengali spelling but often appear in English publications, these are given in square brackets. Other comments are in parenthesis.

bazar	market	'Doing bazar' means 'going grocery shopping'.
bazar-ghat	a public space with characteristics of crowding, gathering, and gossiping	The term is used as a phrase in Bengali and refers to a place with the characteristics typically found at a bazar and at a *ghat*. A bazar is perceived as a public space of selling and buying products, a place of public gathering and a place of endless gossiping at tea stalls. A *ghat* is also perceived as a crowded public space, a meeting place of different people and a gossiping place. The common characteristics are the crowd, the gathering, the meeting opportunities, the gossiping – generally indicating a place where something is happening.
Bepari	professional family title	Refers to a family of businessmen, and to the continuation of the family business through successors.
bhai	brother	Used to address both kin and non-kin.
bhandobi	female friend	-
bhangari dokan	shop trading sorted waste	-
bhit	demarcated shop unit	Refers to a raised place in a variety of contexts, for example in agricultural production (papaya cultivation) or shop units in a market; in the context of this research it refers to demarcated shop units; originates from *bhita* ('rural raised land for housing', i.e. a rural homestead).
bibadi	accused	-
bichar	judgement	Same as *shalish.*
Bideshi(ni)	foreigner (female)	-
bishisto murrobi	notable elder	Knowledge, age, a philanthropist attitude and impartiality, among other things, make a person referred to as *bishisto murobbi* (mostly male). Everyone in a society obeys this person, respects him and relies on him. A *bishisto murobbi* is called upon in all kinds of decision making situations (including *shalish*) and his counselling is generally accepted.
Bisho Ijtema	World Gathering	Specifically refers to an annual Muslim congregation held in Tongi, north of Dhaka, since 1976. It is said to be the second largest Muslim congregation after the *Haj* in Mecca. The congregation is organised by the *Tabligh* movement (see below).
bon-jamai	husband of elder sister	-
borka [burka]	long coat combined with veil worn by women	In Bangladesh *borka* usually refers to the combination of a long coat that is worn by women on top of other clothing with a veil covering the head. This can mean covering the head tightly, with a *hijab* ensuring that no hair is shown, or covering the face completely and only revealing the eyes. Some women in Dhaka also wear only the coat without covering the head with a veil or covering their head only loosely (I have for example seen this with many garment workers). The style chosen depends on the purpose and location, a person's perceptions and the norms and expectations of the (local) society.

bosti [basti, bustee]	neighbourhood, slum area	Originally the term referred to a neighbourhood, however, it has come to be a synonym for slums and squatters and thus carries with it a rather negative connotation.
canvasser	English term used in Bengali	Here refers to a person who advertises something to a crowd with the help of a sound system. The sales pitch most often refers to health related problems, including sexuality. He (seldom she) often sells medicines to the crowd after motivating them to buy his product.
chacha	uncle (father's side)	Used to address both kin and non-kin.
chada	monetary contribution (depending on context: extortion money or 'postitive' monetary contribution)	*Chada* payments can have both a positive and negative connotation. If a leader collects *chada* from local shopkeepers the meaning resembles that of extortion payments, as the leader will supposedly use that money for his own or his party's interest. On the other hand, when a Mosque Committee collects *chada* for development works from the public, or there is a collection of *chada* for a community member who needs support, the term does not have a negative connotation.
chadabaj	extortionist	A *chadabaj* demands *chada* from the public. *Chadabaj*, however, refers only to the negative type of *chada*, and thus if someone is called a *chadabaj* this is meant in a negative sense, denoting someone who collects money from the public unlawfully and forcefully for his personal use, i.e. to 'eat' it.
chadabaji	extortion	The activity of collecting *chada* is referred to as *chadabaji*. Similar to *chadabaj*, this is only used in *chada's* negative sense.
chanachur	spicy snack mixture (Bombay Mix)	-
Chatro League/Dol	Students' Organisation	*Chatro* League: sub-organisation of AL; *Chatro Dol*: sub-organisation of BNP.
chouki	wooden bed	In this research refers to a wooden bed used as a vending table.
dada	elder brother	Used to address both kin and non-kin.
Eid Gah Math (pronounce Id)	Open field for *Eid* prayers	There are Eid Gah Math all over Bangladesh, and they are specifically used during the *Eid-ul-Fitr* and *Eid-ul-Adha* for the morning prayers. In urban areas every community has an Eid Gah Math. These are often under the supervision of a nearby mosque. An Eid Gah Math normally has an *Eid Gah*, a kind of monument, located on the Western side, towards Mecca, from where the prayer will be held.
Eid-ul-Adha		Celebration commemorating the willingness of Abraham/Ibrahim to sacrifice his son Ishmael (also referred to as *Kurbanir Eid*).
Eid-ul-Fitr		Celebration of the end of *Ramadan*.

Ekushe	21st (February)	The day commemorating the martyrs of the language movement in 1952, i.e. the students who demonstrated to keep Bengali the official language and were fired on by the Pakistani army at the *Shohid Minar* (monument).
GD (General Diary)	English term used in Bengali	A documentation or record of an incident that is kept at the police station (*thana*). It does not necessarily trigger legal steps, but it can be drawn upon in future by the party who has filed the GD in case of further incidents.
ghat	boat landing point	Typically, these landing points include stairs leading towards the water and provide access to the water for various activities.
ghomta	covering the head with cloth (women)	Covering of the head with cloth by women, most commonly by the end of the *sari*, but could also mean with an additional cloth; covering by 'giving *ghomta*' means a loose covering of the head, not tightly as with the *hijab*-style common with *borka*.
gunda	gangster	Used in connection to a (criminal) political leader, considered a similar threat to society as *rongbaj* as they can harm the public.
gushti	patrilineal kinship	Patrilineal kinship group of patron-client relations between poor and well-off households.
Hafiz	Islamic scholar	Refers to a scholar who can recite the whole Quran.
Haj	Islamic pilgrimage to Mecca	-
hat	measurement	1 *hat* = 1 foot = 30.48 cm
hijab	head scarf	Refers to an Arabic style head scarf for women, worn tightly around the head without any hair remaining visible.
hijra	transgender	Transgender community of persons born with male characteristics but carrying a female identity; understand themselves as a third gender.
hortal	general strike	Commonly called by the opposition to protest against activities of the ruling party.
iftar	fast-breaking meal at sunset during *Ramadan*	-
ijjot	see *man-ijjot*	-
Jamaat-e-Islami		The largest Islamist political party, going back to the Pakistan period.
jamai	son-in-law, also used for husband	-
Jatiyo Paty (JP)		Party founded by Ershad, the former autocratic ruler of Bangladesh, JP is part of the AL's Grand Alliance (coalition) government.
jomidar [zamidar]	landlord	Used during Mughal and Colonial times.
Jubo League/Dol	Youth Organisation	*Jubo* League: sub-organisation of AL; *Jubo Dol*: sub-organisation of BNP.

kaccha	unripe, nondurable, not stable	In the context of housing this refers to houses made from nondurable materials, i.e. tin or bamboo. Quite often, it is only used to denote the latter, while tin houses are considered a separate category. In the context of road networks and open spaces, the term is used to refer to the quality of the pavement, i.e. a *kaccha* road or space would be covered by sand, clay or mud.
karom	kind of snooker played with fingers on a wooden board	-
katha	1 *katha* = 66.9 m²	-
khomota	power	In most cases refers to political power, but also economic power.
lakh	1 *lakh* = 100,000	-
lojja, lojja-shorom	shyness, uneasiness, shame	Commonly, *lojja-shorom* are used as conjugate words. In everyday language they are used interchangeably and a difference like in the equivalent Hindi words *laj* and *sharum* (Tarlo 2005) could not be observed. *Lojja* and *shorom* refer to feeling shyness, uneasiness and shame and to a certain type of behaviour of women, i.e. a woman who has shyness in her expression, is not arrogant in behaviour, and is known as well-mannered is behaving according to the concept of *lojja*. There is the Bengali proverb '*lojja narir bhushon*' which means 'shyness is the ornament of a girl', and this matches the expectation of society that women are 'born to be shy'.
madrasa	Muslim school	-
mama	uncle (mother's side)	Used to address both kin and non-kin.
man-ijjot, ijjot	respect, prestige, honour, self-pride felt by a person	*Man-ijjot* or *ijjot* refers to the feeling of prestige, honour and self-pride of a person (male and female) or family. A person and/or family seeks to uphold this feeling of being honourable and respected. For example, the feeling of *ijjot*, of being a member of a respected household/of respected status, causes women to not work outdoors because this could impinge on family reputation.
man-shomman	respect, honour, prestige shown to a person by others	*Man-shomman* or *shomman* refers to the respect and status of a person (male and female) or family perceived by others. For example, the members of middle-class households in Manikpara are respected by the society because of their economic status.
mastan	local youth who is employed by powerful leaders	*Mastan* refers to a local youth who is employed/ 'used' by powerful leaders, for example to organise election campaigns. In terms of behaviour a *mastan* is not considered a gentle person and he is not appreciated because he is engaged in unlawful activities, e.g. land grabbing, *chadabaji*, fighting or even murder. In return for his 'services', he is paid by the political leaders. *Mastan* and *gunda* are similar.

matbor	respected person who acts as a judge in *shalish*	Respected person in any settlement who acts as a judge (only outside the legal system, i.e. in *shalish*, *bichar*, *ponchayet*). Generally a *matbor* is a wise, older person in the settlement. All members of the settlement obey him. A *matbor* is a person selected as a judge by other members of the settlement. He is responsible for resolving any problems within the settlement. This is basically a rural concept, but it is also found in the urban lower income settlements. In the urban context, *matbor* refers to the local power holders and they can be involved in misusing their power.
math	playing field	Traditionally, every locality has a *math* for children to play on, both in rural and urban areas. However, this 'concept' is getting lost in urban areas. A *math* can at the same time be used for the *Eid* prayers twice a year (see Eid Gah Math).
maxi	female clothing	Wide gown worn together with an *orna* to cover the chest, mostly worn within the house, but in some confined areas also outside.
mazar	Muslim shrine	Grave of a Muslim saint, people come to the *mazars* to offer a donation so that their wishes are fulfilled.
mela	fair	*Mela* refers to a gathering of people on the eve of a certain occasion, for example *Pohela Boishakh*. In the village context, during a *mela* all traditional items such as jewellery, clothing and pottery are sold. People of all ages and sexes come to the *mela* and buy the items, eat, and have fun.
michil	political demonstration	-
milad	Muslim thanksgiving prayer	This refers to the praising of Allah and the prophet with songs and prayers said on specific occasions. For example, if a person opens a new shop he/she starts operating the shop by arranging *milad*. Some people along with the imam of the mosque are invited. After the *milad*, people eat and say prayers, asking God to help in the person's business.
mohajon	manager, owner	Of a business, but *mohajon* also refers to someone powerful in a locality, someone who is economically successful.
mohalla	unit, neighbourhood	In Dhaka used to refer to the sub-units of a Ward.
mohanogor	megacity	-
Mohila League/Dol	Women's Organisation	*Mohila* League: sub-organisation of AL; *Mohila Dol*: sub-organisation of BNP.
Mollah	religious family title	Refers to a family tradition of Islamic scholars.
Moulana	Muslim scholar	*Moulana* is used as a title with names. It refers to a person who was educated in the *madrasa* system and is engaged as a teacher in a *madrasa* or as a Muslim scholar in a mosque.

Mussolmani	celebration of a boy's circumcision	Celebrated a few weeks after a boy has been circumcised in accordance with Islamic rules.
nagordola	big wheel	A Bengali version of a big wheel commonly constructed out of wood and consisting of four cabins, operated manually and especially popular during the celebration of *Pohela Boishakh* (see below).
Nawab [Nobab]	title for Muslim nobles, family of Old Dhaka	This refers to a title from the Mughal Empire that later on became a high title for Muslim nobles. The *Nobabs* were also one of the most influential families in Old Dhaka in the 19th century.
orna	scarf	Worn by women as part of the *salwar kamij* (see below), or with a *sari* or *borka* to cover the head (can be worn in a loose style, like *ghomta* or tightly like *hijab*).
orosh	religious meeting in honour of a Muslim saint	A religious meeting held annually, where the followers of a Muslim saint meet at the *mazar* (most commonly) and remember the saint/*pir*, listen to some renowned followers' speeches, and celebrate their meeting with food.
paka [pucca]	ripe, stable	In the context of housing this refers to houses made from durable materials, i.e. bricks. In the context of road networks and open spaces, the term is used to refer to the quality of the pavement, i.e. a *paka* road or space would be concrete or tarmac.
pir	spiritual guide	The guidelines of a Muslim saint are taught to the believers of the saint by a *pir*. A *pir* acts as a bridge between the believers and the saint's guidelines.
pitha	cake or bread commonly made of rice flour	-
Pohela Boishakh	Bengali New Year	Celebrated on the 1st day of the Bengali month *Boishakh* (15th April).
ponchayet		The *ponchayet* refers to a committee doing judgements (*shalish/bichar*).
porda [purdah]	seclusion (of women)	*Porda* refers to the social and religious norm of seclusion of women in traditional and religious society; following *porda* includes wearing the *borka* when leaving home, not laughing out loud, and not talking to any men except for family members. The related verb *porda-pushida kora* is related to women's style of dress and the social norms of society (to be well dressed so that no one can say anything bad). Wearing the *borka* whenever leaving the house is one type of *porda-pushida kora*, but these norms also differ depending on regional variations and economic status.
Ramadan		Fasting month of the Islamic calendar.
rongbaj	gangster	Refers to a young boy/man who is a local leader, and who passes his time aimlessly, considered a spoilt youth of a family; may have a political link or be a political leader, but not necessarily; similarly to *gunda* they are considered a threat to society as they can harm the public.

saheb	sir	Honourable way of addressing a man.
salwar kamij	female clothing	Three-piece clothing worn by women consisting of baggy trousers, a long blouse and an *orna* (see above), commonly worn to cover the chest (although the style of wearing the *orna* varies, for example depending on one's socio-economic status and/or religious belief).
sari	saree, sari	South-Asian unstitched garment worn by women, together with a blouse and petticoat (skirt) as undergarments.
semi-paka	half-stable	Refers to a house with brick walls but a CI-sheet roof (or other unstable material).
shalish	Judgement	Carried out by local leaders and elites (*matbors*), outside of the governmental judiciary; concept of conflict resolution especially in conflicts between neighbours in rural Bangladesh, but also common in some urban areas; many local disputes are taken to the *shalish* rather than to the formal judiciary which is especially inaccessible to the poor due to its corruption and lengthy procedures.
Shechhashebok League/Dol	Volunteers' Organisation	*Shechhashebok* League: sub-organisation of AL; *Shechhashebok Dol*: sub-organisation of BNP.
Shohid Minar	monument of *Ekushe*	-
shomaj	society, congregation	In rural areas, *shomaj* refers to a group within a village headed by a respected elder person (male). Every family of a village belongs to one *shomaj*, while there may be several *shomaj* in a village. The head of a *shomaj* is involved in solving disputes (*shalish*) and motivating community initiatives. A family who does not belong to any *shomaj* does not receive any *shalish* verdicts, i.e. traditional judgements. The *shomaj* is an expression of collective interest. In urban areas, *shomaj* more commonly refers to the society of a neighbourhood in general, while it also carries a notion of adhering to Islamic rules. Where there is reference to *shomaj* in urban areas, this means that people in a neighbourhood know and respect each other, help each other and obey the decisions of seniors (although not in all cases), and accept the verdicts from traditional judgements. It is still common in low-income areas and Old Dhaka, but in the new middle and high-income areas, i.e. apartment blocks, neighbourhood relationships are no longer as important. While in rural areas the *shomaj* heads are selected socially, in urban areas the *shomaj* is dominated by political leaders.
shomman	see *man-shomman*	-
shontrashi	armed gangster	Generally armed and has done something unlawful for profit or contrary to the society's choice.
shorkari shomoy	government time	Used to refer to the daylight saving time introduced in June 2009.
shorom	see *lojja*	-

somiti [samiti]	cooperative	Generally based on locality, in the context here mostly involved in saving and loan schemes.
Sromik League/Dol	Labourers' Organisation	*Sromik* League: sub-organisation of AL; *Sromik Dol*: sub-organisation of BNP.
tabij [tabiz]	religious amulet	Amulets containing written verses of the Quran.
Tabligh, Tablighi Jamaat	Islamic group	*Tabligh* is a conservative but pacifist Muslim movement of South Asian origin. The *Bisho Ijtema* (see above) is the annual congregation of the *Tabligh* movement.
Taka	currency of Bangladesh (BDT)	Exchange rate at the time of submitting this thesis: 100 BDT = 0,96 Euro (14.12.2011).
thana	police station	In Dhaka, the *thana* refers to the area of jurisdiction of a police station.
uthan	Courtyard	Originally referring to the courtyard of a rural homestead, also used to denote the courtyard in urban low-income housing compounds.

REFERENCES

Abourahme, Nasser (2011): Spatial Collisions and Discordant Temporalities: Everyday Life between Camp and Checkpoint. In *International Journal of Urban and Regional Research* 35 (2), 453–461.

Adler, Patricia A.; Adler, Peter (1994): Observational Techniques. In Norman K. Denzin, Yvonna S. Lincoln (eds.): Handbook of Qualitative Research. Thousand Oaks, London, New Delhi: SAGE Publications, 377–392.

Ahmed, Sharif Uddin (ed.) (1991): Dhaka – Past Present Future. Dhaka: Asiatic Society of Bangladesh.

Aktinson, Paul; Hammersley, Martyn (1994): Ethnography and Participant Observation. In Norman K. Denzin, Yvonna S. Lincoln (eds.): Handbook of Qualitative Research. Thousand Oaks, London, New Delhi: SAGE Publications, 248–261.

Altrock, Uwe (2012): Conceptualizing Informality: Some Thoughts on the Way towards Generalization. In Colin McFarlane, Michael Waibel (eds.): Urban Informalities: Reflections on the Formal and Informal. Farnham/Surrey: Ashgate, 171–193.

Altvater, Elmar; Mahnkopf, Birgit (2003): Die Informalisierung des urbanen Raums. In Jochen Becker (ed.): Learning from* – Städte von Welt, Phantasmen der Zivilgesellschaft, informelle Organisation. Berlin: NGBK, 17–30.

Andaleeb, Syed Saad; Irwin, Zachary T. (2007): Political leadership and legitimacy among the urban elite in Bangladesh. In Bangladesh Development Initiative (ed.): Political culture in Bangladesh – Perspectives and analysis. Dhaka: The University Press, 101–123.

Anjaria, Jonathan Shapiro (2006a): Street Hawkers and Public Space in Mumbai. In *Economic and Political Weekly*, 2140–2146.

Anjaria, Jonathan Shapiro (2006b): Urban Calamities: A View From Mumbai. In *Space and Culture* 9 (1), 80–82.

Anyamba, Tom J. C. (2006): Diverse Informalities: Spatial transformations in Nairobi. Oslo: Arkitekthøgskolen.

Apentiik, Caesar R.A; Parpart, Jane L. (2006): Working in Different Cultures: Issues of Race, Ethnicity and Identity. In Vandana Desai, Robert B. Potter (eds.): Doing Development Research. London, Thousand Oaks, New Dehli: SAGE Publications, 34–43.

Appadurai, Arjun (2001): Deep democracy: Urban governmentality and the horizon of politics. In *Environment and Urbanization* 13 (2), 23–43.

Bachmann-Medick, Doris (2009): Cultural turns – Neuorientierungen in den Kulturwissenschaften. 3rd edition. Reinbek bei Hamburg: Rowohlt-Taschenbuch-Verlag.

Bal, Ellen Wilhelmina (2007): They ask if we eat frogs – Garo Ethnicity in Bangladesh. Singapore, Leiden: ISEAS Publishing/IIAS.

Banks, Nicola (2008): A tale of two wards: political participation and the urban poor in Dhaka City. In *Environment and Urbanization* 20 (2), 361–376.

Banks, Nicola; Roy, Manoj; Hulme, David (2011): Neglecting the urban poor in Bangladesh: research, policy and action in the context of climate change. BWPI Working Paper 144. Manchester: The University of Manchester.

Batliwala, Srilatha; Dhanraj, Deepa (2007): Gender myths that instrumentalize women: a view from the Indian front line. In Andrea Cornwall, Elizabeth Harrison, Ann Whitehead (eds.): Feminisms in development – Contradictions, contestations and challenges. London: Zed Books, 21–34.

Baud, Isa; Nainan, Navtej (2008): "Negotiated spaces" for representation in Mumbai: ward committees, advanced locality management and the politics of middle-class activism. In *Environment and Urbanization* 20 (2), 483–499.

Baumgart, Sabine; Hackenbroch, Kirsten; Hossain, Shahadat; Kreibich, Volker (2011): Urban Development and Public Health in Dhaka, Bangladesh. In Alexander Krämer, Mobarak Hossain Khan, Frauke Kraas (eds.): Health in Megacities and Urban Areas. Berlin, Heidelberg: Springer-Verlag, 281–300.

Bayat, Asef (2004): Globalization and the Politics of the Informal in the Global South. In Ananya Roy, Nezar AlSayyad (eds.): Urban Informality – Transnational Perspectives from the Middle East, Latin America, and South Asia. Lanham, Maryland: Lexington Books, 79–104.

Bayat, Asef (2010): Life as Politics – How Ordinary People Change the Middle East. Amsterdam: Amsterdam University Press.

Beazley, Harriot (2002): 'Vagrants Wearing Make-up': Negotiating Spaces on the Streets of Yogyakarta, Indonesia. In *Urban Studies* 39 (9), 1665–1683.

Becker, Ruth (2008): Raum: Feministische Kritik an Stadt und Raum. In Ruth Becker, Beate Kortendiek (eds.): Handbuch Frauen- und Geschlechterforschung – Theorie, Methoden, Empirie. 2nd edition. Wiesbaden: VS Verlag für Sozialwissenschaften, 652–664.

Bertuzzo, Elisa T. (2009): Fragmented Dhaka – Analysing everyday life with Henri Lefebvre's Theory of Production of Space. Stuttgart: Franz Steiner Verlag.

Bertuzzo, Elisa T.; Nest, Günter (2008a): Smooth and Striated: City and Water, Dhaka/Berlin. In Elisa T. Bertuzzo, Nazrul Islam, Günter Nest, Salma A. Shafi (eds.): Smooth and Striated: City and Water, Dhaka/Berlin. Dhaka, 14–18.

Bertuzzo, Elisa T.; Nest, Günter (2008b): The Significane of Urban Water Bodies for Public Sphere. In Elisa T. Bertuzzo, Nazrul Islam, Günter Nest, Salma A. Shafi (eds.): Smooth and Striated: City and Water, Dhaka/Berlin. Dhaka, 19–28.

Bode, Brigitta (2002): In Pursuit of Power: Local Elites and Union-level Governance in Rural North-Western Bangladesh. Unpublished research report. Dhaka: CARE Bangladesh.

Bohle, Hans-Georg (2007): Geographies of violence and vulnerability – An actor-oriented analysis of the civil war in Sri Lanka. In *Erdkunde* 61 (2), 129–146.

Bose, Mallika (1998): Surveillance, Circumscriber of Women's Spatial Experience: The Case of Slum Dwellers of Calcutta, India. In Hemalata C. Dandekar (ed.): City, space + globalization – An international perspective. Proceedings of an international symposium, College of Architecture and Urban Planning, the University of Michigan, February 26–28, 1998. Ann Arbor: University of Michigan, 364–382.

Brandt, Carmen (2010): Bangladesch als Fußnote sozialwissenschaftlicher Diskurse? In *Orientalische Literaturzeitung* 105, 7–20.

Breckner, Ingrid; Sturm, Gabriele (2002): Kleiderwechsel – Sackgassen und Perspektiven in patriarchalen Öffentlichkeiten. In Martina Löw (ed.): Differenzierungen des Städtischen. Opladen: Leske + Budrich, 157–186.

Bromley, Ray (2004): Power, Property, and Poverty: Why De Soto's "Mystery of Capital" Cannot Be Solved. In Ananya Roy, Nezar AlSayyad (eds.): Urban Informality – Transnational Perspectives from the Middle East, Latin America, and South Asia. Lanham, Maryland: Lexington Books, 271–288.

Brown, Alison (2006a): Challenging street livelihoods. In Alison Brown (ed.): Contested space-Street trading, public space, and livelihoods in developing cities. Rugby, UK: ITDG Publishing.

Brown, Alison (ed.) (2006b): Contested space – Street trading, public space, and livelihoods in developing cities. Rugby, UK: ITDG Publishing.

Brown, Alison; Lloyd-Jones, Tony (2002): Spatial Planning, Access and Infrastructure. In Carole Rakodi, Tony Lloyd-Jones (eds.): Urban Livelihoods – A People-centred Approach to Reducing Poverty. London, Sterling, VA: Earthscan Publications, 188–204.

Brown, Alison; Lyons, Michal; Dankoco, Ibrahima (2010): Street Traders and the Emerging Spaces for Urban Voice and Citizenship in African Cities. In *Urban Studies* 47 (3), 666–683.

Brydon, Lynne (2006): Ethical Practices in Doing Development Research. In Vandana Desai, Robert B. Potter (eds.): Doing Development Research. London, Thousand Oaks, New Dehli: SAGE Publications, 25–33.

Bujra, Janet (2006): Lost in Translation? The Use of Interpreters in Fieldwork. In Vandana Desai, Robert B. Potter (eds.): Doing Development Research. London, Thousand Oaks, New Dehli: SAGE Publications, 172–179.

Burdett, Ricky; Sudjic, Deyan (2007): The endless city – The urban age project by the London School of Economics and Deutsche Bank's Alfred Herrhausen Society. London: Phaidon.

Butler, Ruth (2001): From where I write: the place of positionality in qualitative writing. In Melanie Limb (ed.): Qualitative methodologies for geographers – Issues and debates. London: Arnold, 264–276.

Centre for Urban Studies (CUS); National Institute of Population Research and Training; MEASURE Evaluation (2006): Slums of Urban Bangladesh: Mapping and Census, 2005. Dhaka.

Chakrabarty, Dipesh (2001): Adda, Calcutta: Dwelling in Modernity. In Dilip Parameshwar Gaonkar (ed.): Alternative modernities. Durham, NC: Duke University Press, 123–164.

Chambers, Robert (1995): The Primacy of the Personal. In Michael Edwards, David Hulme (eds.): Non-Governmental Organisations – Performance and Accountability – Beyond the Magic Bullet. London: Earthscan, 207–218.

Chambers, Robert; Conway, Gordon (1992): Sustainable Rural Livelihoods: Practical Concepts for the 21st Century. Brighton: IDS.

Chatterjee, Partha (2004): The politics of the governed – Reflections on popular politics in most of the world. New York: Columbia University Press.

Chowdhury, A.M; Faruqui, Shabnam (2009): Physical growth of Dhaka City. In Sharif Uddin Ahmed (ed.): Dhaka –Past Present Future. 2nd edition (first published in 1991). Dhaka: Asiatic Society of Bangladesh, 56–76.

Crane, Lucy G.; Lombard, Melanie B.; Tenz, Eric M. (2009): More than just translation: challenges and opportunities in translingual research. In *Social Geography* (4), 39–46.

Dannecker, Petra (2008): Der Aufbruch der Frauen. In *Le monde diplomatique* 4/2008, 92–96.

Davis, Mike (2007): Planet of slums. London: VERSO.

Davy, Benjamin (2008): Plan it without a condom. In *Planning Theory* 7 (3), 301–318.

Davy, Benjamin (2009): The poor and the land: poverty, property, planning. In *Town Planning Review* 80 (3), 227–265.

de Haan, Leo; Zoomers, Annelies (2005): Exploring the Frontier of Livelihoods Research. In *Development and Change* 36 (1), 27–47.

de Smedt, Johan (2009): 'No Raila, No Peace!' Big Man Politics and Election Violence at the Kibera Grassroots. In *African Affairs* 108 (433), 581–598.

De Soto, Hernando (1989): The other path – The invisible revolution in the third world. New York: Harper & Row.

De Soto, Hernando (2000): The mystery of capital – Why capitalism triumphs in the West and fails everywhere else. New York: Basic Books.

Deleuze, Gilles; Guattari, Félix (1987 [1980]): 1440: The Smooth and the Striated. In Gilles Deleuze, Félix Guattari (eds.): A Thousand Plateaus: Capitalism and Schizophrenia. Minneapolis, London: University of Minnesota Press, 474–500.

Denzin, Norman K. (1970): The research act in sociology: a theoretical introduction to sociological methods. London: Butterworth.

Denzin, Norman K. (1989): The Research Act. 3rd edition. Englewood Cliffs, N.J.: Prentice Hall.

Dick, Eva (2008): Residential segregation – stumbling block or stepping stone? A case study on the Mexican population of the West Side of St. Paul, Minnesota, USA. Wien: LIT Verlag.

Dil, Afia (1972): The Hindu and Muslim dialects of Bengali. Dissertation (Diss. phil.) at Stanford University.

Doderer, Yvonne P. (2003): Urbane Praktiken – Strategien und Raumproduktionen feministischer Frauenöffentlichkeit. Münster: Monsenstein & Vannerdat.

Dodman, David R. (2003): Shooting in the city – an autophotographic exploration of the urban environment in Kingston, Jamaica. In *Area* 35 (3), 293–304.

Dünne, Jörg; Günzel, Stephan (eds.) (2006 [1967]): Raumtheorie – Grundlagentexte aus Philosophie und Kulturwissenschaften. Frankfurt am Main: Suhrkamp.

Eckardt, Frank (2009): Die komplexe Stadt – Orientierungen im urbanen Labyrinth. Wiesbaden: VS Verlag für Sozialwissenschaften / GWV Fachverlage.

Eckert, Julia (2006): From subjects to citizens: Legalism from below and the homogenisation of the legal sphere. In *Journal of Legal Pluralism and unofficial law* 38 (53–54), 45–75.

Ellis, Frank (2000): Rural Livelihoods and Diversity in Developing Countries. Oxford: Oxford University Press.

Emerson, Robert M.; Fretz, Rachel I.; Shaw, Linda L. (2010): Writing ethnographic fieldnotes. Chicago: University of Chicago Press.

Erler, Brigitte (1985): Tödliche Hilfe – Bericht von meiner letzten Dienstreise in Sachen Entwicklungshilfe. Kirchzarten: Dreisam-Verlag.

Etzold, Benjamin (2011): Die umkämpfte Stadt – Die alltägliche Aneignung öffentlicher Räume durch Straßenhändler in Dhaka (Bangladesch). In Andrej Holm, Dirk Gebhardt (ed.): Initiativen für ein Recht auf Stadt: Theorie und Praxis städtischer Aneignungen. Hamburg: VSA Verlag, 187–220.

Etzold, Benjamin (2012): Selling in Insecurity – Living with Violence: Eviction Drives against Street Food Vendors in Dhaka and the Informal Politics of Exploitation. In Noa K. Ha, Kristina Graaf (ed.): Urban Street Vending: A Global Perspective on the Practices and Policies of a Marginalized Economy. New York: Berghahn Books, forthcoming.

Etzold, Benjamin; Keck, Markus; Bohle, Hans-Georg; Zingel, Wolfgang Peter (2009): Informality as Agency – Negotiating Food Security in Dhaka. In *DIE ERDE* 140 (1), 3–24.

Fahmi, Wael Salah (2009): Bloggers' street movement and the right to the city – (Re)claiming Cairo's real and virtual "spaces of freedom". In *Environment and Urbanization* 12 (1), 89–107.

Few, Roger (2002): Researching actor power: analyzing mechanisms of interaction in negotiations over space. In *Area* 34 (1), 29–38.

Flick, Uwe (2008): Triangulation – Eine Einführung. 2nd edition. Wiesbaden: VS Verlag für Sozialwissenschaften / GWV Fachverlage.

Flick, Uwe (2009): Qualitative Sozialforschung – Eine Einführung. 2nd edition. Reinbek bei Hamburg: Rowohlt-Taschenbuch-Verlag.

Flyvjerg, Bent (2004): Five Misunderstandings about Case-Study Research. In Clive Seale, Giampietro Gobo, Jaber F. Gubrium, David Silverman (eds.): Qualitative Research Practise. London: SAGE Publications, 420–434.

Forsslund, Annika (1995): "From nobody to somebody" – Women's struggle to achieve dignity and self-reliance in a Bangladeshi village. Umeå: Department of Education University of Umeå.

Foucault, Michel (2006 [1967]): Von anderen Räumen. In Jörg Dünne, Stephan Günzel (eds.): Raumtheorie – Grundlagentexte aus Philosophie und Kulturwissenschaften. Frankfurt am Main: Suhrkamp, 317–327.

Fraser, Nancy (1993): Falsche Gegensätze. In Seyla Benhabib, Judith Butler, Drucilla Cornell, Nancy Fraser (eds.): Der Streit um Differenz. Feminismus und Postmoderne in der Gegenwart. Frankfurt am Main: Fischer, 59–80.

Friedmann, John (1986): The World City Hypothesis. In *Development and Change* 17, 69–83.

Gardner, Katy (1994): Purdah, female power, and cultural change: a Sylheti example. In *The Journal of Social Studies* 64, 1–24.

Gardner, Katy (2000): Global Migrants, Local Lives – Travel and Transformation in Rural Bangladesh. Oxford: Clarendon Press.

Gardner, Katy; Ahmed, Zahir (2009): Degrees of Separation: Informal Social Protection, Relatedness and Migration in Biswanath, Bangladesh. In *Journal of Development Studies* 45 (1), 124–149.

Gehl, Jan (2007): Public Spaces for a Changing Public Life. In *Topos: European Landscape Magazine* 16 (61), 16–22.

Ghafur, Shayer (2005): Socio-Spatial Adaptation for Living and Livelihood: A Post-Occupancy Evaluation of Multi-storey Low-income Housing in Dhaka – Final Research Project Report. Dhaka.

Ghafur, Shayer (2008): Spectre of product fetishism – Reviewing housing development proposal for Dhaka city. In *The Daily Star*, 12.07.2008.

Gilbert, Alan (2007): The Return of the Slum: Does Language Matter? In *International Journal of Urban and Regional Research* 31 (4), 397–713.

Glaser, Barney G. (1978): Theoretical Sensitivity. Mill Valley: University of California.

Glaser, Barney G.; Strauss, Anselm L. (1967): The Discovery of Grounded Theory: Strategies for Qualitative Research. Chicago: Aldine Publishing.

Gregory, Derek (1994): Geographical Imaginations. Cambridge: Wiley-Blackwell.

Ha, Noa K. (2009): Informeller Straßenhandel in Berlin – Urbane Raumproduktion zwischen Störung und Attraktion. Berlin: WVB Wissenschaftlicher Verlag.

Habermas, Jürgen (1981): Theorie des kommunikativen Handelns (Band I) – Handlungsrationalität und gesellschaftliche Rationalisierung. Frankfurt am Main: Suhrkamp.

Hackenbroch, Kirsten (2011): Urban Informality and Negotiated Space – Negotiations of Access to Public Space in Dhaka, Bangladesh. In *disp – The Planning Review* 187 (4/2011), 59–69.

Hackenbroch, Kirsten; Hossain, Mohammad Shafayat; Rahman, Asif (2008): Coping with Forced Evictions – Adaptation Processes of Evicted Slum Dwellers in Dhaka. In *Trialog* 98, 17–23.

Hackenbroch, Kirsten; Hossain, Shahadat (2012): "The organised encroachment of the powerful" – Everyday practices of public space and water supply in Dhaka, Bangladesh. In *Planning Theory and Practice* 13 (3), 397–420.

Hafiz, R. (2007): The Urban Frontiers of Dhaka: Creating Space above the Water. In Sarwar Jahan, K.M. Maniruzzaman (eds.): Urbanization in Bangladesh: Patterns, Issues and Approaches to Planning. Dhaka: Bangladesh Institute of Planners, 55–67.

Hansen, Karen Transberg (2004): Who Rules the Streets? The Politics of Vending Space in Lusaka. In Karen Transberg Hansen, Mariken Vaa (eds.): Reconsidering Informality – Perspectives from Urban Africa: Nordiska Afrikainstitutet, 62–80.

Hart, Keith (1973): Informal income opportunities and urban employment in Ghana. In *Journal of Modern Africa* 11, 61–89.

Harvey, David (2003): A Right to the City. In *International Journal of Urban and Regional Research* 27 (4), 939–941.

Hasan, Arif (2009): Land, CBOs and the Karachi Circular Railway. In *Environment and Urbanization* 21 (2), 331–345.

Hasan, Arif (2010): IIED Density Study: 04 Cases of Housing in Karachi. IIED.

Healey, P. (2002): On creating the 'city' as a collective resource. In *Urban Studies* 39 (10), 1777–1792.

Hernández-Bonilla, Mauricio (2008): Contested Public Space Development: The Case of Low Income Neighbourhoods in Xalapa, Mexico. In *Landscape Research* 33 (4), 389–406.

Holston, James (1998): Spaces of insurgent citizenship. In Leonie Sandercock (ed.): Making the invisible visible – A multicultural planning history. Berkeley: University of California, 37–56.

Holston, James (2009): Dangerous Spaces of Citizenship: Gang Talk, Rights Talk and Rule of Law in Brazil. In *Planning Theory* 8 (1), 12–31.

Hossain, Shahadat (2011): Informal dynamics of a public utility: Rationality of the scene behind a screen. In *Habitat International* 35 (2), 275–285.

Human Rights Watch (2011): World Report 2011.

International Labour Office (ILO) (1972): Employment, Income and Equality: A Strategy for Increasing Productivity in Kenya. Geneva.

Ipsen, Detlev (2002): Die Kultur der Orte. Ein Beitrag zur sozialen Strukturierung des städtischen Raumes. In Martina Löw (ed.): Differenzierungen des Städtischen. Opladen: Leske + Budrich, 233–246.

Islam, Ishrat; Mitra, Sumon Kumar; Shohag, Md Abu Nayem; Rahman, Mohammad Aminur (2007): Land Price in Dhaka City: Distribution, Characteristics and Trend of Changes. In Sarwar Jahan, K.M. Maniruzzaman (eds.): Urbanization in Bangladesh: Patterns, Issues and Approaches to Planning. Dhaka: Bangladesh Institute of Planners, 25–35.

Islam, Nazrul (2005): Dhaka Now – Contemporary Urban Development. Dhaka: Bangladesh Geographical Society.

Islam, Nazrul; Khan, Mohammad Mohabbat; Islam Nazem, Nurul; Rhaman, Mohammad Habibur (2003): Reforming Governance in Dhaka, Bangladesh. In Patricia Louise McCarney (ed.): Governance on the Ground – Innovations and Discontinuities in Cities of the Developing World. Washington DC: Woodrow Wilson Center Press, 194–219.

Islam, Nazrul; Shafi, Salma A. (2008): A Proposal for a Housing Development in Dhaka City. Dhaka.

Israel, Mark; Hay, Iain (2006): Research ethics for social scientists – Between ethical conduct and regulatory compliance. London: SAGE Publications.

Jachnow, Alexander (2003): Die Attraktivität des Informellen – Der große Einfluss der Zivilgesellschaft und andere Fehleinschätzungen. In Jochen Becker (ed.): Learning from* – Städte von Welt, Phantasmen der Zivilgesellschaft, informelle Organisation. Berlin: NGBK, 79–91.

Jahan, Rounaq (2004a): Bangladesh in 2003 – Vibrant Democracy or Destructive Politics? In *Asian Survey* 44 (1), 56–61.

Jahan, Rounaq (2004b): Why are we still continuing with a "viceregal" political system? In *The Daily Star*, 31.01.2004.

Jahan, Rounaq (2007): Bangladesh at a crossroads. In *seminar – the monthly symposium*, Experiments with democracy (576).

Karim, Abdul (2009): Origin and development of Mughal Dhaka. In Sharif Uddin Ahmed (ed.): Dhaka – Past Present Future. 2nd edition (first published in 1991). Dhaka: Asiatic Society of Bangladesh, 34–55.

Kelle, Udo (1994): Empirisch begründete Theoriebildung – Zur Logik und Methodologie interpretativer Sozialforschung. Weinheim: Deutscher Studien Verlag.

Khan, Sharful Islam; Hussain, Mohammed Iftekher; Parveen, Shaila; Bhuiyan, Mahbubul Islam; Gourab, Gorkey; Sarker, Golam Faruk; Arafat, Shohael Mahmud; Sikder, Joya (2009): Living on the Extreme Margin: Social Exclusion of the Transgender Population in Bangladesh. In *Journal of Health, Population and Nutrition* 27 (4), 441–451.

Khatun, Habiba (2009): Pre-Mughal Dhaka. In Sharif Uddin Ahmed (ed.): Dhaka – Past Present Future. 2nd edition (first published in 1991). Dhaka: Asiatic Society of Bangladesh, 674–678.

Kombe, Wilbard J.; Kreibich, Volker (2006): Governance of Informal Urbanisation in Tanzania. Dar es Salaam: Mkuki na Nyota Publishers.

Kreibich, Volker (2012): The Mode of Informal Urbanisation – Reconciling Social and Statutory Regulation in Urban Land Management. In Colin McFarlane, Michael Waibel (eds.): Urban Informalities: Reflections on the Formal and Informal. Farnham/Surrey: Ashgate, 149–170.

Krisch, Richard (2002): Methoden einer sozialräumlichen Lebensweltanalyse. In Ulrich Deinet, Richard Krisch (eds.): Der sozialräumliche Blick der Jugendarbeit. Opladen: Leske + Budrich, 87–154.

Kudva, Neema (2009): The everyday and the episodic: the spatial and political impacts of urban informality. In *Environment and Planning A* 41 (7), 1614–1628.

Kumar, Somesh (2002): Methods for Community Participation – A Complete Guide for Practitioners. Chippenham: Antony Rowe.

Lamnek, Siegfried (2005): Qualitative Sozialforschung. 4th edition. Weinheim, Basel: Beltz-Verlag.
Lefebvre, Henri (1971 [1968]): Everyday Life in the Modern World. New York, Evanston, San Francisco, London: Harper & Row.
Lefebvre, Henri (1991 [1974]): The Production of Space. Oxford: Blackwell.
Lefebvre, Henri (2004 [1992]): Rhythmanalysis – Space, Time and Everyday Life. London, New York: continuum.
Lewis, David; Hossain, Abul (2007): Beyond the net: an analysis of the local power structure in Bangladesh. In David Gellner (ed.): Governance, Conflict, and Civic Action in South Asia. New Delhi: SAGE Publications, 279–300.
Lewis, David; Hossain, Abul (2008): Understanding the Local Power Structure in Bangladesh. SIDA Studies No. 22. Stockholm.
Lewis, Oscar (1959): Five Families: Mexican Case Studies in the Culture of Poverty. New York: Basic Books.
Lo Piccolo, Francesco; Thomas, Huw (2008): Research Ethics in Planning: A Framework for Discussion. In *Planning Theory* 7 (1), 7–23.
Lorch, Jasmin (2008): Politischer Islam in Bangladesch – Wie schwache Staatlichkeit und autoritäre Regierungsführung islamistische Gruppen stärken. Berlin: Stiftung Wissenschaft und Politik.
Lourenço-Lindell, Ilda (2004): Trade and the Politics of Informalization in Bissau, Guinea-Bissau. In Karen Transberg Hansen, Mariken Vaa (eds.): Reconsidering Informality – Perspectives from Urban Africa: Nordiska Afrikainstitutet, 84–98.
Löw, Martina (2001): Raumsoziologie. Frankfurt am Main: Suhrkamp.
Madanipour, Ali (ed.) (2010): Whose public space? – International case studies in urban design and development. London: Routledge.
Marsh, David; Furlong, Paul (2002): A Skin not a Sweater: Ontology and Epistemology in Political Science. In David Marsh, Gerry Stoker (eds.): Theory and Methods in Political Science. 2nd edition. Hampshire, New York: Palgrave Macmillan, 17–41.
Massey, Doreen (1984): Introduction: Geography matters! In Doreen Massey, John Allen (eds.): Geography matters! A Reader. Cambridge: Cambridge University Press, 1–11.
Massey, Doreen (2006): Space, time and political responsibility in the midst of global inequality. In *Erdkunde* 20 (2), 89–95.
Massey, Doreen (2008): Politik und Raum/Zeit. In Bernd Belina, Boris Michel (eds.): Raumproduktionen – Beiträge der Radical Geography. Eine Zwischenbilanz. Münster: Westfälisches Dampfboot, 111–132.
Meagher, Kate (1995): Crisis, Informalization and the Urban Informal Sector in Sub-Saharan Africa. In *Development and Change* 26, 259–284.
Merton, Robert King; Kendall, Patricia L. (1956): The focussed interview. Glencoe, Ill.
Meth, Paula (2010): Unsettling Insurgency: Reflections on Women's Insurgent Practices in South Africa. In *Planning Theory and Practice* 11 (2), 241–263.
Miaji, Abdel Baten (2010): Rural Women in Bangladesh – The Legal Status of Women and the Relationship between NGOs and Religious Groups. Lund: Lund University.
Naher, Ainoon (2010): Defending Islam and women's honour against NGOs in Bangladesh. In *Women's Studies International Forum* 33 (4), 316–324.
Naik, Rineeta (2007): Bangladesh. The Caretaker's Burden. In *Economic and Political Weekly*, 3540–3542.
Neuwirth, Robert (2006): Shadow cities – A billion squatters, a new urban world. New York: Routledge.
North, Douglass C. (1990): Institutions, Institutional Change and Economic Performance. Cambridge: Cambridge University Press.
Patel, Shirish B. (2007): Urban Layouts, Density & the Quality of Urban Life. In *Trialog* 94 (3), 47–50.

Payne, Geoffrey (2002): Tenure and Shelter in Urban Livelihoods. In Carole Rakodi, Tony Lloyd-Jones (eds.): Urban Livelihoods – A People-centred Approach to Reducing Poverty. London, Sterling, VA: Earthscan Publications, 151–164.

Perlman, Janice E. (1976): The Myth of Marginality: Urban Poverty and Politics in Rio de Janeiro. Berkeley, Los Angeles: University of California Press.

Perlman, Janice E. (2004): Marginality: From Myth to Reality in the Favelas of Rio de Janeiro, 1969–2002. In Ananya Roy, Nezar AlSayyad (eds.): Urban Informality – Transnational Perspectives from the Middle East, Latin America, and South Asia. Lanham, Maryland: Lexington Books, 105–146.

Rahman, Mahfuzur (2003): Dhaka – A City of Dirt, Darkness and Deprivation. Dhaka: NewsNetwork.

Rakodi, Carole; Lloyd-Jones, Tony (eds.) (2002): Urban Livelihoods – A People-centred Approach to Reducing Poverty. London, Sterling, VA: Earthscan Publications.

Rashid, Harun-or (2009): The Dhaka Nawab family in Bengal politics. In Sharif Uddin Ahmed (ed.): Dhaka – Past Present Future. 2nd edition (first published in 1991). Dhaka: Asiatic Society of Bangladesh, 153–171.

Rashid, Sabina Faiz (2006): Small Powers, Little Choice: Contextualising Reproductive and Sexual Rights in Slums in Bangladesh. In *IDS Bulletin* 37 (5), 69–76.

Reichertz, Jo (2007): Abduktion, Deduktion und Induktion in der qualitativen Forschung. In Uwe Flick, Ernst von Kardorff, Ines Steinke (eds.): Qualitative Forschung – Ein Handbuch. 5th edition. Reinbek bei Hamburg: Rowohlt-Taschenbuch-Verlag.

Robinson, Jennifer (2006): Ordinary cities – Between modernity and development. London: Routledge.

Rose, Gillian (1993): Feminism and Geography. Cambridge: Polity Press.

Roy, Ananya (2003): City requiem, Calcutta – Gender and the politics of poverty. Minneapolis: University of Minnesota Press.

Roy, Ananya (2004): The Gentlemen's City: Urban Informality in the Calcutta of New Communism. In Ananya Roy, Nezar AlSayyad (eds.): Urban Informality – Transnational Perspectives from the Middle East, Latin America, and South Asia. Lanham, Maryland: Lexington Books, 147–170.

Roy, Ananya (2005): Urban Informality – Toward an Epistemology of Planning. In *Journal of the American Planning Association* 71 (2), 147–158.

Roy, Ananya (2009a): The 21st-Century Metropolis – New Geographies of Theory. In *Regional Studies* 43 (6), 819–830.

Roy, Ananya (2009b): Why India Cannot Plan Its Cities: Informality, Insurgence and the Idiom of Urbanization. In *Planning Theory* 8 (1), 76–87.

Roy, Ananya (2010): Informality and the Politics of Planning. In Jean Hillier, Patsy Healey (eds.): The Ashgate Research Companion to Planning Theory – Conceptual Challenges for Spatial Planning. Farnham, Surrey: Ashgate, 87–108.

Roy, Ananya (2011): Slumdog Cities: Rethinking Subaltern Urbanism. In *International Journal of Urban and Regional Research* 35 (2), 223–238.

Roy, Ananya (2012): Urban Informality: The Production of Space and the Practice of Planning. In Randall Crane, Rachel Weber (eds.): The Oxford Handbook of Urban Planning. New York: Oxford University Press, 691–705.

Roy, Ananya; AlSayyad, Nezar (2004a): Prologue/Dialogue. Urban informality: Crossing Borders. In Ananya Roy, Nezar AlSayyad (eds.): Urban Informality – Transnational Perspectives from the Middle East, Latin America, and South Asia. Lanham, Maryland: Lexington Books, 1–6.

Roy, Ananya; AlSayyad, Nezar (eds.) (2004b): Urban Informality – Transnational Perspectives from the Middle East, Latin America, and South Asia. Lanham, Maryland: Lexington Books.

Rozario, Santi (2001): Claiming the campus for female students in Bangladesh. In *Women's Studies International Forum* 24 (2), 157–166.

Rozario, Santi (2006): The new burqa in Bangladesh: empowerment or violation of women's rights? In *Women's Studies International Forum* 29 (4), 368–380.
Sahityo Academy (ed.) (1988 [1966]): Bongiyo Shobdokosh: A Bengali-Bengali Lexicon. 3rd edition. New Dehli.
Salway, Sarah; Jesmin, Sonia; Rahman, Shahana (2005): Women's Employment in Urban Bangladesh: A Challenge to Gender Identity? In *Development and Change* 36 (2), 317–349.
Sassen, Saskia (2005): The Global City: Strategic Site, New Frontier. In Marco Keiner, Martina Koll-Schretzenmayr, Willy A. Schmid (eds.): Managing Urban Futures – Sustainability and Urban Growth in Developing Countries. Aldershot, Hampshire: Ashgate, 73–88.
Satterthwaite, David (2007): The transition to a predominantly urban world and its underpinnings. Human Settlements Discussion Paper Series. London: IIED.
Schmid, Christian (2005): Stadt, Raum und Gesellschaft – Henri Lefebvre und die Theorie der Produktion des Raumes. Stuttgart: Franz Steiner Verlag.
Scott, James C. (1985): Weapons of the Weak – Everyday Forms of Peasant Resistance. New Haven: Yale University Press.
Sharp, Joanne P.; Routledge, Paul; Philo, Chris; Paddison, Ronan (eds.) (2000): Entanglements of power: geographies of domination/resistance. London: Routledge.
Sheuya, Shaaban (2004): Housing Transformations and Urban Livelihoods in Informal Settlements. SPRING Research Series 45. Dortmund.
Siddiqui, Kamal (ed.) (2004): Megacity governance in South Asia – A comparative study. Dhaka: The University Press.
Siddiqui, Kamal; Ahmed, Jamshed; Awal, Abdul; Ahmed, Mustaque (2000): Overcoming the governance crisis in Dhaka City. Dhaka: University Press.
Siddiqui, Kamal; Ahmed, Jamshed; Siddique, Kaniz; Huq, Sayeedul; Hossain, Abu; Nazimud-Doula, Shah; Rezawana, Nahid (2010): Social Formation in Dhaka, 1985–2005 – A Longitudinal Study of Society in a Third World Megacity. Farnham, Surrey: Ashgate.
Simmel, Georg (1903): Die Großstädte und das Geistesleben. In Theodor Petermann (ed.): Die Großstadt. Vorträge und Aufsätze zur Städteausstellung, Vol. 9 (9), 185–206.
Simone, Abdoumaliq (2006): Pirate Towns: Reworking Social and Symbolic Infrastructures in Johannesburg and Douala. In *Urban Studies* 43 (2), 357–370.
Simone, Abdoumaliq (2010): City life from Jakarta to Dakar – Movements at the Crossroads. New York: Routledge.
Soja, Edward W. (1996): Thirdspace – Journeys to Los Angeles and other real-and-imagined places. Cambridge, Massachusetts: Blackwell Publishers.
Soja, Edward W. (2003): Thirdspace – Die Erweiterung des Geographischen Blicks. In Hans Gebhardt, Paul Reuber, Günter Wolkersdorfer (eds.): Kulturgeographie – Aktuelle Ansätze und Entwicklungen. Heidelberg, Neckar: Spektrum Akademischer Verlag.
Soja, Edward W. (2010): Seeking Spatial Justice. Minneapolis, London: University of Minnesota Press.
Staffeld, Ronny; Kulke, Elmar (2011): Informal Employment and Health Conditions in Dhaka's Plastic Recycling and Processing Industry. In Alexander Krämer, Mobarak Hossain Khan, Frauke Kraas (eds.): Health in Megacities and Urban Areas. Berlin, Heidelberg: Springer-Verlag, 211–221.
Strauss, Anselm L.; Corbin, Juliet (1994): Grounded theory methodology. An overview. In Norman K. Denzin, Yvonna S. Lincoln (eds.): Handbook of Qualitative Research. Thousand Oaks, London, New Delhi: SAGE Publications, 273–285.
Tarlo, Emma (1996): Clothing matters – Dress and identity in India. Chicago IL: University of Chicago Press.
Taylor, Peter J. (2004): World City Network: A Global Urban Analysis. London: Routledge.
Terlinden, Ulla (2002): Räumliche Definitionsmacht und weibliche Überschreitungen. Öffentlichkeit, Privatheit und Geschlechterdifferenzierung im städtischen Raum. In Martina Löw (ed.): Differenzierungen des Städtischen. Opladen: Leske + Budrich, 141–156.

The Daily Star (2009a): Dhaka gone crazy – Heat, outage, water crisis cripple city life. In *The Daily Star*, 27.04.2009.
The Daily Star (2009b): Death of a lifeline. In *The Daily Star*, 01.06.2009.
The Daily Star (2010a): Sadarghat Launch Terminal – Leasing to be withdrawn to lessen passengers' suffering. In *The Daily Star*, 19.04.2010.
The Daily Star (2010b): DAP now official – All recommendations find place on gazette notification; fate depends on implementation, 24.10.2010.
The Daily Star (2011): 13 sued for disgracing devotees of Lalon, 11.04.2011.
Transparency International Bangladesh (2010): Corruption in the Service Sectors: National Household Survey 2010. Dhaka.
Turner, Barry A. (1981): Some practical aspects of qualitative data analysis: One way of organising the cognitive processes associated with the generation of grounded theory. In *Quality and Quantity* 15 (3), 225–247.
Turner, Ralph L. (1966): A Comparative Dictionary of the Indo-Aryan Languages. London: Oxford University Press.
UN-Habitat (2003): Slums of the World: The face of urban poverty in the new millennium? Nairobi.
UN-Habitat (2008): State of the World's Cities 2008/2009 – Harmonious Cities. London, Sterling, VA: Earthscan.
van Schendel, Willem (2009): A History of Bangladesh. Cambridge: Cambridge University Press.
Wedel, Heidi (2004): Alltagsleben und politische Partizipation – Gecekondu-Viertel als gesellschaftlicher Ort. In *European Journal of Turkish Studies* 1 (1).
Weinberg, Merlinda (2002): Biting the Hand That Feeds You, and Other Feminist Dilemmas in Fieldwork. In Will C. van Den Hoonaard (ed.): Walking the Tightrope: Ethical Issues for Qualitative Researchers. Toronto: University of Toronto Press, 79–94.
Woiwode, Christoph (2007): Urban Risk Communication in Ahmedabad – India – between Slum Dwellers and the Municipal Corporation. Unpublished dissertation (Diss. phil.), Development Planning Unit, London.
World Bank (2007): Dhaka: Improving living conditions for the urban poor. Dhaka: The World Bank Office.
World Bank (2008): Whispers to Voices – Gender and Social Transformation in Bangladesh. Dhaka: The World Bank Office.
Yiftachel, Oren (2009): Theoretical Notes On 'Gray Cities': the Coming of Urban Apartheid? In *Planning Theory* 8 (1), 88–100.
Yin, Robert K. (2003): Case Study Research – Design and Methods. 3rd edition. Thousand Oaks, London, New Delhi: SAGE Publications.
Zakaria, Fareed (2003): The Future of Freedom: Illiberal Democracy at Home and Abroad. New York: W.W. Norton.

Websites

Website ICG (International Crisis Group)
http://www.crisisgroup.org/en/publication-type/crisiswatch/crisiswatch-database.aspx?CountryIDs={C3F2C698-4DFE-4548-B5EA-B78C8D53B2BF}#results, last access on 21.04.2011

Website RAJUK (Rajdhani Unnayon Kortripokkho)
www.rajukdhaka.gov.bd, last access on 21.04.2011

Website RAJUK-DAP (RAJUK-Detailed Area Plan)
http://www.rajukdhaka.gov.bd/rajuk/dapHome, last access on 19.11.2011

Website TI (Transparency International)
http://www.transparency.org/policy_research/surveys_indices/cpi/2010/results, last access on 21.04.2011

Website Urban Density Project
http://www.urbandensity.org/, last access on 01.12.2011

List of interviews

Below the interviews conducted are listed by the two study settlements and the three categories:

- interviews with participants of solicited photography (including interviews conducted after the solicited photography),
- key informant interviews, and
- focus group discussions with women.

A list of expert interviews is included separately at the end.

Manikpara: interviews with participants of solicited photography

Name (fictive)	*Sex*	*Main occupation (at time of interview)*	*Age*[1]
Abbas Ali	m	employee in a plastic sorting shop on embankment slopes	~30
Fahima	f	employee in a plastic sorting shop along Embankment Road (see also focus group 1)	~35
Hortem	m	employee in a plastic sorting shop, sometimes washes/dries plastic at the Bagan for his employer	23
Kabin Mia	m	manager of a rickshaw garage and operates a rickshaw repair shop along Embankment Road (his wife operates a *pitha* stall in public space next to Embankment Road), later on he had to shift to a different location	60
Mesbah Uddin	m	owner of a plastic sorting business, when he received the camera he was operating a tea stall (see also key informants)	47
Rashida	f	employee of a plastic sorting shop along Embankment Road	21
Riaz	m	employee of a plastic shredding shop along Embankment Road	20
Rokib, Momena	m, f	owners of a business washing and drying plastic on embankment slopes on commission from plastics businessmen (*mohajon*), married	50 30
Roxana	f	changes jobs almost every month, mostly works in factories in the surrounding areas	~30
Tariq	m	owns a plastic sorting business, involved in construction of the garden at Dokkin Ghat (see also key informant interviews)	~25

1 The age is based on the age of interviewees in 2010. Many of the interviewees did not know their exact age or only gave a possible range.

Manikpara: key informant interviews

Name (fictive)	*Sex*	*Function, reason to conduct key informant interview*	*Age*
Akkas	m	member of the committee responsible for the Bagan	~45
Foyez	m	operated the leasehold of Manikghat for two years, became the leaseholder himself in 2010	~35
Jahangir	m	previous leaseholder of Manikghat, respected person of the locality with distinct knowledge about its development, owns a plastics business	~45
Mesbah Uddin	m	respected 'brother' of the locality with distinct knowledge about its development, owns a plastics business (also participant of solicited photography)	46
Nabin	m	AL leader of Ward level	~40
Ripon	m	community police, owner of plastics business,	~40
Ward Commissioner	m	since 2001, BNP politician	~50
Tariq, his friends	m	together have established a small garden on the embankment slopes at Dokkin Ghat	~25

Nasimgaon: interviews with participants of solicited photography

Name (fictive)	*Sex*	*Main occupation*	*Age*
Abdul	m	works as a rickshaw (van) puller, sometimes also does construction work, or sorts plastics	~35
Aklima	f	operates a stall/shop on Nasimgaon Math	~30
Assad	m	works as a carrier for garment factories	~25
Dipa	f	operated a stall/shop on Khalabazar, later on opened a shop selling and sewing dresses	~39
Hamena Ramisa	f, f	works in a garment factory (daughter), sells rice from Khalabazar (mother)	~20 ~50
Hamidul	m	works as a boatman	~40
Khalil	m	owns a shop at Shaonbazar	~55
Meher	f	owns a shop at Amlabazar (with her husband)	~38
Mokbul Hossain	m	owns a shop at Nasimgaon Math	~45
Nasrin	f	owned a tea stall at Nasimgaon Math together with her husband, later on transformed into a restaurant	~23
Rabeya	f	young girl living at Amlabazar, sometimes helps in her father's business	13

Rehana	f	operates a stall/shop at Khalabazar	~40
Riad	m	sells puffed rice from a mobile stall	~40
Rina	f	works in a garment factory	~22
Rohima	f	operates a stall/shop at Khalabazar together with her husband	~25
Shihab	m	works in a shop on Nasimgaon Math	~45
Shoma	f	operates a stall/shop at Khalabazar	~30
Taslima	f	operates a stall/shop on Nasimgaon Math	~40
Zakir, Sania	m, f	operate a restaurant at Nasimgaon Math, married	~39 ~45

Nasimgaon: key informant interviews

Name (fictive)	*Sex*	*Function, reason to conduct key informant interview*	*Age*
Afsana	f	involved in Dhaka Mohanogor Committee (AL) and several NGO committees, shopkeeper at Khalabazar	~35
Arif	m	president of Mosque Committee	~50
Azad	m	owner of a rickshaw garage	~35
Baharul	m	chairperson of AL Sub-area Committee, Khalabazar, involved in several other political and NGO committees	~60
Faruk	m	involved in AL Chatro League (Ward level), son of a local *pir*	~20
Harez	m	president and secretary of Sub-area Committee	~45
Hatem	m	used to live in Nasimgaon, holds several positions in AL committees in Nasimgaon and outside	~50
Kamal	m	night guard, employed by AL office	~35
M. Hossain	m	shop owner, Nasimgaon Math (see sol. photogr.)	~45
Rabiul	m	Imam of a local mosque, AL supporter	~50
Rizia	f	used to live in Nasimgaon, AL politician (*thana* level), involved in NGOs	~40
Sayed	m	member of Amlabazar Committee, AL supporter, local water supplier	30
Shahidul	m	night guard, employed by Amlabazar Committee	25
Shahin	m	operates a shop and tea stall at Khalabazar (rented), with Rohima (see above)	~30
Wahab	m	AL-supporter (Ward level)	50

Manikpara and Nasimgaon: focus group discussions with women

Name (fictive)	*Occupation*	*Age*
Manikpara FG1:	with working women	
Fahima, Gulshana (S.'s daughter) Salma (G.'s mother)	- all three work in plastic sorting shops along Embankment Road, all are married	~40 ~35 ~50
Manikpara FG2:	with women mainly staying at home	
Mashrufa (M.U.'s wife),	- home-based plastic sorting (sometimes), married	~30
Subina,	- home-based sewing/stitching (sometimes), widow,	~35
Azufa (H.'s daughter),	- home-based sewing/stitching (sometimes), married,	~35
Hasna (A.'s mother)	- only visiting her daughter	~55
Nasimgaon FG:		
Meher,	- owns a shop and works there, married,	~37
Ishrat,	- currently does not work, previously worked as a housemaid, married but her husband left,	~42
Firoza,	- about to start work in a garments factory, married,	~22
Niger	- works in a garments factory, unmarried	~20

Expert interviews

Function/name	*Affiliation*
BIWTA officer	-
RAJUK planner	RAJUK, Department of Planning
Roxana Hafiz, Salma Shafi	BUET, Department of Urban and Regional Planning, Centre for Urban Studies (CUS)

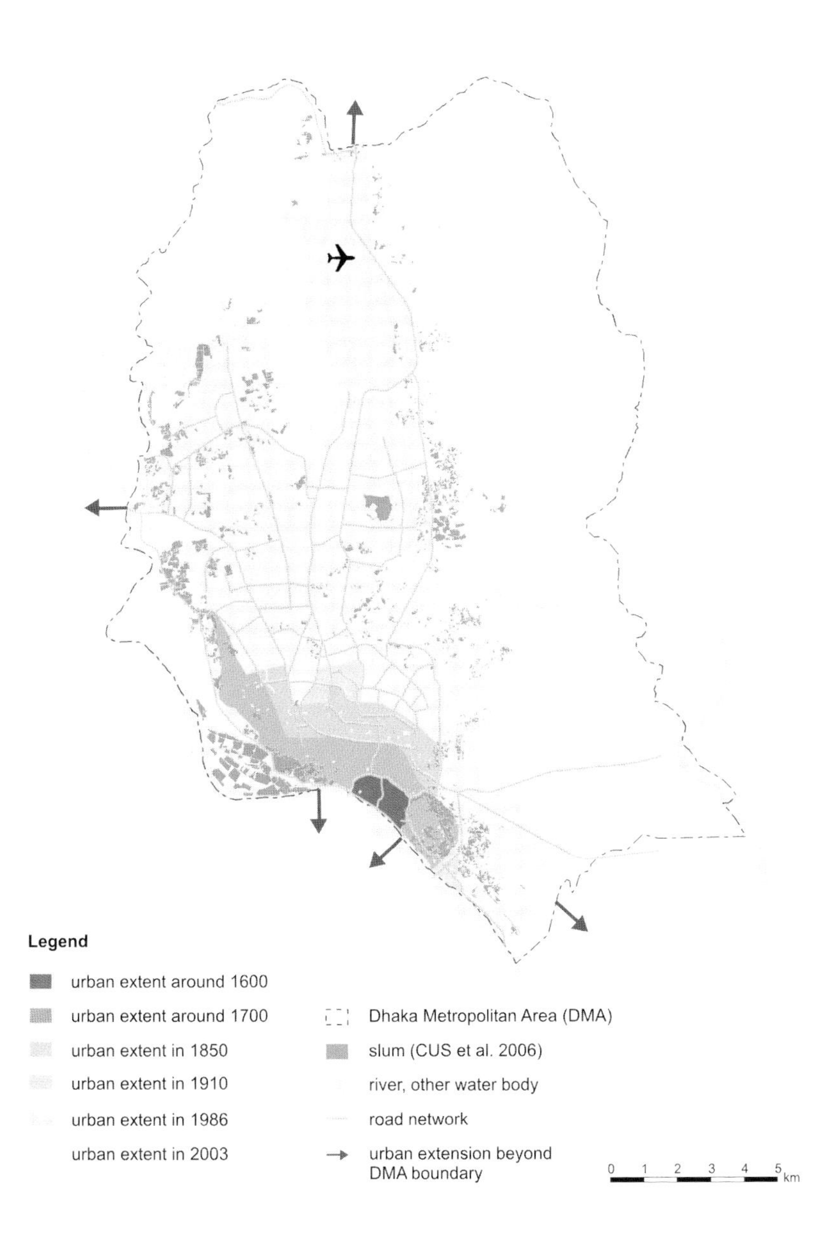

Figure A-1: Urban development and 'slums' in Dhaka Metropolitan Area (own construct, urban extent based on Ahmed 1991, Google Earth Imagery of 22.01.2006; 'slums' based on CUS et al. 2006)

Photo A-1: Panorama of Manikpara

Photo A-2: The embankment slopes in Manikpara

Photo A-3: Khalabazar in February 2008

In 2008 the bazar was still an open field which vendors occupied by selling their materials from *chouki*. (photo: Juliane Hagen)

Photo A-4: Khalabazar in March 2010

This picture is taken from almost the same perspective. On the right the space has been transformed into stable shops, while on the left vendors sell from *paka* platforms instead of *chouki* from the ground.

Photo A-5: Typical courtyard and corridor in Nasimgaon

Photo A-6: Typical courtyard and corridor in Nasimgaon

Photo A-7: Beggars selling rice at Khalabazar

Photo A-8: Canvasser in Nasimgaon

This picture was taken by Rohima in April 2009 as part of the solicited photography.

Photo A-9: Semi-permanent vending unit

The shop encircled is selling coconuts and Bengali plums from a *chouki*. It is not permanently fixed to the ground, thus the vending unit is classified as semi-permanent.

Photo A-10: Permanent vending unit

The two shops selling plastic toys are permanent. Although they also use *chouki* for selling, the roof structures supported by bamboos make these shops less easily removable.

Photo A-11: Stable vending unit, Manikpara

The first shop in the picture sells plastic bags, the second shop is a tea stall. Both are not as stable as the surrounding, but less easily dismantable than the permanent structures.

Photo A-12: Stable vending unit, Nasimgaon

The vendors now selling from the encircled structure used to sell from *chouki*, but since the beginning of 2009 the structures transformed into what now can be classified as stable.

Photo A-13: Plastic storage and sorting in public space, Manikpara

The photo shows how storage of plastic bags and sorting activities occupy much of the road during a working day. In front of most of the shops on the left side of the road, both women and men sit outside sorting the plastic stored in the bags.

Photo A-14: Plastic sorting on Fridays, Manikpara

This photo shows the same place (left side of the road) on a Friday. Most of the sorting shops are closed. Only in the foreground one of these shops is operating and the two people sitting on the ground are sorting plastic.

Photo A-15: Embankment slopes beginning of August 2009

The photo shows the water table which has increased considerably compared to dry season. However, there is still just enough space for plastic washing and drying to be conducted.

Photo A-16: Embankment slopes end of August 2009

The photo shows the same place as in Photo A-15 about four weeks later. The height of the water table no longer allows plastic drying. Instead, the slopes are used for domestic activities (doing laundry, bathing).

Photo A-17: Production and storage of firewood on Nasimgaon Math

Photo A-18: Parking and repairing of rickshaws on Nasimgaon Math

Photo A-19: The Buriganga River

This picture was taken in May 2009 by Hortem, male participant in solicited photography, in Manikpara while on a boat ride. He often went on such boat rides with his friends and in the rainy season, when they considered the water to be cleaner, they also jumped from the gauge tower as shown in the pictures. (Hortem, 29.05.2009)

Photo A-20: Lake east of Nasimgaon

This picture was taken in April 2009 by Abdul, male participant in solicited photography in Nasimgaon, because he liked the new bridge (background). He mainly went to this place to sit there and relax late in the evening or at night, and especially if he felt hurt or not in a good mood. (Abdul, 26.04.2009)

Photo A-21: Bird cages, Bagan

This picture was taken in April 2009 by Roxana, female participant in solicited photography, in Manikpara Bagan because she liked the garden with its birds and trees, which gave her 'pleasure of mind'. (Roxana, 08.05.2009)

Photo A-22: Open space in Nasimgaon

This picture was taken in April 2009 by Nasrin, female participant in solicited photography, in Nasimgaon because she liked to see the children playing kites. She visited this place sometimes with her children (who then were aged one, five and seven). (Nasrin, 11.05.2009)

Photo A-23: Women's *cinema* at a private room

Photo A-24: Men's *cinema* at a tea stall

Photo A-25: Cricket game in Manikpara

This picture, taken in January 2010, shows children playing cricket on the extension site for the school, which was sand-filled in 2009 and was previously a waste dump.

Photo A-26: Children playing kites in Nasimgaon

This picture was taken in 2009 by Nasrin, female participant in solicited photography, in Nasimgaon because she liked to see the children playing with kites on the Nasimgaon Math. (Nasrin, 11.05.2009)

Photo A-27: *Nagordola* on the Math

Photo A-28: Men gathering around a *karom* board

Photo A-29: Selling of mangos along the road

This picture was taken by Nasrin in April 2009. She wanted to show a vendor selling the first (green) mangos of the season who she passed on her way to the bazar. (Nasrin, 11.05.2009)

Photo A-30: Open Market Sale of rice

This picture was taken by Rehana in April 2010. She wanted to show the government subsidised Open Market Sale of rice from where she buys the rice for her rice shop at the bazar. (Rehana, 23.04.2010)

Photo A-31: Bazar area in Nasimgaon during *Ramadan* 2009

Photo A-32: Tea stalls along Embankment Road during *Ramadan* 2009

Photo A-33: Temporary shop selling *Eid* greeting cards in Nasimgaon

This shop had only been set up towards the end of *Ramadan* 2009 (August/September) to sell *Eid* greeting cards. The vending unit can be considered permanent in terms of mobility, but it was removed after the *Eid* day.

Photo A-34: Shop extensions selling *iftar* items

On the right, a row of selling tables has been put up in front of the stable shops to sell *iftar* items. Some of these are operated by the owners of the shops for an additional income. Others are set-up by vendors with the permission of the shop owners.

Photo A-35: *Eid* prayers on the Math during *Eid-ul-Adha*

This picture was taken by Mokbul Hossain's son on *Eid-ul-Adha* during the morning prayers (4th camera for the occasion of *Eid*). A tent had been put up for the *Eid* prayers by DCC, as this is an enlisted Eid Gah Math (according to Rabiul, 23.03.2010).

Photo A-36: *Orosh* at Jahangir's place on the embankment slopes

Photo A-37: *Mussolmani* on open space

Photo A-38: *Pohela Boishakh* celebration in Manikpara Bagan

Photo A-39: Crowd at mobile vendors' stalls on *Pohela Boishakh* on Nasimgaon Math

Photo A-40: Tent put up on the Math for the winter *mela*

Photo A-41: Playing equipment on Nasimgaon Math for the winter *mela*

Photo A-42: *Michil* on Victory Day

This photo was taken by Shihab during solicited photography (second camera) on 16th December 2009, showing a Victory Day procession.

Photo A-43: Victory day decoration at a *ghat*

The banner erected at a *ghat* gives Victory Day greetings by different political persons, among them the local MP.

Photo A-44: Embankment Road before reconstruction, February 2009

Photo A-45: Embankment Road during reconstruction, April 2009

Photo A-46: Barrier to slow down traffic on Embankment Road, erected in April 2010

Photo A-47: Kabin Mia's rickshaw garage before improvement of Embankment Road

Photo A-48: A village home in Komilla

This picture was taken by Khalil in May 2009. It shows the neighbouring house at his village housing compound in Komilla. Khalil did not take any pictures in Nasimgaon (his son took a few) but took the camera to the village, indicating how he felt he belonged to the village while Dhaka was solely his workplace (see also below on mobility patterns).

Photo A-49: Agricultural field in Sherpur

This picture of a field in his village was taken by Shihab in May 2009 because he liked to see the view. About village life and the city he expressed: "Village is good. But I stay here [in Dhaka] to earn my livelihood. [...] Dhaka is good; there is no scarcity of rice in Dhaka. But the village is good for living." (Shihab, 07.06.2009)

maxi worn with *orna* (head covered with *orna*)	*salwar kamij* with *orna*	*salwar kamij* with *orna* (head covered with *orna*)	*sari* (head covered with end of *sari* - *ghomta*)	*borka* (head covered with veil in *hijab* style)

Figure A-2: Female clothing styles

Photo A-50: Bagan on *Pohela Boishakh*

Photo A-51: Bagan on an afternoon in May 2009

Photo A-52: Taking possession of Nasimgaon Math after the water tank conflict

In the foreground, the photo shows bamboo poles (green ellipses) connected by thin ropes that were put on the Math symbolising the possession-taking. In the background, a crowd gathers at the place of the water tank.

Photo A-53: Concrete pillars for fencing of Nasimgaon Math (September 2009)

The concrete pillars were erected on two sides of the Math in preparation for fencing. However, before the end of 2010 the pillars had deteriorated and no fencing attempt had been made.

Photo A-54: Fenced area designated for community purposes

The photo shows the fence that was put up by AL leaders to demarcate a ground for a school, hospital and community centre to be built on the Math. The reconstructed rickshaw garages are located left of the picture (the bamboo fence runs parallel to the northern wall of the garages, facing the Math).

Photo A-55: Blocked access to the garages under reconstruction

The photo shows the blocked access (bamboo pillars) to the rickshaw garages. In the afternoon, the signboard mentioned in the narrative was attached to this fence.

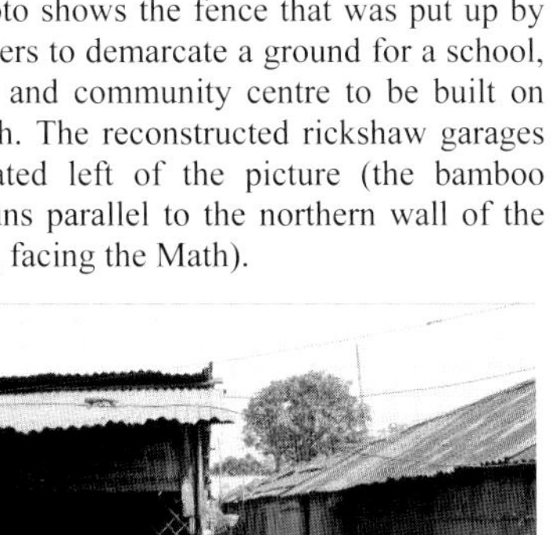

Photo A-56: New access road to the rickshaw garages

On the left side, the photo shows the community police office – one of the three units that were decreased in width in March 2010 in order to make space for the road providing access to the rickshaw garages.

Photo A-57: Tea stall and new rickshaw garages in November 2010

In the foreground the photo shows the new tea stall north of the rickshaw garages. Behind are the new rickshaw garages with the new roof structures. The CI-sheet walls in the background are part of the rickshaw garages that had been reconstructed in March 2010.

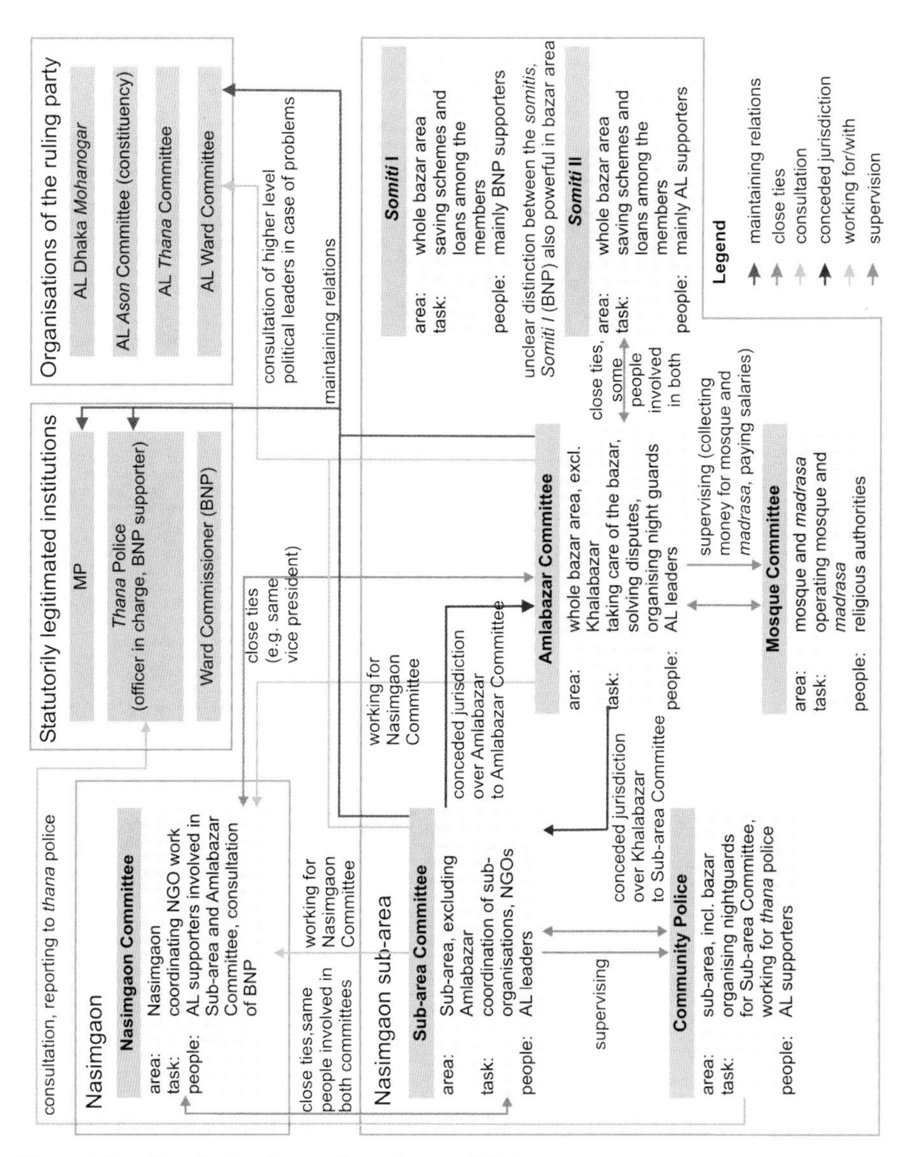

Figure A-3: Organisational set-up in a sub-area of Nasimgaon

The lower part of the figure displays the organisations most active within one sub-area of Nasimgaon. Above this the left box displays an organisation responsible for the whole area of Nasimgaon; the box in the middle displays the statutorily legitimated actors (local government, organs of the executive and Member of Parliament); the box on the right displays the corresponding AL party organisations. The figure presents one possible reading of the institutional set-up in Nasimgaon, with a focus on one sub-area. It cannot be considered to fully represent reality as perceived by all actors, as the interviews revealed no 'shared reality' but a representation dependent on individual perspectives and preferences. Despite the difficulty of presenting a definite organisational set-up, the main message transported by the figure of the dominance of political organisations and the competing claims within one sub-area remains, independent of inconsistent representations.